BODY MAPS

body maps

improvising meridians and nerves in global chinese medicine

Lan A. Li

JOHNS HOPKINS UNIVERSITY PRESS | *Baltimore*

Printed in the United States of America on acid-free paper
2 4 6 8 9 7 5 3 1

Johns Hopkins University Press
2715 North Charles Street
Baltimore, Maryland 21218
www.press.jhu.edu

Library of Congress Cataloging-in-Publication Data
Names: Li, Lan A., 1988– author.
Title: Body maps : improvising meridians and nerves in global Chinese Medicine / Lan A. Li.
Description: Baltimore : Johns Hopkins University Press, 2025. | Includes bibliographical references and index.
Identifiers: LCCN 2024021293 | ISBN 9781421450964 (hardcover) | ISBN 9781421450971 (ebook) | ISBN 9781421450988 (epub open access)
Subjects: LCSH: Human anatomy—China—History. | Human anatomy—History. | Medicine, Chinese. | Medical illustration—History. | Acupuncture points.
Classification: LCC QM11 .L55 2025 | DDC 611—dc23/eng/20240508
LC record available at https://lccn.loc.gov/2024021293

A catalog record for this book is available from the British Library.

Special discounts are available for bulk purchases of this book.
For more information, please contact Special Sales at specialsales@jh.edu.

To my family

To access the digital companion to this book, visit **www.bodymaps-book.com.**

CONTENTS

PREFACE

Anatomy has often relied on images to be seen. Without images, corporeal objects escaped the eye—whether trained or untrained, believing or unbelieving. Images mattered. They allowed illustrators and readers to make sense of the body. Images were technologies for sense-making and, more importantly, for making common sense.

There is a long history to anatomical images as problematic and poetic objects.[1] Many excellent books on wax-model makers, engravers, and photographers have shown how anatomical images relied on specific material conditions and, further, commingled with historical, social, and political concerns. Images from the fringe to the famous came with strings attached. They derived from a close study of stolen corpses, gorilla parts, and trees. Although not all anatomical images were art objects, they operated within aesthetic constraints to move the mind.

I first came to the history of science and medicine wondering why there were so few books on the history of anatomy in Asia. When I asked medical practitioners about their views on anatomy, some lamented that Asian medical practices "did not have anatomy."[2] How strange, I thought, especially when anatomy was everywhere. Medieval and early modern images showed bodies packed with organs. The gut was the key to health. It was filled with oily membranes, clouds, spirits, and channels stretching from the head to the hands and feet. Now most recognizably associated with styles of acupuncture and moxibustion, these channels facilitated the movement of fluids like Blood (血) and internal Qi (氣). They were central to life, death, and longevity.

This book focuses on images of anatomical channels, called *jingluo*

(經絡), and how they became entangled with representations of neurophysiology.[3] *Jingluo* were corporeal objects that included larger structures (*luomai*) and smaller branches (*jingmai*). They were both material and invisible—felt in the body through pulse-taking practices and seen in hand-drawn images. These images have an impressively long history. For instance, one image that appears throughout this history is **Figure P.1**. You might notice paths twisting around each other, bending at sharp angles above the eyebrows and cascading down the cheeks. You might notice a knot of text crowding around the eyes, flanked by circles and lines clasping the eyes to the ears. Looking even more closely, you might even see how details in the image make it asymmetrical. The paths forming shapes around the cheeks are different on the right than on the left. The inscriptions on the right of the image are also different from the text on the left. For instance, *er* (耳) appears on the right ear lobe, and *men* (門) on the left ear lobe.

I will return to this image in the introduction. What you need to know now is that this image of *jingluo* had origins in the second, seventh, and fourteenth centuries. Its long history was complex, multiple, and not entirely coherent.[4] Throughout this broad and fragmented history, classical corpora described how *jingluo* moved at different depths. They could rise to the surface, twist, and sink into the body. They had ephemeral qualities, not all of which were illustrated on paper. And the paths that did appear on paper were usually asymmetrical. Asymmetrical images were different from right to left and were meant to be read from right to left, such as the site named *ermen* (耳門) on lower part of the ear.[5]

Images of *jingluo* existed in specific graphic genres. In this book, I call illustrations of *jingluo* "meridians" to conceptually separate the

Figure P.1. (opposite) Meridian man, seventeenth century. Unknown illustrator, woodblock print. This image is titled *renzheng mingling tu* (正人明堂图), or "facing up, *mingtang tu*." The vertical text flanking the meridian man describes the beginning and ending sites of four meridians on the front and back of the body. The rest of the image shows the 12 primary meridians. From "Chen Chiu Ta Chheng, Charts," Needham Research Institute

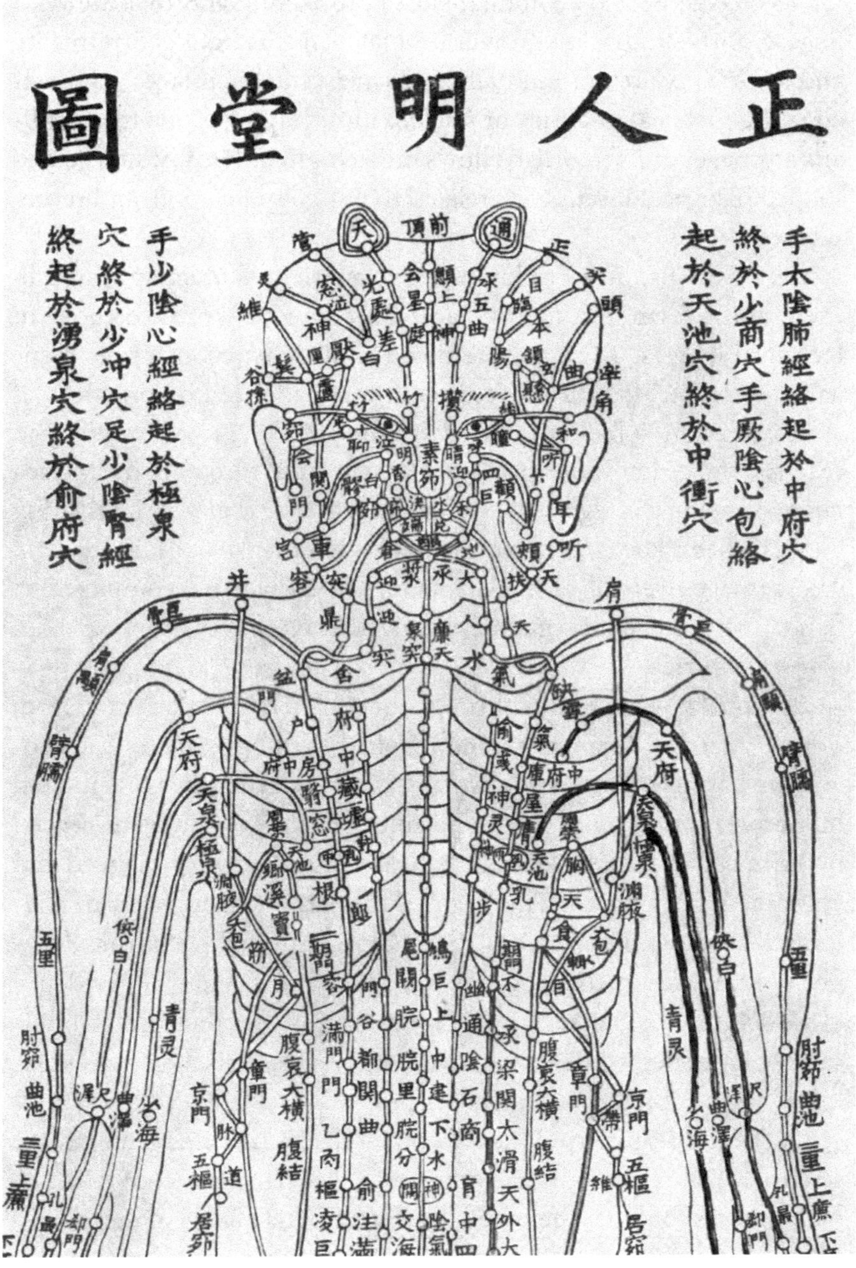

正人明堂圖

手太陰肺經絡起於中府穴終於少商穴手厥陰心包絡起於天池穴終於中衝穴

手少陰心經絡起於極泉穴終於少冲穴足少陰腎經終起於湧泉穴絡於俞府穴

image of corporeal things from the corporeal body. This also allows me to engage with the simultaneous corporeality and invisibility of *jingluo* in the body. As you will find later in this book, corporeality and invisibility are not mutually exclusive. Graphic images acknowledge the material qualities of *jingluo*. Limiting my scope to meridians as images of *jingluo* also allows me to highlight how visual genres shaped historical discussions related to ontology, location, and representation.

What you also need to know is that meridians often appeared in the graphic form of a *tu* (圖), an East Asian historical category of technical images. *Tu* as a genre demanded a particular mode of engagement. Meridian *tu* moved the mind. *Tu* facilitated practices of approximation, which grounded cultures of accuracy. Healers who read the image knew how to locate the site on a human body. Meridian *tu* were not a category of fine art but rather a means for cultivating craft. This is a key concept in the book: that meridians existed as *tu*. These images were illustrated by hand, carved by hand, completed by hand. The hand-made qualities of meridian *tu* have sustained them across centuries.

As global powers shifted throughout the late early-modern period, practitioners of Asian medicine invoked the young field of neuroanatomy to explain meridian *tu*. Early modern nerves facilitated human sensation, cognition, and finer spirits. Nerves were moved by an invisible mind. Like meridians, the mind was experienced and embodied, but unseen. This seemed to match well with Asian anatomical images filled with oily membranes, clouds, and spirits. European and Eurasian conceptions of the mind and soul were evidence of something divine. Yet theories of sensation remained philosophically open-ended while meridian *tu* persisted as a discrete and evolving set of corporeal anatomical images.

By the modern period, new political discourses merged colonial science with national modernity, and physicians felt compelled to make claims for what they thought was the true nature of meridians. They used meridian *tu* to represent an indigenous kind of practice that was rational, scientific, and "Chinese." Twentieth-century physicians debated over whether fluids like Blood (血) and internal Qi (氣) were only acting on nerves. Perhaps meridians were nerves, they thought.[6] As debates continued, neurophysiologists risked ascribing all physiological features to nerves, replacing one black box for another. But being a black box, nervous anatomy and physiology became an unreliable object of modernity. It could not redefine or reimagine the broad taxonomy of *jingluo*.

For neuroscientists, nerves left a lot to be desired. Mechanosensory input via nerves did not square with the broad therapeutic practices of acupuncture, massage, mental choreography, *materia medica*, and moxibustion that acted on *jingluo*.[7] Each practice involved different kinds of external input that implicated different kinds of cells and triggered different kinds of physiological pathways.[8] These cells and pathways could have been lymphatic, vascular, fascial, or something else entirely.[9] At the same time, not all cells or signals were nervous. In other words, nerves were not always at the center of sensory perception and therapeutic action. As a result, the juxtaposition of *jingluo* and nerves has remained awkward, asymmetrical, and incommensurable.

I contend that *tu* was at the heart of this friction. It facilitated and inhibited how physiologists compared, overlapped, and translated meridians as nerves.[10] It was from the visually absolute inscription of a line and a dot that physiologists grew to question whether knowing *where* meridians lay defined *what* constituted them. Place somehow presumed essence. Meridian *tu* manifested the uneasy relationship between aesthetic practice and ontological judgment. Conflating *what* meridians were to *where* they appeared on paper gave anatomical images power. Reading meridian *tu* as body maps pinned meridians to a place on the page. It confused the sign with the signified.

This book takes a dynamic disciplinary approach to telling this long history of anatomical representation. I draw on analytical tools from science studies, Chinese studies, art history, medical anthropology, and critical cartography to consider images that scholars of religion might call "subtle" entities. Representations of meridians and neurophysiology guided and legitimated practices meant to preserve the body, such as Qigong, *neigong*, moxibustion, acupuncture, and drug formulas. This book centers on the images that have led scholars and illustrators to conflate ontological certainty with topographical uncertainty. It does not aim to be comprehensive but clarifying.

Although historians and science studies scholars have considered the social and political dimensions of observation and objectivity in the production of modern anatomical images, we have not yet considered nerves and meridians together at length. I take inspiration from works on the expressiveness of the body and respond to political histories of medicine through a cultural history of images.[11] Reading technical images with care requires that viewers take images on their own terms. It requires us to withhold assumptions of what the images showed. By slowing down, we can understand why these images have posed some of the greatest challenges to empiricists in the last two centuries.

Introduction

Global Chinese Medicine as Method

"I don't read Chinese," Nilza said to me.

It was a cloudy afternoon in March. We sat on the back patio of her clinic in São Paolo, Brazil, speaking through a translator.[1] At that time, Nilza was running her own community clinic, where she saw patients and trained new physicians.[2] On the walls of her examining room hung certificates of the conferences that she had attended in China. Facing the examining table were charts of the eyes, ears, and full body filled with lines and dots. These body maps guided two kinds of practices: *acupuntura*, or the application of needles, and *moxabustão*, the application of burning moxa, made of dried and compressed mugwort leaves.[3]

Nilza struck a match and lit what looked like a small cigarette. She placed the roll of moxa close to the back of her hand and said, "Watch how the skin absorbs the smoke." As I sat across her desk, Nilza pulled out an embroidered box that was inlaid with golden fabric. A set of thick copper needles the size of hairpins lined the interior. They were facsimiles of late nineteenth-century bloodletting tools.

"A gift from my colleagues in Beijing," she smiled.[4]

That day, Nilza wore a bright red satin shirt with a high collar adorned with golden phoenixes and dragons. She had learned *acupuntura* and *moxabustão* from French textbooks and understood herself to be preserving ancient medical practices of needling and cauterizing the body. When Nilza began her training, she thought she was carrying on a tradition that no longer existed in China. Then, she went to China. "I was so surprised to learn how similar our perspectives were," she recalled. Her idea of tradition resonated with her colleagues in Beijing. They collectively contributed to maintaining an ancient practice even as, or especially as, Nilza drew on other techniques. She used *acupuntura* and *moxabustão* alongside chromotherapy and radiesthesia tools. In her clinic, I watched her swap color filters in a battery-powered torch and spin it over a client's back. When I asked how she squared recent technologies in light therapy and energy healing with her identity as a traditional practitioner, she said, "We are Brazilian! We blend all cultures, all influences."[5]

Scholars are familiar with contemporary practitioners who identify as traditional healers that actively modify their practice.[6] It is not that they were never modern. They are all modern, if modernity were so distinct.[7] When they invoke Chinese medicine, they engage in modern, postcolonial identity politics. When they express attitudes that align with the state, they participate in modern, postcolonial national discourses. When they manage their own acu-moxa clinic, they exist as subjects to modern, postcolonial, capitalist demands.[8] The result is that there is no single "Chinese" medicine but rather many kinds around the world. As anthropologist Volker Scheid put it, Chinese medicine does not exist in "zones of purity."[9] It is—and has been—multiple, complex, evolving.

Before Chinese medicine was Chinese, it was global *and* local. Medical practices in East Asia have enjoyed a long global history because trade and diplomacy have a long global history.[10] Travelers made their way along the many Silk Routes on land and sea.[11] They carried with them objects, ideas, and skills. These journeys were not always easy; knowledge transmissions were never straightforward.[12] On these many missions, images of the body toured the world, arriving at seaports and royal palaces. They moved across empires in the hands of itinerant monks in the thirteenth century and medical missionaries in the eighteenth century.

Single-origin stories of "Chinese" medicine have sustained an uneasy sense of cohesion. They have sustained the myths that separate tradition from modernity. These same myths were useful for campaigns in building a modern nation-state.[13] Early twentieth-century political reformers in Republican China politicized national practices and named them "Chinese" so that they belonged to China. This name offered a cultural legacy. For the first time, physicians were compelled to identify themselves as uniquely Chinese medical practitioners. By the mid-twentieth century, *acupuntura* and *moxabustão*, or acu-moxa, became recognizably Chinese.

The idea of a singular Chinese medicine also traveled.[14] It traveled with the sale of moxa dust, bloodletting tools, anatomical images, textbooks, handbooks, research articles, and peripatetic practitioners.[15] Buying and selling these objects of Chinese medicine meant that they belonged to the Chinese diaspora and its sympathizers.[16]

As a global object, Chinese medicine could belong to physicians like Nilza. It belonged to them as they reinforced the dichotomies of East and West, of tradition and modernity. These dichotomies entangled global Chinese medicine within postcolonial legacies. It connected Chinese medical practitioners in Singapore to Chinese medical practitioners in Berlin, Germany. It related acu-moxa clinics in San Francisco, California, to acu-moxa clinics in São Paolo, Brazil. And among these connections, identity politics felt like splitting hairs. They were absolute and imagined, national and ethnic, local and universal.

This book engages with the identity politics of global Chinese medicine from a different perspective. It takes a long view, from the early modern period to the recent past through a series of case studies. It engages with visual sources originating in different times in different places. I focus on images with origins in tenth-century East Asia and fourteenth-century Eurasia/Europe that facilitated knowledge about the human body to explore how conceptions of the body shifted when these images traveled and changed.

Specifically, I study representations of channels known as *jingluo* (經絡) to understand their influence on contemporary neurophysiology. *Jingluo* were corporeal and ephemeral objects that facilitated therapeutic practices to guard and enhance the body.[17] One could access *jingluo* through the mind, the skin, and the gut. As fluid entities, *jingluo* responded to food, medicine, poison, meditation, ritual, and alchemical practices. Practitioners engaged with *jingluo* by palpating the wrist to reach the organs.[18] These were stable, corporeal, and permanent paths.[19] Etymologically, the characters *jing* (經) and *luo* (絡) referred to weaving. *Jing* was the warp that held the weft in place.[20] *Jingluo* included larger structures known as *luomai* and smaller branches called *jingmai*.[21] Larger *jing* appeared on paper; smaller *jing* did not.

The images I present throughout the book are meant to appeal to the reader.[22] They are meant to perform graphic symmetry so that the images within each case study can speak to, past, and alongside one another. Doing so offers insight into why representations of *jingluo* and nerves have not and might never cohere in the way we expect. To this end, I take inspiration from feminist and decolonial science studies to inform my intellectual approach and aesthetic practice. My training in the history of science and medicine allows me to inspect images as objects of knowledge production; medical anthropology challenges me to engage with issues of embodiment and power; critical cartography allows me to conceptually situate graphic sources. Although I am not trained in art history or literary studies, I have found scholarship on genre and performance particularly important in reconceptualizing Asian sources so that I can take on comparative analysis with care.

Though the historical actors considered here often relied on the dichotomous categories of West/East, Asia/Europe, and North/South, I try to resist dualities that reinforce the essentialist categories of "traditional" or "Chinese," "Western" or "biomedical." Instead, I operate on a different axis by offering new names. Throughout the book, you will meet "meridian man," "sensation man," "man with goatee," "man with hat," "man with flaps," and "man with robes." Renaming these images acknowledges the ways in which they represent a basic body that was explicitly gendered and mostly male.[23] Both "meridian man" and "sensation man" extend from complex histories of gendered and androgynous bodies. Although some women appeared in medical texts, they did not stand in for a universal anatomy or universal body.

This book thus aims to help you to approach unfamiliar images with patience and regard familiar images with caution. As art historian James Elkins put it, seeing anything involves persistent attention for the images to speak.[24] Regarding images with new terms and new names allows them to say something new.

Meeting Meridian Men

Jingluo (經絡) and their associated types were old objects.[25] They were even older than the texts that described them. Most of these sources include silk and bamboo manuscripts from the second century BCE.[26] For instance, archaeological artifacts like small figurines recently excavated from the Laoguanshan and Tianhui tombs in Sichuan already featured paths in the body that engaged with smaller and larger cosmological movements.[27] They controlled physiological processes that manifested in a variety of internal and external Qi. By the first century BCE, the famous corpora called the *Huangdi neijing* already described *jingluo* and the movement of Qi through rising and sinking pulses.[28]

Although current scholarship across a number of academic fields is working to understand the empirical origins of *jingluo*, this book begins with a later part of this history in the late medieval and early modern period. By this time, *jingluo* were already commonplace. Their structures were already well established. Images of *jingluo* sustained particular kinds of epistemic practices. Specifically, I call illustrations of *jingluo* "meridians" to provide an analytical category that engages with both the corporeal and invisible qualities of *jingluo*.[29] I borrow "meridians," a Middle English word, to describe hand-drawn lines that represented invisible corporeal structures. To be clear, I do not mean to suggest that *jingluo* themselves were literally astronom-

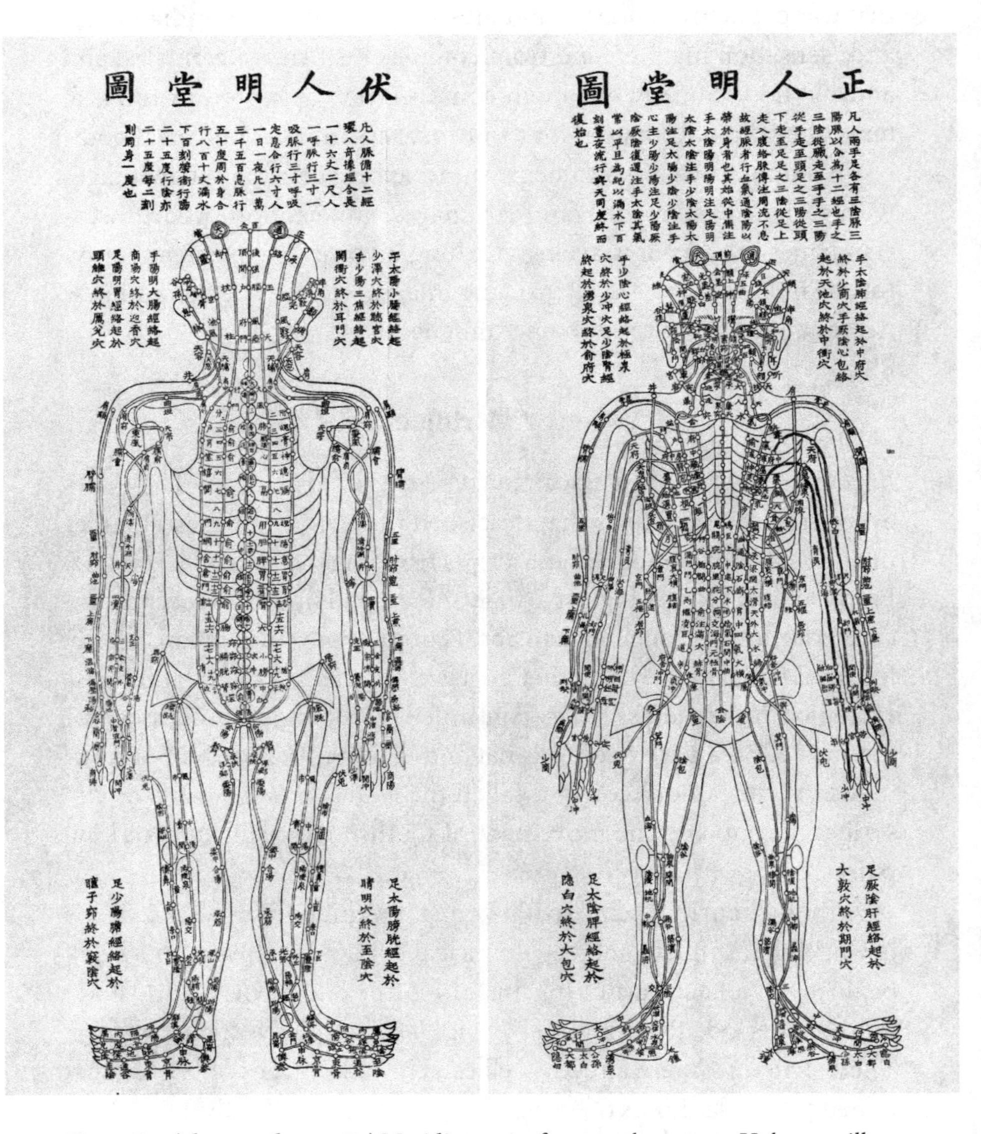

Figure I.1. (above and opposite) Meridian men, fourteenth century. Unknown illustrator(s), woodblock print reproductions. These three panels reproduce images of primary meridians from a fourteenth-century text. From a 1956 reprint of *Zhenjiu jicheng* (Collected Works on Acupuncture and Moxibustion)

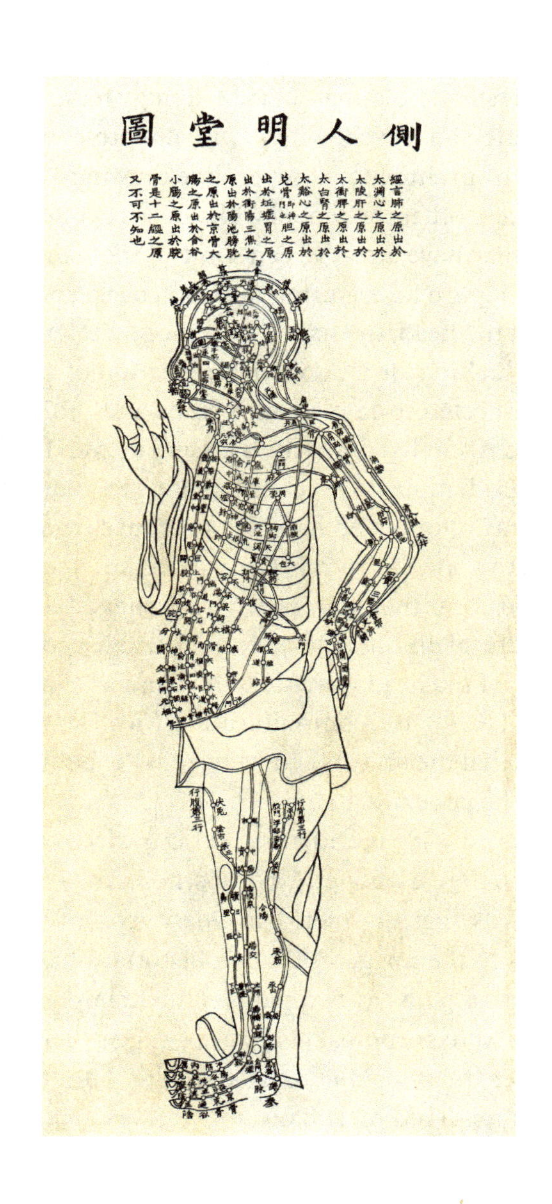
側人明堂圖
經言肺之原出於
太淵心之原出於
太陵肝之原出於
太衝脾之原出於
太白腎之原出於
太谿心之原出於
兌骨即神門也胆之原
出於坵墟胃之原
出於衝陽三焦之
原出於陽池膀胱
之原出於京骨大
腸之原出於合谷
小腸之原出於腕
骨是十二經之原
又不可不知也

ical meridians. Instead, considering their representations as meridian lines creates distance between the sign (lines) and the signified (*jingluo*). It allows me to understand how anatomical ontology manifested *through* visual representation and, eventually, *as* visual representation.

Meridian men ranged from simple to complex line drawings. Some images showed a single path, others brimmed with dozens of lines and hundreds of labels. You may have already seen some of these images, like the meridian men in **Figure I.1**, featuring the front, back, and side of the body. Text flanked the head, the hands, and the feet to explain the movement of individual *jing*. It described the location of 12 primary meridians that connected to 12 primary organs and indicated two core meridian paths that circled down the back and up the chest.[30]

This three-column meridian man was called a *mingtang tu* (明堂圖), which roughly meant "illuminated hall." It was a puzzling name that technically referred to a specific room within palace architecture where emperors would call on their advisors.[31] This suggested that the *mingtang tu* was a cultural object, a didactic object, and a cosmological object. At the top of the head were two tiny banners that featured the words *daotian* (道天), or "communicating with Heaven."[32] These banners suggested that the universe converged in the body, that this man was more than he appeared to be.

Mingtang tu were packed with meridians. Text crowded around parallel lines and displaced the ears, bones, eyes, ribs, nose, and mouth. This displacement suggests that the meridian lines were inscribed first. They were essential to the image. Although historians have argued that drawing meridian paths as two parallel lines simply reinforced the idea that they were corporeal channels, I suggest another possibility. If we consider the *mingtang tu* as a graphic object, it is possible that illustrators rendered meridians as two lines for clarity of vision. The clarity of vision perhaps reflected the "illuminating" purpose of *mingtang tu*. Maybe they made the many meridians easier to see.

These images were modern and medieval. For instance, **Figure I.1** is a 1956 reproduction of an fourteenth-century image. It had origins in the seventh century during the Sui and Tang dynasties with connections to the legendary physician Sun Simiao (d. 682) and the lesser-known Wang Tao (670–755).[33] Different versions of *mingtang tu* belonged to different medical lineages. They were associated with medical families that thrived beyond direct government control.[34] They did not represent a single, cohesive practice but instead many types of practices. As a category, they cast a wide net that captured diverse ideas that operated roughly within similar medical cosmologies and medical therapeutics.[35] For instance, they referenced Yin and Yang orientations, which were old and common motifs.[36] They invoked entities like spirits, demons, and ghosts that affected disease and disorder in broadly Daoist, Buddhist, and Confucian frameworks.[37]

Over six centuries, *mingtang tu* have changed.[38] They appeared as a recognizable graphic genre even as illustrators made their own modifications when they traced the images by hand or scaled them onto larger or smaller woodblocks. Paths have altered course; sites have vanished and reappeared.[39] In some early modern cases, illustrators who did not read Chinese characters reproduced *mingtang tu* with bodies that looked uncanny and warped. Men faced in the opposite direction with slightly distorted features. Their torsos were too long, their legs too short.[40] *Mingtang tu* with their many modifications roughly resembled other *mingtang tu*, until they did not. They had similar faces, similar bodies, similar blocks of text.[41] The images themselves never appeared alone and free-standing, but as a set.

Not the Same

In 2015, I attended a lecture in Beijing with members of the China Academy of Chinese Medical Sciences.[42] The keynote speaker was Dr. Zhu Bing, one of the leading investigators of acupuncture and moxibustion (acu-moxa, or *zhenjiu*). In the title of his talk, he promised to reveal the "origins" of this global Chinese practice.

But over the course of two hours, Zhu could not quite pin down a single origin story. He tried to rationalize forms of medicine in China, arguing that they were innovative and ordinary. They were innovative because meridians had a long history in the Sinographic world. They were also ordinary because the methods of intervention—needling and heating—were not unique to East Asia.

In the middle of his lecture, Zhu paused to compare two images. One image was a fourteenth-century *mingtang tu*; the other was a nineteenth-century illustration of disturbed sensation. Speaking to the audience, Zhu announced that these two images were "basically identical." To be the same, they had to reflect a universal vision of the body. They had to be read the same way. They needed to be the same, even when Zhu confessed that he did not know what to make of the fourteenth-century *mingtang tu* and had not read the nineteenth-century thesis on disturbed sensation. And even if he had known them, he was not entirely sure how his own research on cellular morphology related to these figures.

To the untrained eye, the two images were obviously different, obviously unidentical. They had come from different methods of production, followed different aesthetic principles, made different historical inferences, and expressed different modes of representation. Despite his earnest effort, Zhu's images failed to meet.

Meeting Sensation Men

The nineteenth-century sensation men that Zhu referenced in his talk first appeared in an 1893 thesis by the British physician Henry Head (1861–1940) (**Figure I.2**). As a young physician, Head was puzzled by how disturbances in the gut registered on the skin. Neuroanatomists illustrated elaborate images of nervous structures but could not explain how sensations changed when the structures supposedly remained stable. A single branch that innervated a discrete area on the skin could somehow be both sensitive and insensitive to

Figure I.2. (opposite and verso) Sensation men, nineteenth century. J. Griffith (active in late nineteenth century) engravings. These two engraved plates by Griffith illustrate Henry Head's mapping of sensitive areas on the skin in the late 1800s. Head collected numerous patient cases to create this composite map that left many questions unanswered. From Head, "On Disturbances of Sensation with Especial Reference to the Pain of Visceral Disease," 1893, 16

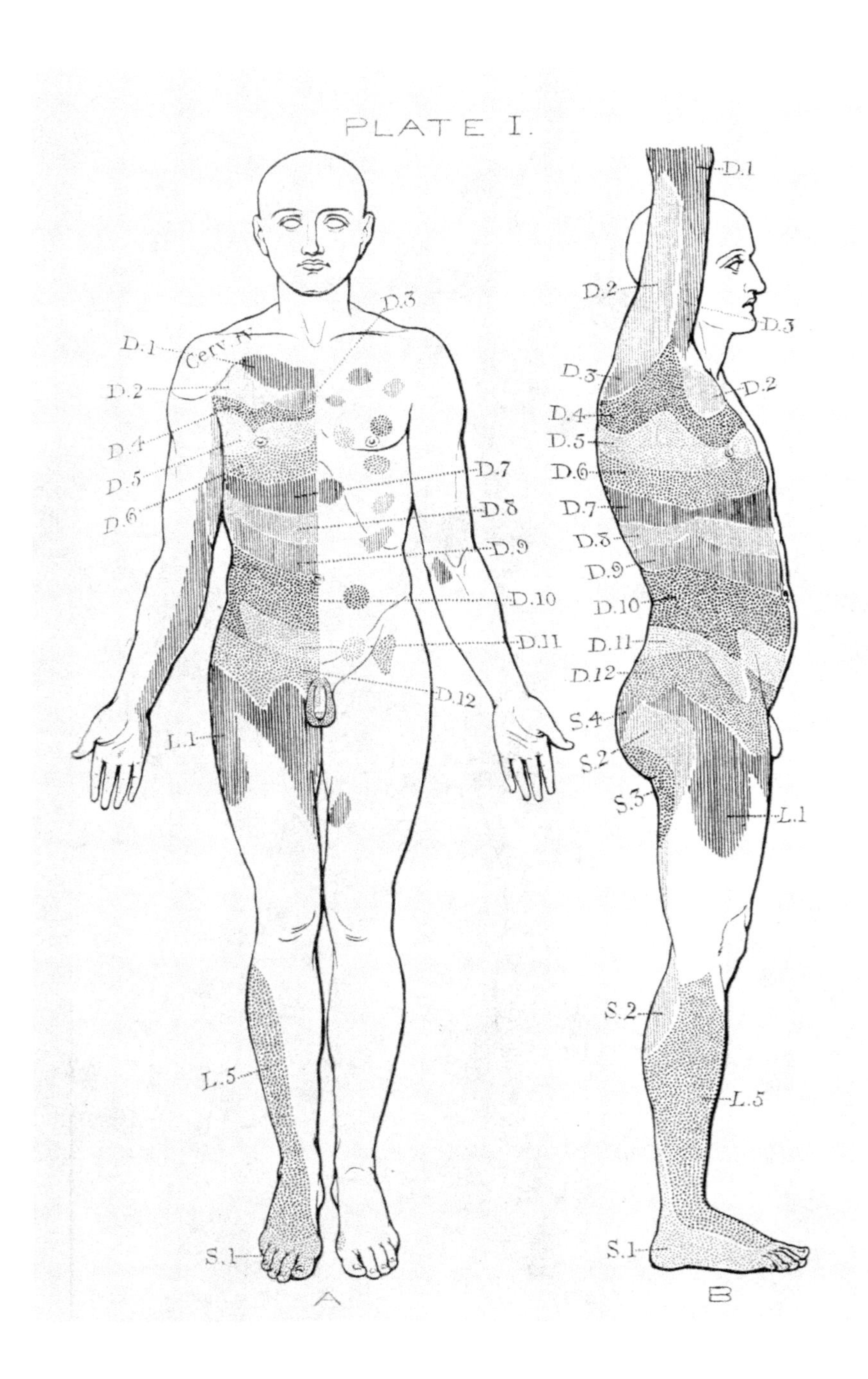
PLATE I.
D.3
D.1
Cerv. IV
D.2
D.4
D.5
D.6
D.7
D.8
D.9
D.10
D.11
D.12
L.1
L.5
S.1
A
D.1
D.2
D.3
D.3
D.2
D.4
D.5
D.6
D.7
D.8
D.9
D.10
D.11
D.12
S.4
S.2
S.3
L.1
S.2
L.5
S.1
B

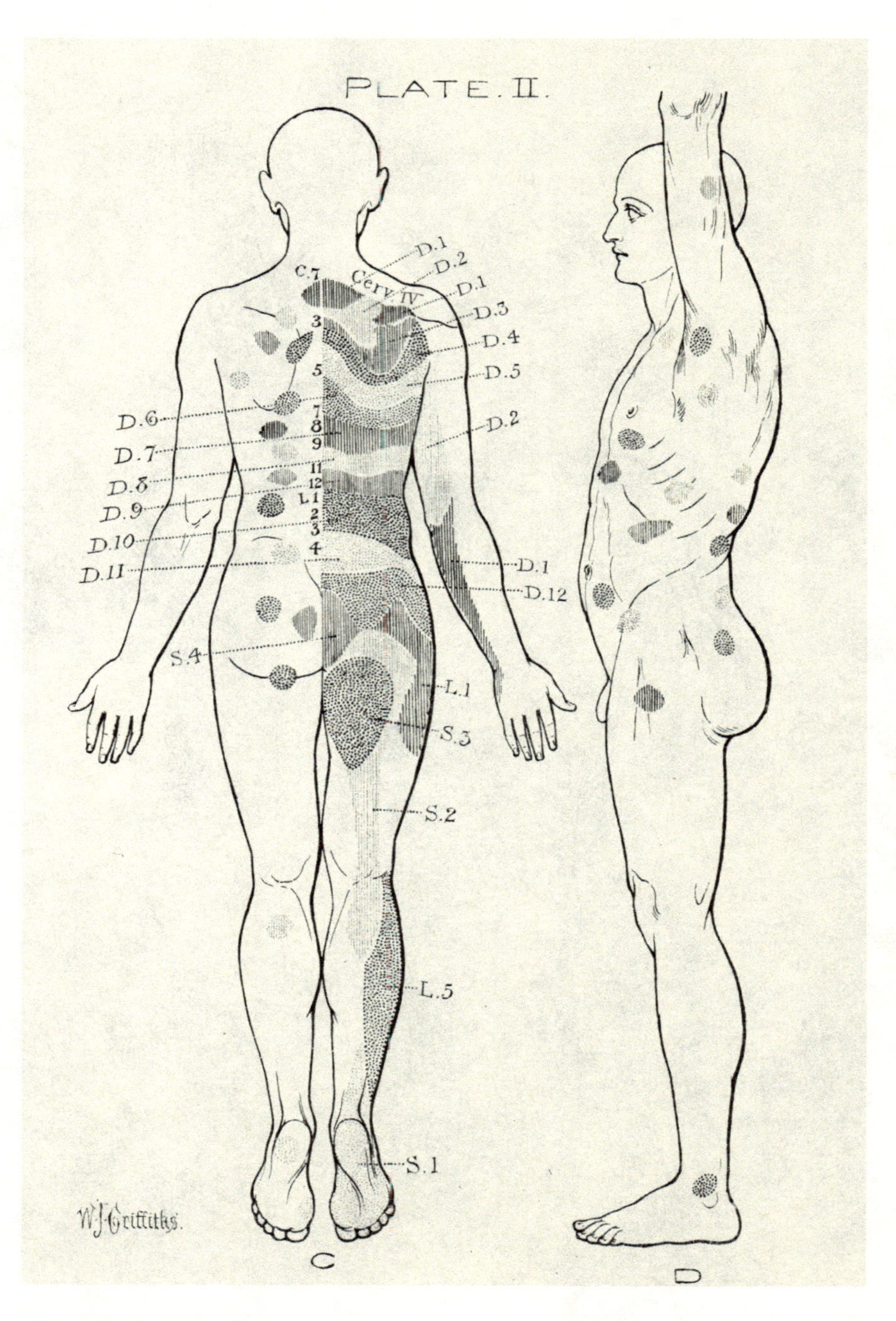

Figure I.2. (continued)

pain. Puzzled, Head tried to track these contradicting patterns on a bare and bald body with the help of the illustrator W. J. Griffiths. The result a composite template.

Looking closely at sensation man, we see a masculine figure with textured bands extending across his torso, arms, and legs (**Figure I.3**). He was an asymmetrical figure. His left side featured discrete areas of hatching and stippling. His right side displayed patches on the elbows, breast, and ribs. According to the image caption, these areas were sensitive to touch. Some areas appeared more articulated, other areas seemed less articulated, curving, peaking, and trailing off the page.

The thin halo of lines assigned labels that read as follows: Cerv. IV, D.1, D.3, D.4, D.5, D.6, D.7, D.8, D.9. These seemed to indicate an early version of spinal levels, although they were not spinal levels. According to the text, these areas represented hypersensitivity. They were illustrated based on the experiences of patients who had described issues in the bladder that registered in the thigh, issues of the liver that registered on the stomach, and issues in the lung sensed on the shoulder. Sensation man was not an anatomical figure but a physiological one. Although researchers like Zhu Bing assumed that sensation men represented nervous anatomy, sensation men did not show nerves. They were simply the preliminary findings of a preliminary investigation.

Meridian men and sensation men were not the same. They displayed two kinds of male bodies. Meridian men showed a full set of meridians. Sensation men showed partial expressions of pathology. These two sets of images were not identical, even when a chief researcher at the largest government-funded institute on medical science in China claimed that they were "basically" the same. Of course, there were similarities. Both images depicted ephemeral impressions in the body. Both were drawn by hand. Both showed structures that could not be captured in full through photography, magnetic resonance imaging, computed tomography, fluorescent dyes, or radioactive isotopes. Meridian men and sensation men appeared as a complete set only when they were manually inscribed.

It was not so simple as that, of course. Nerves also had a contested

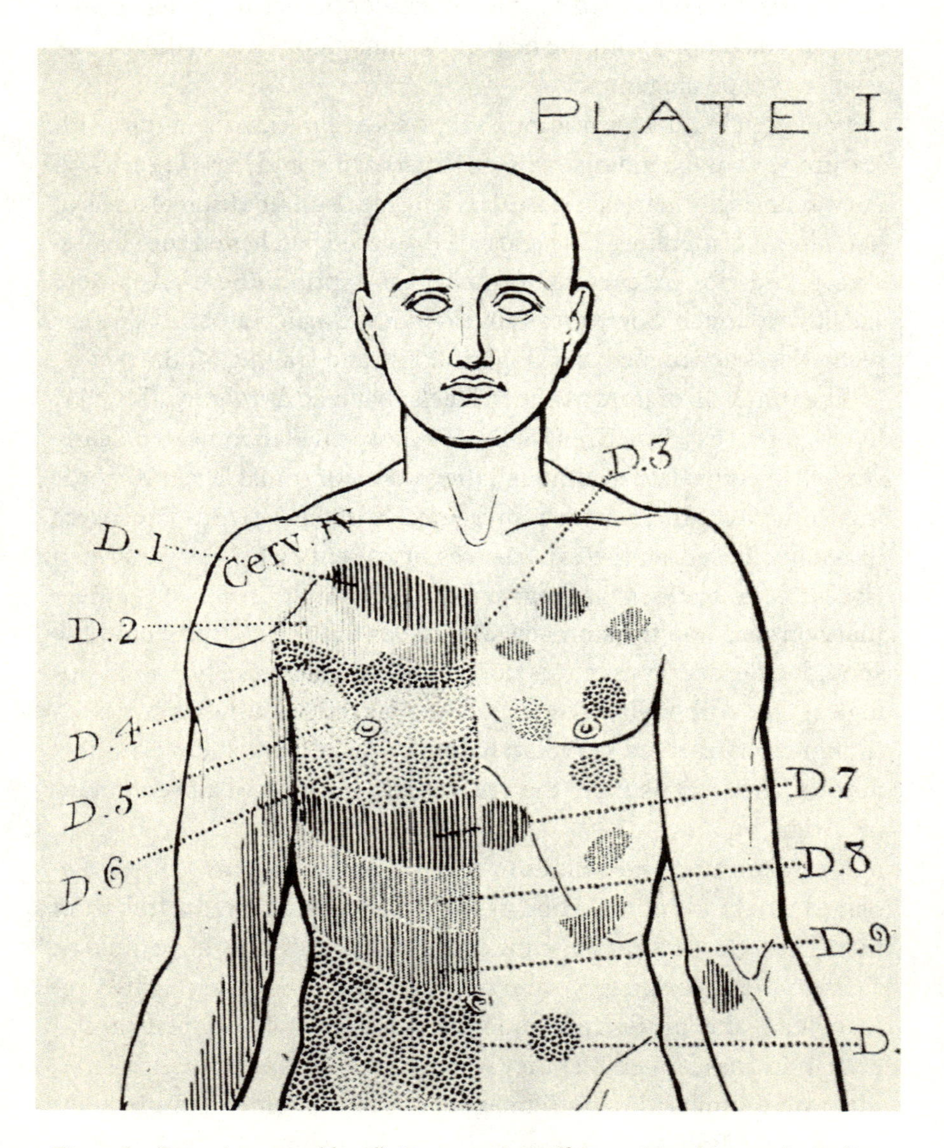

Figure I.3. Sensation man (detail). Engraver J. Griffith used hatching and stippling to represent areas of skin sensitivity. Henry Head found that cutaneous sensation was distributed asymmetrically on the body, seen in the distribution of "maximum spots." From Head, "On Disturbances of Sensation with Especial Reference to the Pain of Visceral Disease," 1893, 16

history. Early modern natural philosophers described nerves as structures that defied anatomy. They were objects that registered the mind and the soul. On paper, they imitated anatomical structures like blood vessels and bones, which were easier to see. Illustrators often copied anatomical images when they had not performed dissections of their own—as if dissections on a rotting corpse could offer a clear vision of fluid bodies. Whether living or dead, bodies and their extensive network of nerves were too fine to figure out.[43]

Furthermore, the images that Zhu Bing compared were the copies of copies that had warped in the hands of the copier. Despite the enduring names of original authors and their original texts, the dozens of missionaries, translators, scribes, illustrators, engravers, practitioners, printers, and editors contributed their own flourishes. Images shrank. Text became garbled. Details became embellished. Images of meridian men and sensation men were linguistically specific, temporally specific, and cosmologically specific. They were products of a particular time and place that entered new contexts, new times, and new places. They acted as the mediators of knowledge, what Susan Leigh Star and James R. Griesemer have famously called "boundary objects."

The boundaries that defined bodies as objects, sites, and relics made knowing them hard work.[44] Images moved and changed. They could have absorbed new modifications in the style of Bruno Latour's "immutable mobiles." They lived in the creolized space of Peter Galison's "trading zones," where ideas spoken by new tongues painted new pictures. As Projit Mukharji has observed, medical multiplicity was full of friction and aspiration. Images of bodies braided knowledge within, beyond, and across national boundaries.[45] Similarities made to diminish difference were politically constituted, even when social and cultural circumstances steadily fostered further multiplicity.

Theorizing Body Maps

Although the images of bodies that appear here might take the body for granted, attempts to locate its exact materiality and function have rendered the body—or "a" body as Caroline Bynum reminds us—less known and less obvious.[46] Aparecida Vilaça has described bodies as "chronically unstable" across forms, relations, natures, and appearances.[47] For instance, meridian men existed within a cosmology of bodies that did not look like human bodies and sat within a broad spectrum of material things like valleys, hills, eyes, ears, trigrams, streams.[48] These references coexisted in an ongoing state of transformation.[49]

Body maps referenced bodies on the page and bodies in the world.[50] Some body maps were of no one in particular. Other maps showed individual people, characters, or composites. Idealized models served as templates that defined injection sites; highlighted joints vulnerable to injury; marked sections of flesh to cut, peel, and press; and articulated symptoms when symptoms manifested pathology. They allowed experts to perform their expertise—to explain that the problem was *here*. They allowed patients to register their experiences—that they felt discomfort *there*. Like maps, bodies were physical and symbolic artifacts. Like landscapes, bodies were natural and cultural products.

In this book, I describe body maps as maps insofar as they mimicked cartographic virtues of objectification and objectivity. Maps as a graphic genre were self-consciously un-objective, however.[51] As cartographer Mark Mononier put it, maps lie.[52] They lie because they are made to persuade. They have served as objects of political authority and epistemic superiority. They naturalized familiar objects and froze them from constant motion—shifting borders, vanishing lakes, growing cities.[53] They purposely presented too much information for readers to believe they were seeing everything at once.[54] Maps were active constructions. They represented the map maker's own imagination and dreams of accuracy. Maps made scientific images possible. And they were made possible through a process of distortion where illustrators composed, corrected, and compounded inscription upon inscription, line upon line.[55]

Critical cartographers who work with social theory have critiqued the epistemic nature of maps.[56] Scholars like Jeremy Crampton have characterized the power of maps—however limited and contradictory—as an expression of control and as a site for contesting control.[57]

Maps were tools for power. They were instruments of sovereignty.[58] They were political modes of resource allocation.[59] They were sites for both reclaiming control and subverting control.[60]

Maps can be also described as what they were *not*. They were not universal; they were not transcultural; they were not timeless.[61] Readers who expected body maps to represent a real body would find themselves constantly disappointed. These expectations intensified as body maps became medicalized tools meant for performing expertise. Like cartographic maps, maps of bodies followed rules of universal geometry even when bodies in the world and on paper did not.[62] They were expected to be fixed within photography, confined to the virtues of a "mechanical objectivity" in Lorraine Daston's and Peter Galison's words, even when they could not be so.[63] Body maps as images were bound by their material construction, their political circumstances, and the demands of graphic conventions.

Anatomical illustrations as body maps, then, come with the same caveats for using maps as points of reference. They allow us to take seriously images as participating in and allowing for performances of topographic specificity. Maps also allow us to think about scale. Bodies could be any size as long as they implicated the reader.[64] This intensified the ways in which an idiosyncratic body was at odds with the body map. In Maurice Merleau-Ponty's words, encountering a map could become "interiorized" in a way that separated the individual from the subject of perception and the object it perceived.[65] These experiences rendered one exposed and vulnerable.[66] In other words, bodies and embodiments were not isolated but were mutually influencing.[67]

The Importance of *Tu* (圖)

Historically, meridian men were not quite maps. While maps presented an illusion of accuracy, meridian men could not lie. They were made with different expectations. Instead of existing within the symbolic restrictions of absolute accuracy, meridian men operated within flexible rules of approximation.

Many of the images in this book existed within the graphic category of a *tu* (圖). Whereas maps performed with accuracy, *tu* (roughly translated as a "graphic image-text") did not. *Tu* gave way to the imagination. It was a specialist term, a style of representation. It could include maps, although not all maps were *tu*.[68] Some maps could be paintings filled with affective detail, but *tu* were neither paintings (*hua*) nor likenesses (*xiang*).[69] Although *tu* looked schematic and cartoonlike, they were not the kinds of cartoons that Scott McCloud would recognize, however broad McCloud's definition.[70] Instead, *tu* communicated knowledge about music theory, law, ritual, linguistics, ethics, geography, cosmology, architecture, astronomy, mathematics, and more. As Francesca Bray put it, *tu* offered a template for action and a template for enacting objects in the world.[71] It was a functional category that expressed technical knowledge by doing work and provoking work.[72]

Tu were not quite scientific and not quite artistic, although they followed general aesthetic rules.[73] Historically, topographical images in East Asia shared the same aesthetic principles as painting and poetry.[74] Map-like *tu* were made to be studied and enjoyed.[75] They had the ability to elicit a precise physical, emotional, or visceral response. And the physical response was guided through the *tu*. As a template for action, it initiated gesture, practice, and performance. It was already embedded within embodied rituals of looking, doing, and knowing.

All the while, *tu* registered recognizably cartoonish features that I speculate was further rooted in the material practice of artisanal woodblock carving in China, which differed from woodblock carving practices in Japan.[76] Whereas Renaissance printmaking and image reproduction flourished as an expression of artistic virtuosity in the early modern period, woodblock prints remained simple and simplistic in many areas of the Asian empire. This, I argue, was in part due to the graphic genre of *tu*. Print images did not need to be an expression of aesthetic virtuosity. In East Asia, scholar-elites were scholarly; woodblock artisans were not. They did not need to be because *tu* engaged the imagination. *Tu* were already animated by text,

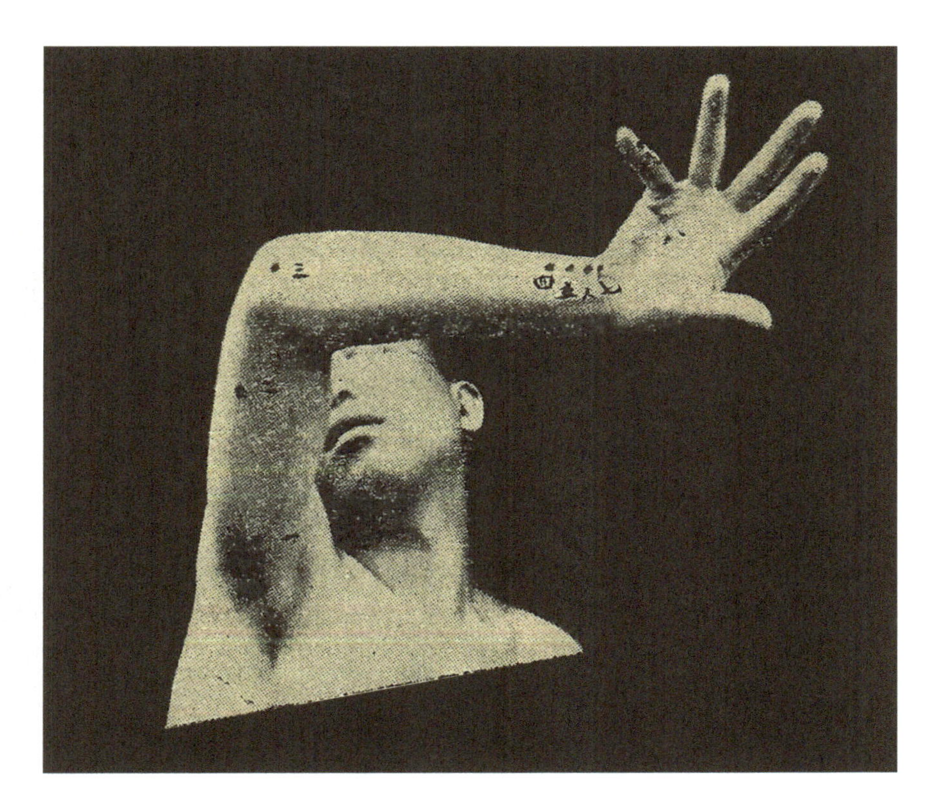

Figure I.4. Photograph with hand-drawn additions, twentieth century. Cheng Dan'an (1899–1957) manually inscribed needling sites onto the human model, took a photograph, then re-inscribed and numbered the points on the photograph. From Cheng, *Zengding zhongguo zhenjiu zhiliao xue*, 1933, 121

which the scholar-elite provided. In other words, the scholar-elite performed expertise through the artisan's *tu*.

More importantly, meridian men lived in *tu*. Even when physicians photographed the body, they still marked meridian sites by hand (**Figure I.4**). They lived in the form of a *tu* with its many dots and lines (recall that *mingtang tu* is a *tu*). Like Shigehisa Kuriyama's expressive bodies, lines indicated expressive structures.[77] Lines—whether bolded, colored, or broken—represented objects that were animate and animated. At the same time, lines were ontologically problematic. They lacked inherent texture and tangible quality.[78] As psychologist

James Gibson has remarked, lines were "ghostly."[79] Lines as symbolic entities were distinct from the things on which they rested and the things that they represented.[80] Although lines and dots, in Tim Ingold's words, participated in the "sway of modernity," they manifested differently in different discourses of what objects could be or should look like.[81]

Lines in *tu* behaved differently. The genre of *tu* did not care for the language of objectivity. They did not correspond to what Hans-Jörg Rheinberger has described as the "economy of the scribble" that characterized biomedical charts.[82] It did not perform the same kind of work as line diagrams measuring heartbeats that marked modern imaginations.[83] Meridian lines inscribed on a *tu* invited specific and broad interpretations. They were not specifically energetic, economic, or anatomical. Yet they were open to those possibilities. They were open to many possibilities because they were always made by hand. The hand-made reality of these images pointed to the ways in which meridians continue to resist digital and mechanical technologies of objectivity and objectification.

Although cartographers have long recognized the complex work of maps, science studies scholars have not yet considered *tu* as a meaningful epistemic object.[84] Indeed, maps and *tu* shared important similarities. Maps captured things in constant motion; they constructed lived spaces.[85] So did *tu*. Maps graphically carved out boundaries that contained a constellation of cosmological meaning.[86] Maps represented the imaginary and the real; they communicated accuracy and approximation.[87] So did *tu*. But whereas maps could lie, *tu* could not. *Tu* intentionally made room for the imagination, whereas maps constrained the imagination. *Tu* could not lie until the image graphically mimicked Renaissance realism. Thus, closely reading *tu* reveals how illustrators composed, corrected, and compounded inscriptions.[88]

Considering images as maps and as *tu* allows me to take on the "provincializing" work that Dipesh Chakrabarty, John Law, Lin Wenyuan, Suman Seth, and others have invoked.[89] I focus on select case studies to offer a new history of body maps.[90] In doing so, this book, again, reveals how physiologists have conflated *what* meridians and nerves were to *where* they were located, further pinning epistemic confusion to the problem of incommensurable graphic genres. Connecting the nature of an object to its location assumed that maps revealed objective truths. Far from objective, maps represented imperfect objects; cartographers have already said as much. It is time to look elsewhere, toward *tu*.

Notes on Genre and Improvisation

When I use the term "genre," I draw directly from literary studies to focus my attention to the relationship among form, action, and ontology.[91] I take after Carolyn Miller's groundbreaking work on genre as social action to Anis Bawarshi's expansion on genre function to consider the provocative responses that images command.[92] Genres operated as ideological stereotypes, as generic types, and as conduits for organizing social action.[93]

In this book, genre manifests in two ways. It appears as a "graphic" genre to articulate the stylistic demands and specific elements of a *tu* and maps. Both *tu* and maps render a particular engagement with the body.[94] They mediate one's experience of the body. Graphic genres operated like literary genres in that they manifested what Wittgenstein has famously called "life forms."[95] Life forms reflected a logic of classification. Generic body maps presented stereotypical forms that registered as didactic anatomical atlases and required one to act. It manifested action. They took on what Heather Bastian has called the "genre effect," or what Charles Bazerman has described as "frames of social action."[96] Body maps as graphic genres were forms of life that expressed and determined ways of being. They guided gaze, touch, and intention.

My second use of "genre" operates on a more specific register to describe specific sets of generic images. This mode of classification focused on genealogies of meridian men, like *mingtang tu*, that were recognizable, stereotyped, and named. In both cases, genre commanded social action. For instance, researchers like Zhu Bing debated the origin and ontology of *jingluo* through *mingtang tu*. The graphic genre of a *tu*, along with recognizable cultural objects like *mingtang tu*, served as powerful framing devices. What would happen in the case of Zhu Bing, for instance, was that the idea of a map became conflated the idea of the thing. The paths on paper became the paths in the body, a projection that made finding the paths impossible.

The graphic genre of meridian men and sensation men furthermore invoked a particular kind of improvisatory subjectivity.[97] Hand-drawn images, like stylized jazz performances, possessed "expression marks."[98] They were subject to intentional and accidental embellishments. They involved in a kind of performance that Judith Butler described as practices "within a scene of constraint."[99] These constraints required images to be legible. In other words, improvisation was not completely free and autonomous activity.[100] It relied on fundamental element of routines, rituals, and activities. Such fundamental elements are particularly relevant in this book because full sets of meridian paths and nervous expression were illustrated by hand. They did not appear as a full set through digital technologies like computed tomography, positron-emission tomography, electroencephalography, X-rays, and more.[101]

To emphasize, I do not use "improvisation" in a colloquial sense of simple invention or fantasy, but as a rigorous mode of representation and knowledge production. As Andrew Goldman has urged, improvisation represents different ways of knowing that facilitate improvisatory behaviors.[102] Improvisation as empiricism and empiricism as improvisation require attention to contextual, historical, material, and individual differences among practitioners, artists, scholars, and writers.

Graphic improvisation through hand-drawn images furthermore connected ethical value and aesthetic value.[103] It joined different forms of artisanal practice and illustration. This further explains why meridian *tu* were highly stylized, numerous, plural, and inconsistent. Many maps over many centuries presented many variations. Some sites that had names were unmarked on the page; in other cases, multiple sites shared the same name.[104] In the twentieth century, the more practitioners tried to clinically define, locate, and localize a single needling site, the more they found clusters of sites rather than an individual point. Hundreds of acupuncture sites erupted into thousands of locations.

Yet, as sites proliferated in the thousands, the number of primary meridians remained relatively stable. There were only ever 14 primary meridians in the *mingtang tu* set.[105] Historically, this created one of the main constraints that shaped the improvisation of, or modes for making knowledge about, meridians and nerves.

Notes on Transliteration and Translation

The transliterations and translations I offer here are consistent with my larger argument on the relationship between graphic images and anatomical ontology. In this book I transliterate common words to serve as points of difference.[106] Some transliterations are meant to emphasize a new analytical category. Other transliterations emphasize historical change. For instance, I transliterate words like *tu* (圖) to remind the reader of the graphic genre in form and name.[107] Doing so brings forth *tu* as an analytical category that enacts the book's theoretical thrust. It directs attention to the idea that meridian paths only ever existed in the form of a *tu*.

Of course, the meaning of *tu* has changed over time. Any image in a textbook—whether a photograph or a painting—is now labeled as a *tu*. In other words, *tu* is the contemporary translation of a "figure." Although I recognize this contemporary use, describing images as *tu* allow me to maintain its theoretical utility.

Other words are transliterated to emphasize them as historical objects. For instance, I do not use the word *guan/kan* (管) interchangeably with the word "vessels." This translation exists, but I resist it. Later in this book, I discuss the history of how Edo physicians came to articulate *guan/kan* as an anatomical object. Before it appeared as "vessels," this *guan/kan* (管) referred to bamboo pipes. They were firm, fibrous organic structures. The graphic word itself even features the "bamboo" radical in the form of *zhu* (⺮) at the top of the character. They did not exist as anatomical objects in early antiquity. From a textual survey, the ancient treatise and early modern reproduction of the *Cold Damages* (*Shanghan lun*) treatise only ever featured the word 管 in reference to pipes, straws, and tubes. Rather than being an ordinary device, *guan/kan* as bamboo pipes further referred to musical instruments used for constructing chromatic scales. These were known objects. In other words, *guan/kan* existed as a kind of empirical instrument.[108] It was a mathematical tool. It was defined by its hollowness.

Transliteration thus serves two different purposes. In the case of *tu*, transliteration reinforces an analytical category. This allows me to highlight the fact that, again, meridians only ever lived in *tu*. The transliteration of *tu* creates distance from its contemporary use of generically labeling images as "figures" alongside text. In contrast, words like *guan/kan* are not given English equivalents to maintain their integrity as historical objects. By presenting *guan/kan* without its English translation, I bring attention to the historical moments where it transformed into "vessels" and, more specifically, "blood vessels." These words may appear foreign to the uninitiated, but they will seem less abstract with consistent exposure.

I have also discussed that I describe illustrations of *jingluo* as "meridians." This is not a translation exactly but narrows my analytical frame to focus on representation while also attending to *jingluo* as corporeal and ephemeral objects. While *jingluo* also manifested in the pulses, they were not etymologically tied to flesh.[109] Individually, the characters *jing* (經) and *luo* (絡) referenced threads, webs, or networks. *Jing* were stable and permanent entities like vertical warps pinned by horizontal wefts. For instance, the idiomatic phrase "*luoyi bu jue*" described a "endless stream" or an "endless thread."

In contrast, I interpret *jingmai* (經脈) as "flesh meridians" in part because the character *mai* (脈) often related to flesh and *xue* (血), a substance that resembled blood. The character *mai* itself featured a flesh radical on the left (月) and a network of paths on the right. This visually represented a corporeal structure in the character. Even the word *mai* has appeared as the "pulses" as just seen in the title of the classical treatise on pulses *Mai Jing*. This book explores the implications of *mai* and flesh in the second chapter.

You will also find that I maintain the historical transliterations and scripts, such as traditional and simplified, used by my actors as objects of study. Whether early modern or twentieth-century Romanizations, Wade-Giles or Pinyin, I preserve the Romanizations to conceptually analyze how historical actors grappled with the nature of illustrated anatomy.[110] Similarly, I do not use the 361 standardized names established by the World Health Organization in their 1991 *Proposed Standard International Acupuncture Nomenclature Report*.[111] Instead, I historically contextualize how the names of acu-moxa sites have changed over time. As you will see in this book, their representation and translation have been closely aligned with discourses and debates on modernizing and standardizing anatomical and physiological structures.

Furthermore, I have translated some book titles into English for the sake of convenience and comprehensibility for readers unfamiliar with classical Chinese medical texts. Given the extensive history of medical books and translations, in cases of compendiums, references to other books, or texts with multiple commentaries, revisions, editions, and name changes, I have opted to use the transliterated names rather than attempt new translations. Other book titles, given that they pertain to areas of scholarly debate, will be addressed in each chapter. For the reader's convenience, I have included the English translations of titles in the footnotes and bibliography where applicable. Readers will also find two glossaries in the back matter. I recommend reviewing glossary A for a summary of the book's key concepts. Others who will use this book for further research can refer to the list of other Sinographic characters in glossary B.

Reading This Book

Historically, meridian men existed unencumbered in the form of a *tu*. At times they still do. But when paired with maps of nerves, they took on new anxieties of accuracy. The ontological tensions that emerged from ideas *about* meridians and nerves and representations *of* meridians and nerves further reveal larger issues in practices of measurement, approximation, variation, accuracy, and difference.

Chapter 1 begins with early modern histories of materiality long before the intersection of meridian *tu* and sensation maps. It compares classical representations of meridians to classical representations of the mind and the brain by discussing similarities and differences between Qi, which animated meridians, and the rational soul, which animated the brain and nerves. Whereas Qi moved beyond the body, the rational soul was constrained within the body. I then examine how these ontological differences manifested within graphic representations and practices of approximation and accuracy.

Chapter 2 follows the translation of meridians and nerves as corpo-real structures in the seventeenth and eighteenth centuries. In this period, physicians redefined and speculated on the content and composition of meridians on paper alongside blood vessels and nerves. Translation invited debate on the nature of anatomical structures, which generated and sustained uncertainty towards fleshy objects, now known as *kei* (經) and *myaku* (脈) in Japanese. This process of transformation resulted in a mixing of identities where metaphors *became* the objects that they were meant to represent.

Chapter 3 explores the technological limits of representing anatomy in the nineteenth and early twentieth centuries. It follows physicians who used photography to map nervous sensations *onto* anatomy, and in turn use nerves to turn meridians *into* anatomy. The discursive and material technologies established a contradiction between what structures looked *like* and what they *did*. Despite the empirical promises of photography, it highlighted the challenges of describing nerves as anatomical objects, and the hazards of using nerves as an ontological framework for anatomizing meridians.

Chapter 4 examines the politics of materiality in the mid-twentieth century, where physiologists began folding their anxieties of localization into meridian *tu*. This chapter interrogates the comparative, conflicting, and diverging conceptions of meridians and nerves that manifested in dialectical-materialist discourses sponsored by the Chinese Communist Party. It argues that despite the unifying aspirations of transnational Communist discourses among Chinese and Russian scientists, dialectical materialism facilitated theoretical plurality in understanding medical therapeutics and its connections with the mind, behavior, and the body.

Chapter 5 considers the relationship between meridians and nerves as they were used to explain each other in during the second half of the twentieth century. This chapter centers on the rising popularity of acupuncture analgesia, or *zhenjiu mazui*, that became popular in the 1970s and undergirded the rise of "gate control" theory. Researchers in Berlin, Germany; London, United Kingdom; Singapore; Michigan, United States; and Colombo, Sri Lanka, speculated on the mechanisms of acupuncture analgesia, amused and bewildered by its effects on chronic and acute pain. Under these circumstances, meridians and neurophysiology encountered a close, almost intimate, ontological proximity, as researchers with a deep interest in acupuncture analgesia remained unsettled by their findings.

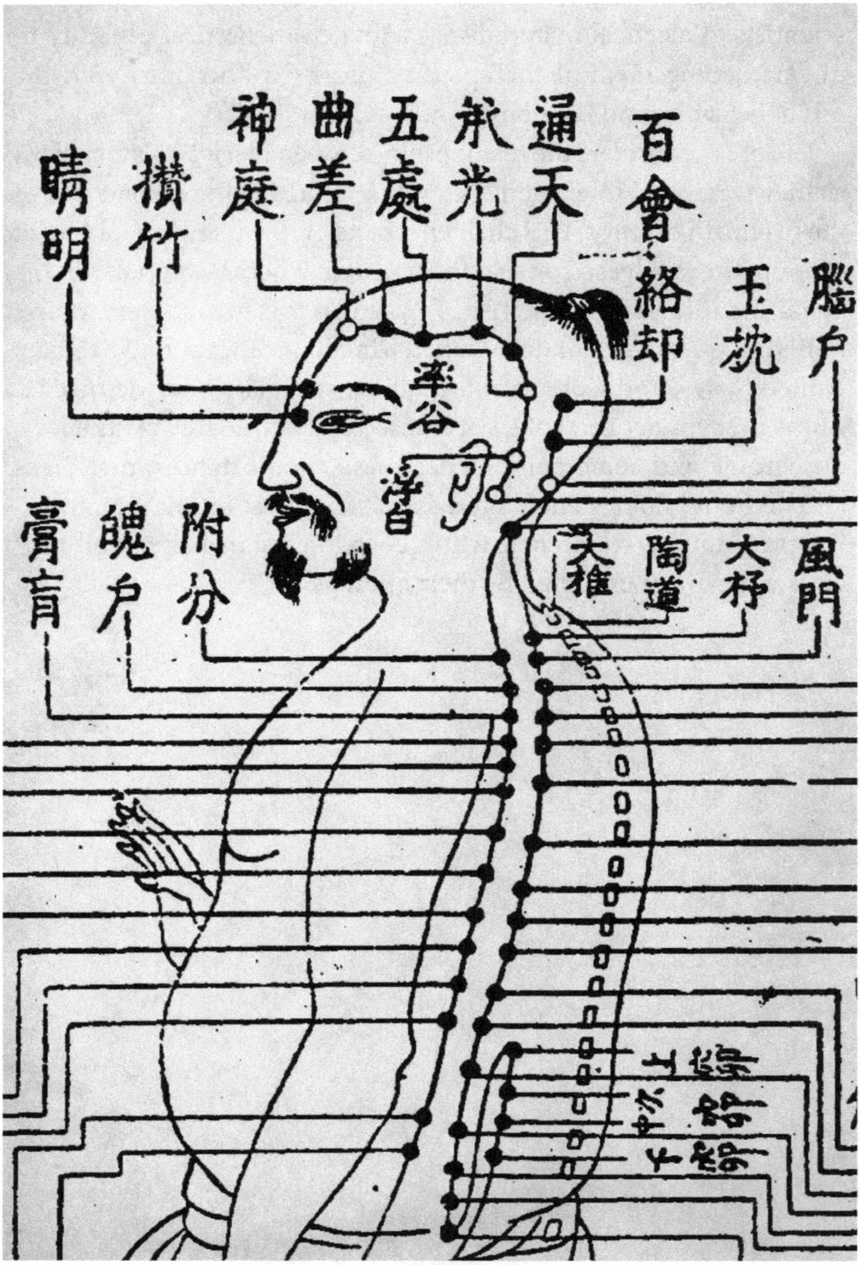

睛明
攢竹
神庭
曲差
五處
承光
通天
百會
絡却
玉枕
腦户
率谷
膏肓
魄户
附分
大椎
陶道
大杼
風門

CHAPTER ONE

Representing Meridians and the Mind

Meeting Manchu Man

He faced left (**Figure 1.1**). Dots connected on the side of his head, curving with his skull and splitting at his neck. The man appeared unaware of the dotted constellation gathering between his eyes. His image represented a single meridian path called the major-Yang-foot-bladder-meridian, or the *zu tai yang pangguang jing* (足太陽膀胱經). The major-Yang-foot-bladder-meridian always appeared on a man facing left.

Published in 1956, this bladder meridian man was a modern version of an old trope. His form invoked a series of meridian men that dated to the fourteenth century, attributed to the scholar-physician Hua Shou (1304–1386). But his face suggested that he had originated during the Qing dynasty (1644–1912). This version of the bladder me-

Figure 1.1. (opposite) Partial Manchu man (detail), Qing dynasty (1644–1912). Unknown illustrator, woodblock print. This detail of the bladder meridian shows the path known as "major-Yang-foot-bladder-meridian," or *zu tai yang pangguang jing* (足太陽膀胱經). The figure's Manchu hair and dress reflect styles from the Qing dynasty, even though this print referred to an older image. From Hua, *Guben jiaozhu, Shisi jing fahui*, 1969, 39

ridian man most likely appeared during the reign of the Manchu empire that initiated "modern" China.[1] His mustache and patterned goatee did not look ethnically Han. He was not a typical Confucian scholar. The shaved head and mini-queue likely reflected Manchu male fashion around the eighteenth century, a marked shift from earlier meridian men who sported full heads of hair.[2] He was a modern man.

The Manchu bladder man joined a long lineage of meridian men who displayed different versions of Chinese masculinities and ethnicities. Some had chest hair, many did not.[3] Some carried Daoist, Confucian, and Buddhist ritual objects and talismans like flywhisks or feathered fans, resembling adolescent gods, old sages, scholar-officials, or monks who lounged, conversed, and contemplated from afar.[4] New political regimes introduced new faces and new bodies. As the bodies changed, the paths changed with them. The bladder man transformed from a Daoist sage to a Confucian scholar to a Manchu official. He typically displayed one continuous path that began at the head, moved down the back, and ended in the big toe (**Figure 1.2**). Classical images were in constant reconfiguration as dots changed position on a smaller face or a larger belly. Over time, the path fractured into two, three, or four segments. Some versions abruptly diverted the path on the lower back while others continued the flow uninterrupted.

Despite these changes, meridian men remained consistent as a visual genre, reinforced by numerous authors, illustrators, and editors. For instance, within the genre of 14 meridians, the images often showed 14 individual meridian men.[5] In nearly every reproduction of the series, these men appeared in the same order—with bladder man always in the seventh position. Each meridian connected to an organ and was classified as either Yin or Yang. As a standardized image, bladder Manchu man embodied cultural continuity and change. His form and features reflected shifting political regimes, which manifested in many versions of the bladder meridian across centuries of reproduction. He embodied multiple modernities, if modernity was so divisible.

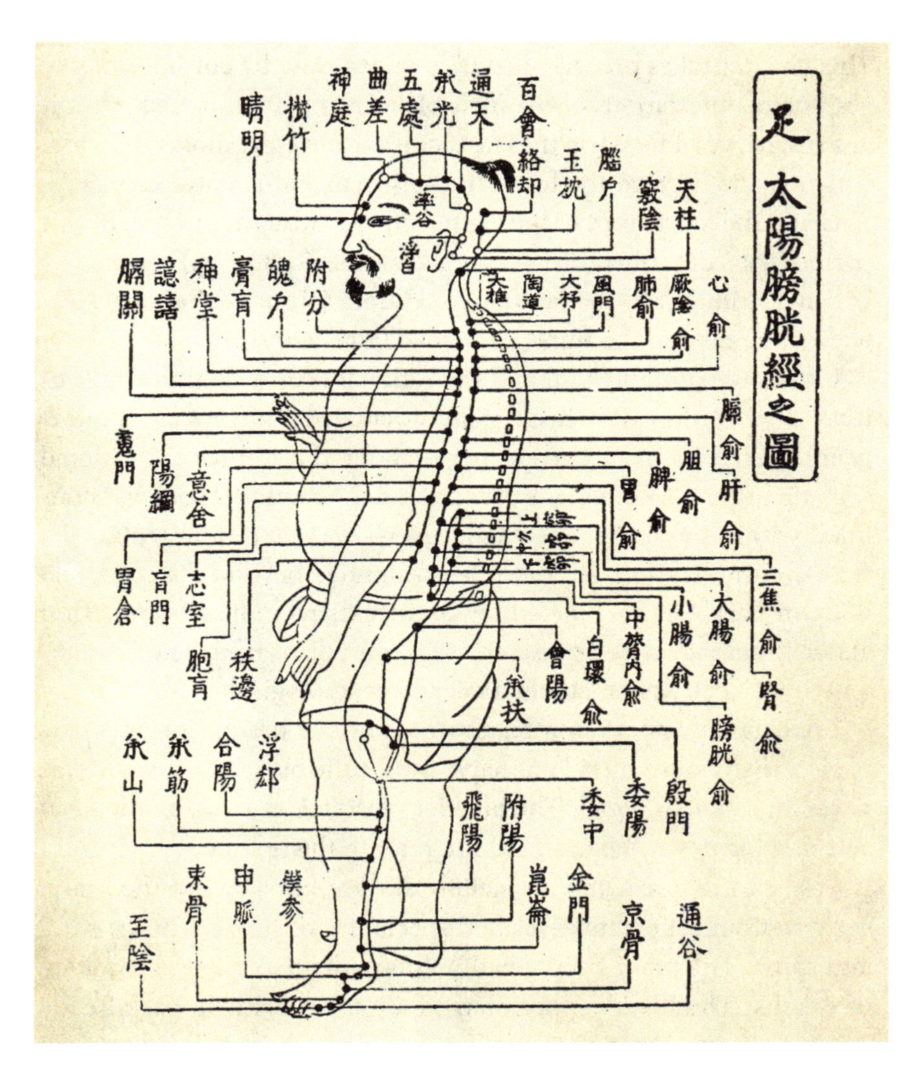

Figure 1.2. Full Manchu man, Qing dynasty (1644–1912). Unknown illustrator, woodblock print. Meridian men largely remained consistent as a visual genre. They established a graphic lineage to which authors, illustrators, and editors understood themselves as contributing. The 14 meridian series always showed 14 individual meridian men in the same order, with bladder man appearing seventh. From Hua, *Guben jiaozhu, Shisi jing fahui*, 1969, 39

Introduction

This chapter takes on early histories of anatomy by considering the graphic manifestations of the invisible agents that animated meridians and nerves. I focus on the late medieval and early modern period, which spanned from the tenth to the seventeenth centuries in East Asia and the fourteenth to the eighteenth centuries in Europe. In this period, texts described meridians as conduits for Qi (氣) and nerves as being animated by the rational soul.[6] Both Qi and the rational soul, like meridians and the mind, remained largely unseen.

I do not assume that Qi and the rational soul were universal objects.[7] Rather, they were historically situated and historically contingent. In early modern Europe, nerves offered a seat for the rational soul, defining the human body and mind. The rational soul was confined within the brain, heart, and liver—it existed only within the human form. In contrast, meridians extended beyond the corporeal body in East Asia. Animated by Qi, meridians moved both within and without the body. As discrete units, meridians measured distance and time. They were mathematical and cosmological.

Importantly, meridian men never featured a brain. This was true of all early modern meridian maps. Unlike the bladder, the brain was never an essential organ. The bladder meridian starting on the head and moving down the back did not refer to the brain or spinal cord. It was not that pre-Qing physicians did not know about the brain; early anatomical prints described special fluids flowing in the skull and spine. The brain simply had no place in an existing cosmology of organs. The bladder meridian as a Yang meridian was already associated with other Yang organs. The Yang organs already paired with Yin counterparts. They existed within an anatomical cosmology without the brain, a negative ontology of sorts.

Furthermore, placing these invisible agents—whether corporeal, mathematical, or cosmological—into graphic representations required imagination and improvisation. Images articulated Qi through meridians and the mind through nerves. Illustrators drawing and naming tangled networks of nerves in the limbs invented anatomical detail on the page. Similarly, the dots, lines, and landscapes in meridian *tu* (圖) took many configurations and forms that approximated the location of Qi. Measuring meridians involved flexible units, like the adjustable "same-body inch." However, not all products of imagination and imprecision yielded the same results. As I will later show, these practices had divergent effects.

Anatomical images involved interpretation through improvisation. Improvisation grounded speculative illustrations of mutable contours, composures, and configurations. It served as a way of knowing that allowed for practices of accuracy. In other words, the history of anatomical measurement was a history of improvised aesthetics where approximation on the page grounded accuracy on the body. Imprecision and imagination underpinned the corporeal mapping of invisible things like Qi and the rational soul.[8] Approximation tentatively located these agents, conceptually locking them through robust anatomical articulations like Manchu bladder man.

Containing the Soul

Her brain burst from her head (**Figure 1.3**). The individual layers separated neatly from each other, filling the skull like an oversized hat. Her long hair hung from her scalp. She was weeping. Thick swirls of breath curled from her mouth, and she parted her lips where spit turned into air. Like a grotesque *écorchée* with a gutted throat, she looked mechanical, her arteries rigid and lungs peeled apart. Her swelling heart perched on her chest.

She stood in the classic *pudica* pose, with one hand covering her genitals and another outstretched. Unlike the standard *pudica*, her arm did not cover her breasts (**Figure 1.4**). For the initiated, she resembled Sandro Botticelli's *Birth of Venus* from the fifteenth century. Her tears evoked a weeping Virgin Mary or a shameful Eve after the Fall. She could have been a goddess, a holy woman, or a renegade.

Printed in 1678 and illustrated by the French physician Amé Bourdon (1638–1706), the weeping *écorchée* Venus was a curious image—not only because she was the only female body to appear in Bourdon's

Figure 1.3. Weeping woman (detail), seventeenth century. Unknown artist, engraving. This detail shows a solitary weeping figure, referencing Christian iconography of the Virgin Mary and Eve's shame. Her tears follow the path of the lacrimal ducts. From Bourdon, *Nouvelles Tables Anatomiques*, 1678, fig. 5

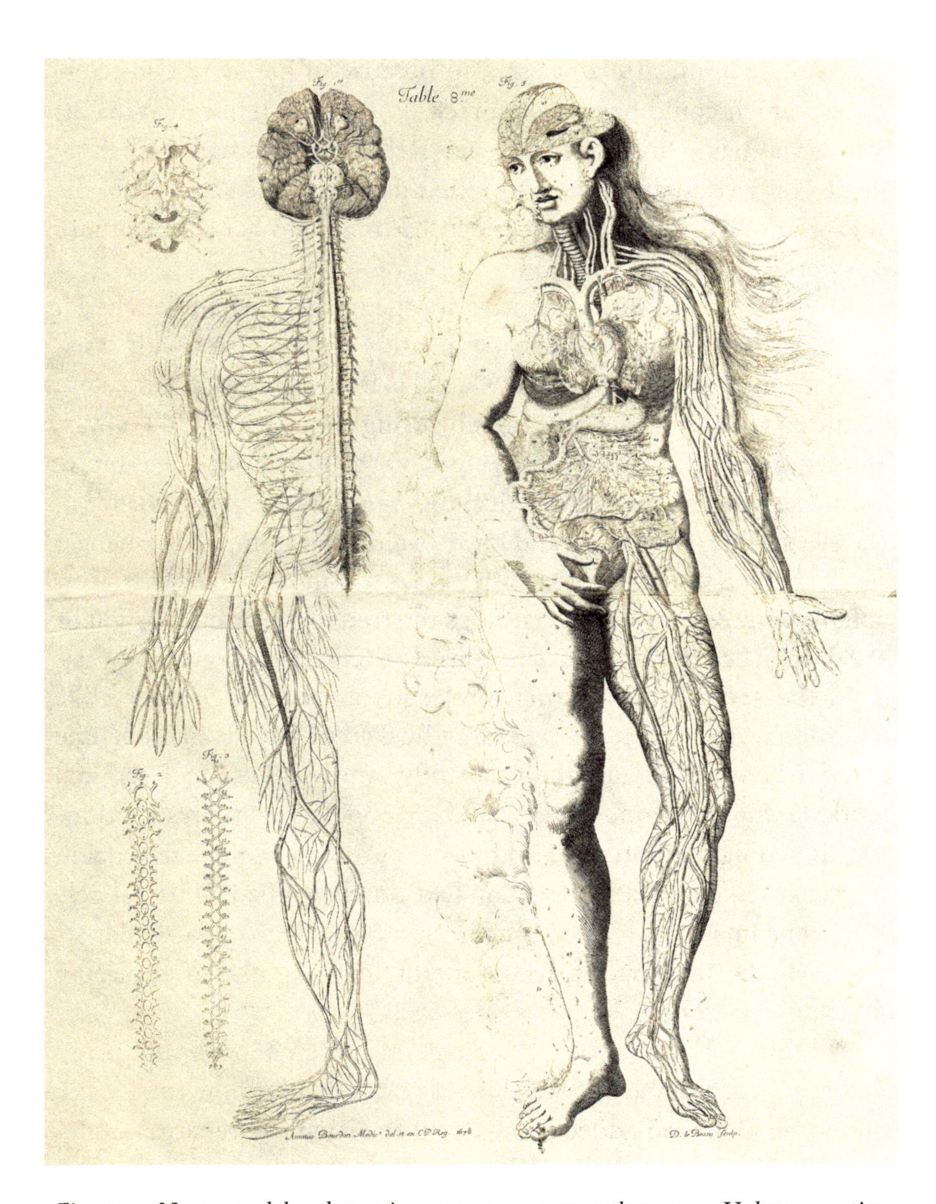

Figure 1.4. Nerve model and weeping woman, seventeenth century. Unknown artist, engraving. *Left*: a nerve model copied from the final plate in Andreas Vesalius's anatomical treatise *De humani corporis fabrica* (1543), showing the arteries and nerves. *Right*: an image of a weeping Venus, the only female body depicted in a series of anatomical plates by Amé Bourdon (1638?–1706). She is shown standing in the classic *pudica* pose, making symbolic references to Greek, Roman, and biblical tropes. From Bourdon, *Nouvelles Tables Anatomiques*, 1678, fig. 5

series of anatomical plates but also because of her symbolic references, which could have been Greek, Roman, or biblical. Whereas Venus was being born from the sea, the weeping woman was nearly dead. Her wet organs calcified as she exhaled. At the same time, she was not quite dead, as vapor condensed back onto her skin like rain or sweat.

Bourdon offered no explanation for the image. He had sketched it himself as a personal study, managing to print his plates with assistance from the French engraver Daniel le Bossu. Bourdon paid close attention to individual body parts by using characters—f, I, t, x, y, z, 15, 14, 56, 40, 34, 35, 54—to label areas on the throat, face, and brain. Doing so suggested an ontological difference between the corner of the eye, bridge of the nose, and lower eyelid. It also implied sameness for branching vessels labeled with the same number, even when they split in four. Perhaps these markings referred to the artery's origin or the diminished importance of its frayed endings. Bourdon did not say.

In the seventeenth century, grotesque anatomical imagery like Bourdon's Venus *écorchée* (meaning husked or flayed) had become standard both as art and medical study. Andreas Vesalius's seminal work, *De humani corporis fabrica* (1543), popularized representations of skinned figures and exposed musculature, captivating subsequent physicians like Bourdon. Bourdon had closely studied *Fabrica*, likely tracing its images to create his own variations. He paid particular attention to the final plate on nerves, modifying it slightly before placing it next to the *écorchée* Venus (**Figure 1.4**).

The image showed another bulging brain, a pair of protruding eyes, and a ribbed spinal cord. Wiry vessels extended into the torso, legs, and arms. Bourdon added more detail to articulate the areas under the brain. He included more labels marking the many parts with further flourishing, finer cross-hatching, and smaller print. He rounded out the eyeballs and added more dimension to individual levels on the spine to make it look more like a ladder and less like a tube. Bourdon's practice of modifying anatomical images, rounding out eyeballs and dissecting a Venus, articulated his vision of the body. He had applied numerous labels to new appendages. He drew the reader's attention to details on the spinal cord, arteries, veins, nerves, bones, eyeballs, skin, nose, skull, and hair. Adding more names would grant the parts more meaning to Bourdon and his contemporaries.

Yet naming body parts did not reveal their function. Form carried no inherent meaning. Historically, anatomical knowledge was inferior to physiology, which involved more than grasping the simple function of parts. Since antiquity, physiology entailed reading the movement of fluid humors—an exercise in interpreting the humoral body. This required rigor in observation, experimentation, and logic.[9] Observation attended to the hot, moist red blood; cold, moist white phlegm; cold, dry black bile; and hot, dry yellow bile.[10] Whereas anatomy categorized structures, physiology read flows and qualities. Parts were generally static, but the humors were made to move. Physiology demanded sustained attention and discernment to track these changes.

The weeping *écorchée* Venus could have been a humoral body, inviting the reader to observe moving fluids. She was surrounded by her own breath, which carried meaning in a humoral sense. Breath related to the soul, mind, and organs, connecting to older concepts of animation. Physicians referred to these invisible objects as *hēgemonikon*, *pneuma*, and *organon*—each term suggesting different things in different contexts. *Hēgemonikon* could denote the governing soul, rational soul, mind, or simply the soul. By late antiquity, it was primarily expressed through the brain, although not equivalent to it.[11] *Pneuma* invoked breath or air but also had technical meanings in philosophy, theology, and medicine.[12] Embodied *pneuma* moved from the heart through the body to the brain, where it turned into something like the soul.[13] The weeping figure's vaporous exhalations could have prompted considerations of these ephemeral, animating essences even though *pneuma* was not quite the soul.

These physiological flows enabled cognition, sensation, and movement.[14] As *hēgemonikon* and *pneuma* inhabited and traversed the body, the brain and heart functioned as *organon*—"instruments" housing the vessels and vesicles.[15] They were the containers that captured moving elements, and the tools that gave objects like *hēgemonikon* and *pneuma* shape and sound. For instance, earlier Greek texts used the lyre metaphor to articulate forms of *organon*. Silent when untouched, the lyre's parts activated when its strings were plucked. Similarly, *hēgemonikon* and *pneuma* did not reside in ventricles but relied on them, invigorating the body into song.[16] Organs had to be expressive to function.

The organ metaphor became the organ itself. As instruments, the material construction of the organs mattered. Like the lyre, an organ's texture and expressiveness related directly to its physical substance. Late antiquity writers like Galen of Pergamon (129–210 CE) who interpreted Platonic and Aristotelian cosmologies, expressed concern for structural materiality. Particularly, organic matter required distinct qualities to fulfill a distinct purpose. "Now the substance of the spinal cord and the nerves is in every respect the same as that of the brain," Galen wrote, "just as that of the casements which contain them is the same as that of the membranes which contain the brain."[17]

Brain stuff and nerve stuff had to be identical. Furthermore, their outer wrappings or "casements" needed to be the same for nerves to categorically differ from arteries and veins.[18] The brain contained no blood; it was just brain. Nerves had no blood; they were just nerves. Similarly, flesh lacked blood; it was just flesh.[19]

Each organ was a distinct instrument made of unique matter, influencing the expressions of *hēgemonikon* and *pneuma*—the governing soul, rational soul, mind, soul, air, breath. Organs were fully formed and discrete. As Galen considered, "It is also most probable that once the parts are constructed, and have achieved their final perfection, they begin to act with the functions belonging to their own particular substances; that the kidneys, for example, have no need of any other organ for their proper functioning—and the same applies to womb, spleen, intestines and in general to every organ of nature."[20] To Galen, organs needed to exist independently from one another. Their "final perfection" established them as individual instruments and parts of an ensemble. Again, the metaphor of the organ—of the lyre—became the organ itself. As independent units, the brain could then monitor the systole and diastole of the heart; organs and organ parts could be assigned names and functions.

Galen's focus on individual parts, perhaps similar to Bourdon's preoccupation with excessive labeling, intensified as philosophers expanded on the function of separate sections. Enumerating the individual nerves of the nervous system—distinct from the heart and liver—enabled the counting of hundreds of minute expressions. Specifically, over 300 integrated muscles enabled nervous and vascular movements, Galen explained. These 300 individual muscles engaged in 1,000 movements, together expressing over 3,000 functions; 400 bones each engaged in 10 functions, together expressing over 4,000 functions; 6,000 nerves and arteries each engaged in 10 functions, together expressing 12,000 functions. These totals, Galen amended, quantified the "purposes" of functioning parts that expressed particular details of the soul.[21] The compulsion to enumerate reflected a view of bodies as assemblages of discrete yet interrelated parts.

Bourdon's weeping Venus presented a curious juxtaposition of form and formlessness. She exhaled something reminiscent of *hēgemonikon* and *pneuma*—yet outside the body, they lost function and form. There, they did not animate hundreds of organs and thousands of functions.[22] Like any bodily fluid, they resembled blood, which took shape within illustrated veins.[23] Of course, they differed from blood, which connected body and soul.[24] Blood demarcated interior and exterior spaces; blood contained the soul within the body's innermost vessels.[25] This is to say that anatomical images depicted mere humoral "casements." To physicians like Bourdon, organs were ideal structures for restricting the movements of the soul. Absent a soul, the brain, spine, heart, and liver lost all interest. They became unremarkable containers. Expressionless organs. The weeping figure's breath hinted at these elusive, animating essences.

Ignoring the Brain

Anatomical illustrations associated with meridian *tu* did not feature the brain. Meridian maps did not need the brain. Already embedded within an articulated medical cosmology, meridians connected 12 primary paths extending from the hands and feet to the 12 organs.[26] These

12 organs were called the *zang* and *fu*, which scholars have described as also called "orbs," "storehouses," or "agents." Specifically, the five *zang* and six *fu* represented hollow and solid organs, including the kidneys, spleen, liver, lungs, and heart.[27] They were not strictly fleshy masses. They functioned as vessels, orbs, storehouses, or agents because they contained the emotions and housed minor gods.[28]

Early modern Chinese texts closely studied the organs, although full-body anatomical images were rare. Some sixth- and seventh-century writers did stress the importance of illustrations, even when internal structures and functions were well-represented through text, metaphor, and symbolic representation.[29] Detailed descriptions of organs in the famous *Huangdi neijing* corpus likely derived from dissections.[30] Meanwhile, *zang* and *fu* were further associated with sounds, colors, and qualities, which reflected deeper teleological orientations.[31] Beyond housing the emotions, individual organs also connected to creatures or "animal spirits" from early Daoist canons. A white lion, vermilion bird, dragon, phoenix, and two-headed deer occupied the Yin organs—lungs, heart, liver, spleen, and kidneys.[32]

One of the earliest printed anatomical compendiums, credited to Yang Jie (1002–1063), appeared during the Song dynasty.[33] Yang's famous *Illustrations of Internal Organs and Circulatory Vessels* compiled 10 images based on the remains of a criminal.[34] The illustrations combined a view of the organs from the front, back, and side.[35] In over half of Yang's images, a curving spine cradled a densely packed gut. Some images prominently featured arterial branches extending from the heart; some displayed a cross-section of the larynx leading to the lungs; some featured a nested diaphragm; others showed a kidney tucked beneath the corrugated bulk of the small intestine.[36]

From reading the text, some organs seemed to be more straightforward than others. "The throat is critical for swallowing," one author later explained.[37] But truly understanding the *zang* and *fu* required understanding their cosmological placements. Organs took on Yin or Yang orientations, reflecting certain climates, directions, seasons, and phases. Lungs were metal, kidneys were water, liver was wood, heart was fire, spleen was earth.[38] These multitudes of meanings demanded experience and discernment to grasp. There was a lot at stake. The containers that housed invisible spirits, Qi, and emotions, were all equally concrete. Organs were more than mere flesh.

Rather than discrete parts, Yang's images represented interdependent systems. His anatomical prints addressed a cosmological framework. For instance, the image featuring the heart did not show the heart (*xin*) on its own but embedded within the membranes of other organs (**Figure 1.5**). The labeled anatomical heart sat at the center of the page, with the lungs, esophagus, liver, spleen, and stomach. These were all Yin organs that anchored Yin meridians. Per the text, what appeared on the page were not discrete entities but "systems," or *xi* (系). Below the heart was the liver *xi*, and below that was the spleen-stomach *xi*. Heart extensions reached up, down, and out, even connecting to Heart Qi (心氣), which unproblematically placed Heart Qi as a corporeal object alongside the organ systems.[39]

So important were the organs that meridian paths were illustrated alongside the *zang* and *fu*. For instance, the *mingtang tu* set often included a fourth image, known as the *neijing*, which could have meant the inner canon or the internal view of the body (**Figure 1.6**). In this image, the body appeared as a legless, armless bulge filled with folded, curved, and stacked viscera.[40]

The brain, again, was not important in this view of anatomy. Although texts described it as the seat of pain, the brain and the spine had no assigned function. Classical texts like the *Huangdi neijing* described the "ocean of bone marrow" as one of the Four Seas, which included the Sea of Blood, the Sea of Qi, and the Sea of Water-Millet.[41] Meridians moved with, not through, these oceans.[42] The brain as a *nao* (脑) did not feature in moving fluids or have a place in the hierarchy of organs. It was not a vessel, agent, or vital storehouse. Nor was it described as a *zang* and *fu*. Instead, what would have been a brain existed as a fluid mass. It may have looked like a brain, but it was not the brain (**Figure 1.7**). The inscription instead instructed, "The ocean of Yin bone marrow penetrates all the way down."[43] In other words, inside the head was an ocean of bone marrow. It flowed and aligned with the Yin organs.[44]

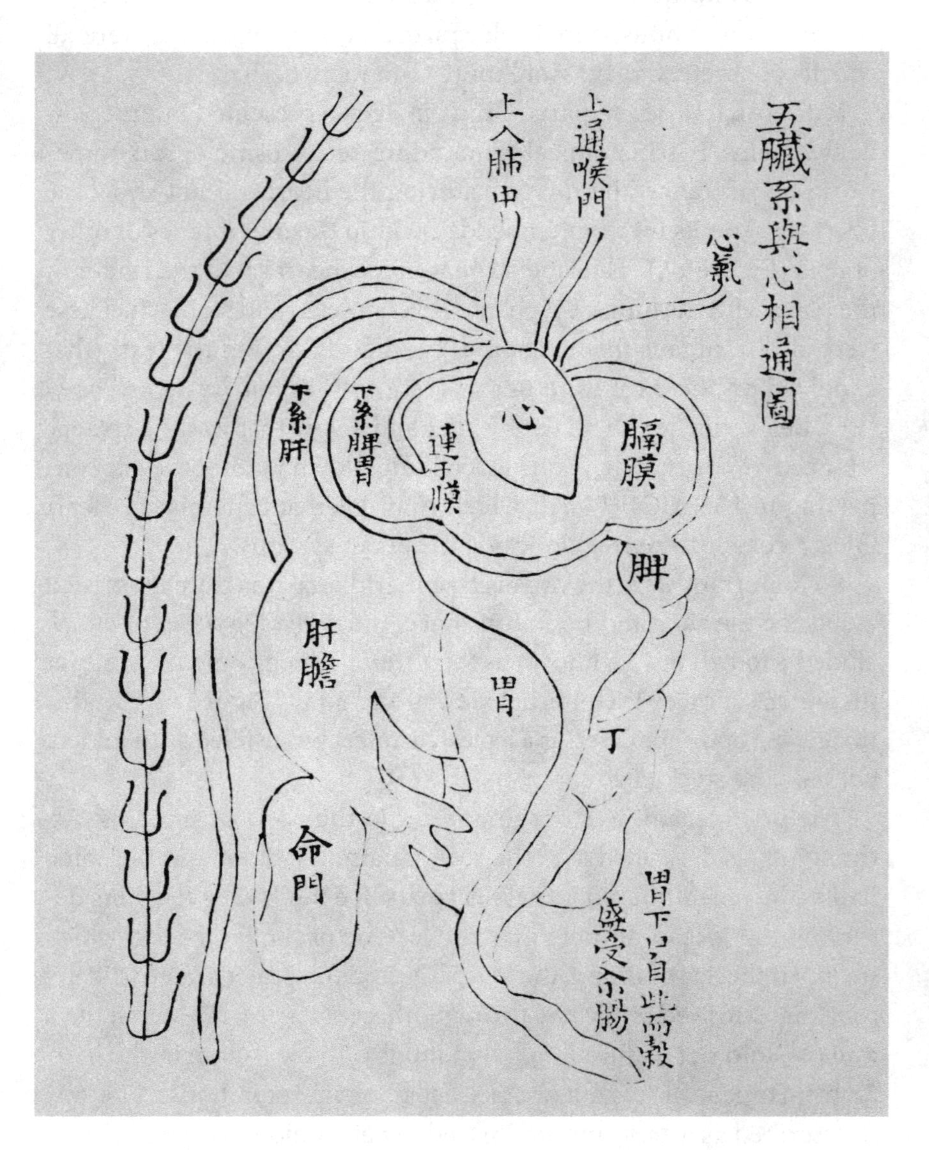

Figure 1.5. Heart with organ systems, thirteenth century. Unknown artist, hand-drawn woodblock reproduction. Images attributed to Yang Jie (1060–1113) show the heart *xin* (心) alongside other Yin organs like the lungs, liver, spleen, and kidneys. From Wang Haogu, *Yi yin tang ye zhong jing guang wei dafa*, 1234, 13

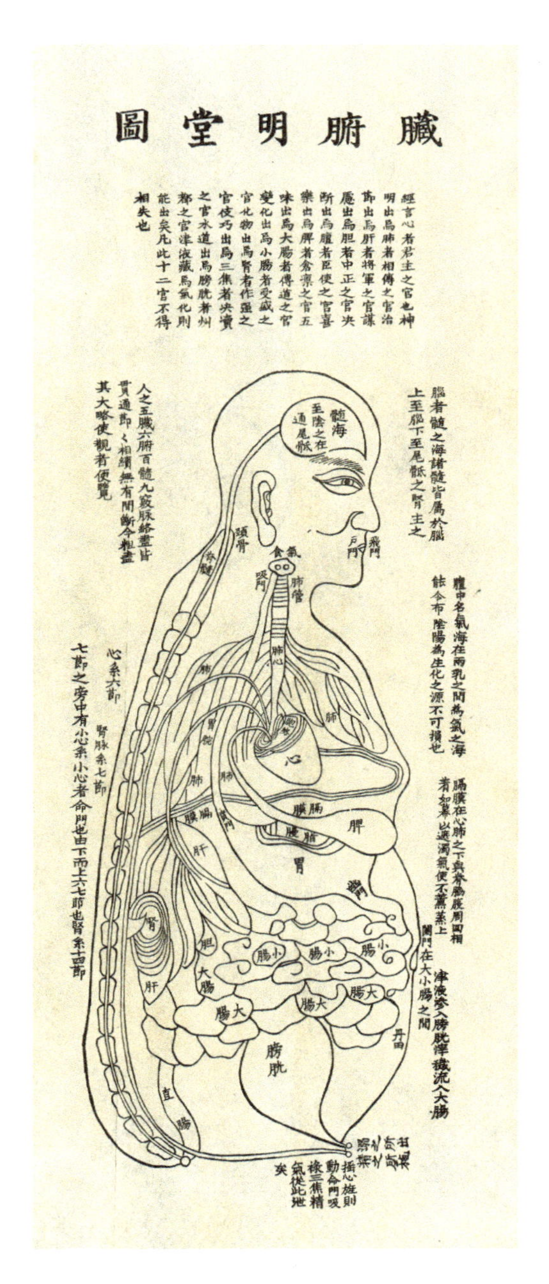

Figure 1.6. Orb man, fourteenth century. Unknown illustrator, woodblock reproduction. This "inner vision," or *neijing tu* (內景圖), diagram shows the *zang* (臟) organs, translated as "viscera" or "orbs." The orb man was part of a classic four-image set. From a reprint of *Zhenjiu jicheng*, 1956

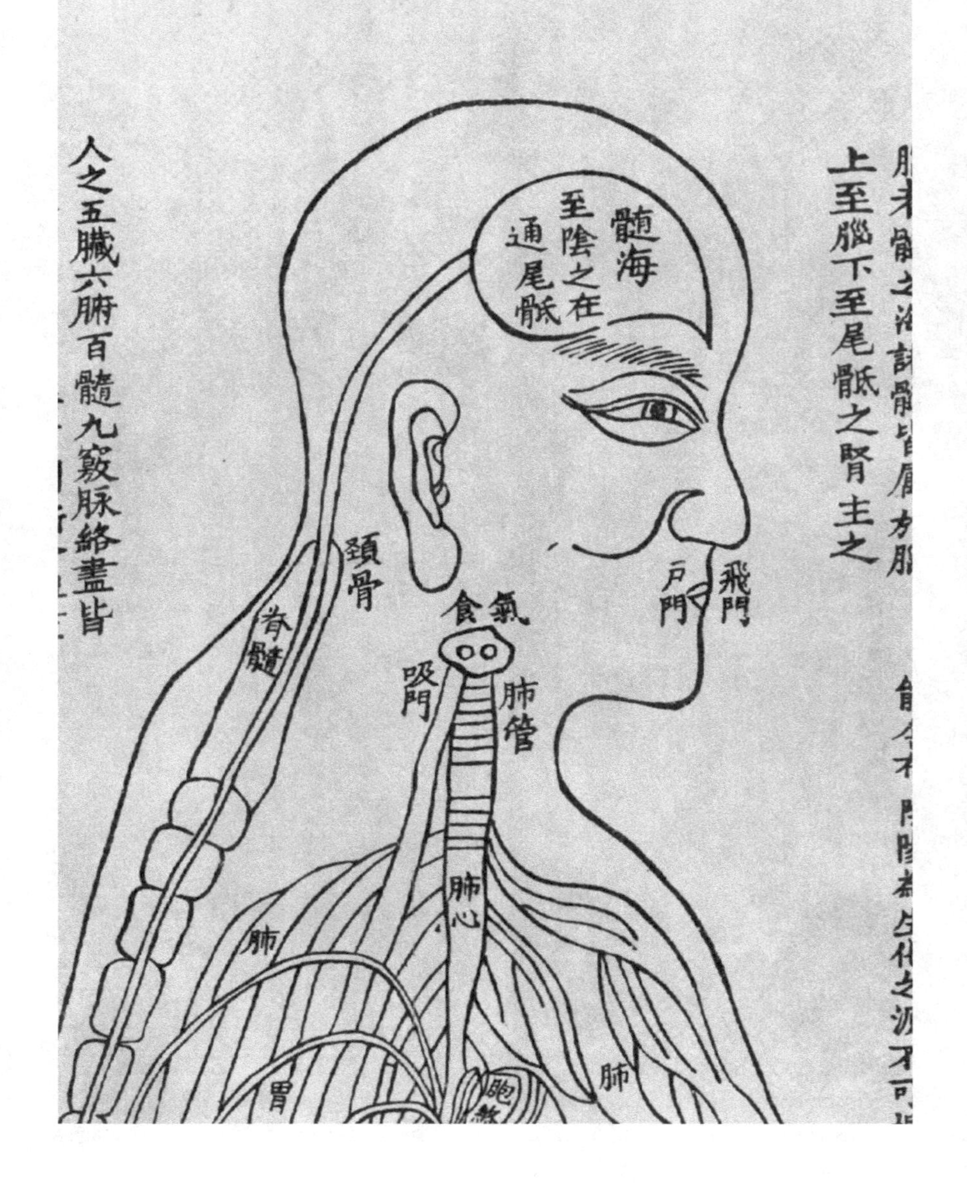

官也神
之官治
之官謀
之官决
之官喜
之官五
道之官
受盛之
作強之
者决瀆
胱者州
氣化則
官不得
上至腦下至尾骶之腎主之
人之五臟六腑百髓九竅脉絡盡皆
髓海
至陰之在
通尾骶
頸骨
脊髓
食
氣
吸門
肺管
戶門
飛門
肺心
肺
肺
胃

Qi Taxonomies

Then, there was Qi (氣). The many types of Qi existed as material things that enmeshed in worldly and other-worldly environments. Historically, Qi appeared as a taxonomy of invisible things operating in distinct spaces.[45] Etymologically, the common character 氣 depicted wind (气) over rice (米)—a material object with material effects. A more esoteric Qi appeared as 炁, nothing (旡) over fire (火) or water (氵).[46] Despite being nothing, 旡 still suggested something fiery or aqueous. Qi was like steam yet not steam. It was both the object and the effects of the object.

Fundamentally, the image of Qi in the characters 炁 and 氣 pointed to a material essence—an epistemic object known through worldly effects. At one point, seventeenth-century philosophers linked Qi to fire, describing it metaphorically as a flame traveling by boat in the body. To this end, the physician Wu Youxing (1580–1660) articulated Qi as "a vessel with oars that moved and transported fire."[47] This boat-flame metaphor gave Qi spatial qualities. It was like fire, but unlike fire, it occupying vessels without traveling in them. Imagining Qi as fire materialized an invisible thing. Conceptualizing it meant it was no longer unseen.[48]

Qi had many manifestations and bore many names. It existed as "Heavenly Qi," "Earthly Qi," "Essential Qi," "Ordering Qi," "True Qi," "Five Qi," "Six Qi," "*Yang* Qi," "*Zong* Qi," "*Yingwei* Qi," and "Four Period Qi," among others. Each of these manifestations had directional and temporal qualities.[49] The "Five Qi," for instance, referenced the five seasons and agents of wood, fire, earth, metal and water. In the body, Qi also proliferated: "*Yuan* Qi" indicated fundamental Qi; "*Zang* (Viscera) Qi" filled the organs; "*Zong* Qi" pooled in the chest. "*Jingluo* Qi" occupied the *jingluo*; "*Ying* Qi" traveled through the *jingluo*; "*Wei* Qi" moved outside the *jingluo*.[50] Qi also divided into Yin

Figure 1.7. (opposite) Ocean of "bone marrow" (detail), fourteenth century. Unknown illustrator, woodblock reproduction. The inscription in the head reads, "The ocean of Yin bone marrow penetrates all the way down" [*sui hai zhi yin zhi zai tong wei di* (髓海至陰之在通尾骶)]. From a reprint of *Zhenjiu jicheng*, 1956

and Yang, which were relational categories rather than ontological categories that classified bodily things like blood, bones, flesh, *jingluo*, organs, and diseases.[51] Like Qi, Yin and Yang linked the body to its environment, situating disease patterns in time, place and season.[52]

Qi represented a taxonomy of objects and effects. It existed within and outside the body, its shifting characters—氣, 炁, and 汽—suggesting wind, flame, and steam. Different historical conditions determined different references to the materiality of Qi. In the twentieth century, sinologists attempted to situate Qi in the brain, like *pneumata*, perhaps to bridge invisible forces across cosmologies. But Qi did not fit in the brain. As its historical categories indicated, Qi moved with *jingluo*, protected them, resided in the chest, inhabited the whole body, and fluctuated with the seasons. Qi embodied a broad, mutable ontology uninhibited by cerebral structures.

Illustrating the organs also involved illustrating different varieties of Qi. Organs that housed the emotions and linked the material effects of Qi to the material effects of the *xin* (心), which poorly translated as a cogitating-heart-and-mind.[53] Both Qi and the *xin* were vulnerable to the emotions. The *xin* could perish under the weight of extreme joy, anger, sadness, fear, and anxiety.[54] Unrelenting joy compromised Qi in the kidneys. Excess anger damaged the Qi in the liver.[55] The organs that localized the emotions also materialized Qi.[56]

If the broad taxonomy of Qi was poetic and confusing, the names of meridian paths were dull and direct. Yin organs gave rise to the Yin meridians (**Figure 1.8**). Yang organs gave rise to the Yang meridians[57] (**Figure 1.9**). Together, six Yin and six Yang meridians constituted 12 primary paths that initiated at the hands or feet. Twelve organs oriented 12 channels, totaling 24 primary organ-based primary meridians. This Yin and Yang orientation of viscera from the hands and feet

Figure 1.8. (opposite) Yin meridian men, Qing dynasty (1644–1912). Unknown illustrator, woodblock print reproduction. Six of the 12 primary meridian men show paths associated with the Yin organs, including the spleen, kidneys, liver, lungs, heart, and pericardium. New political regimes introduced new faces and new bodies. For instance, the bladder man has changed from a Daoist sage to a Confucian scholar to a Manchu official. From Hua, *Guben jiaozhu, Shisi jing fahui*, 1969

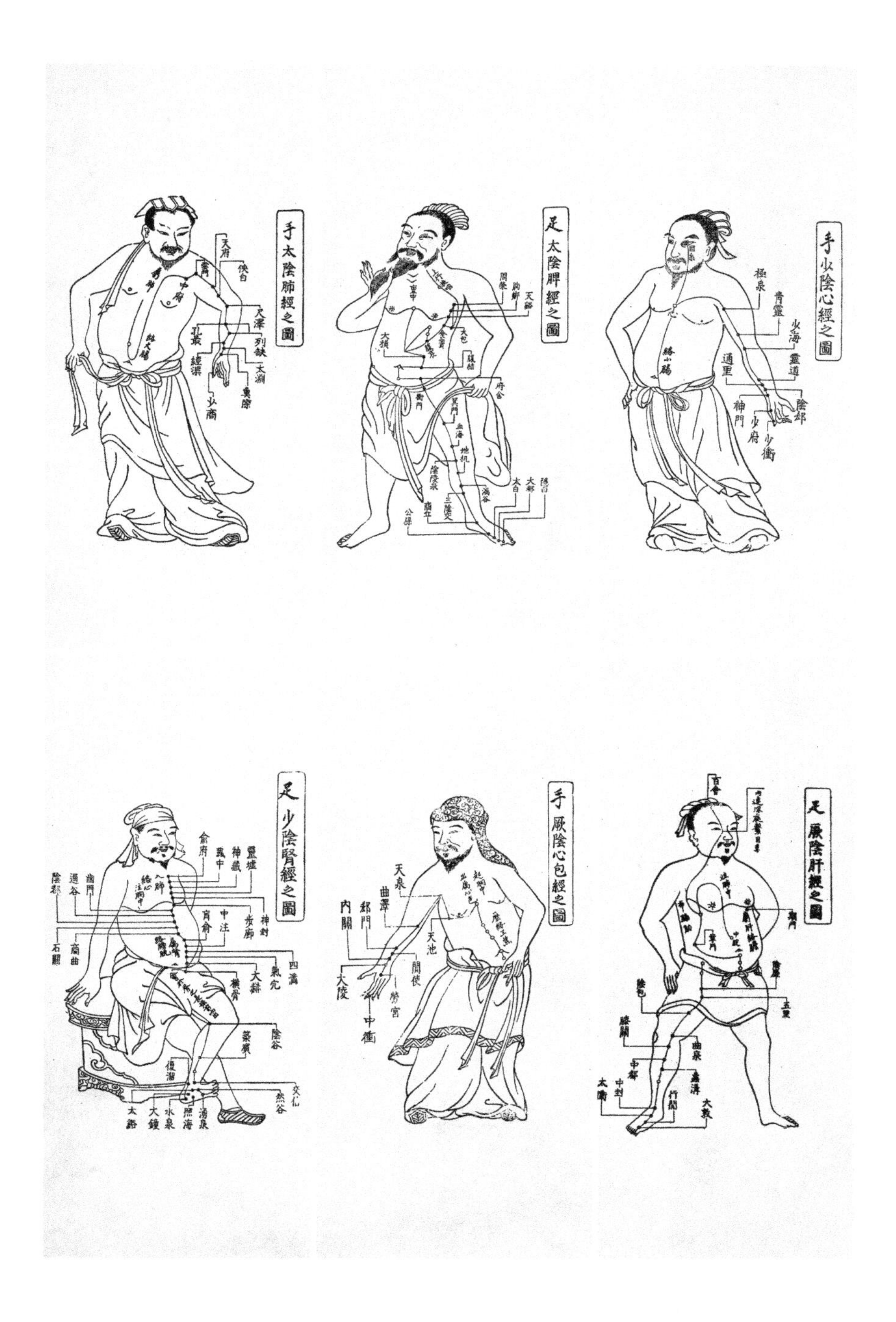
手太陰肺經之圖
足太陰脾經之圖
手少陰心經之圖
足少陰腎經之圖
手厥陰心包經之圖
足厥陰肝經之圖

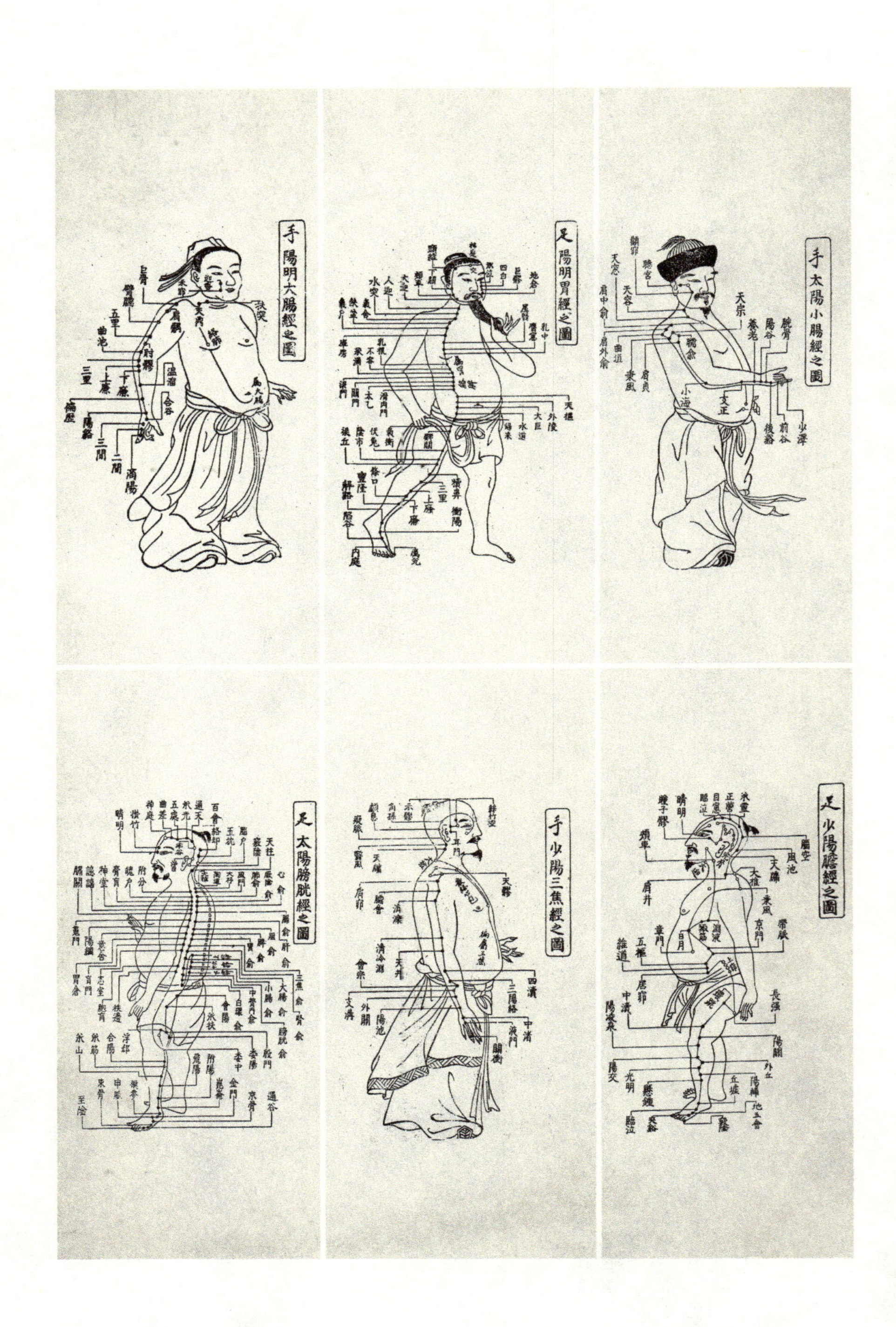
手陽明大腸經之圖
足陽明胃經之圖
手太陽小腸經之圖
足太陽膀胱經之圖
手少陽三焦經之圖
足少陽膽經之圖

determined the names of meridian paths. For instance, the lung meridian was called the hand-major-Yin-lung meridian (*shou taiyin fei jing*). The heart meridian was called the hand-minor-Yin-heart meridian (*shou shao yin xinjing*). The bladder meridian was called the foot-major-Yang-bladder meridian (*zu taiyang pangguang jing*). Names functioned mnemonically, denoting origins in the hands and feet, orientations of Yin and Yang, and the status of being major or minor.

Meridian men were simple images. They showed generic gods, sages, scholars and monks with paths on the chest, arm, and belly. This now-popular genre often dated to the fourteenth-century physician Hua Boren and his influential *Treatise on the Fourteen Meridians* (*Shisi jing fahui*).[58] *Fourteen Meridians* presented straightforward images and text that circulated broadly in Tokugawa Japan as authors continued modifying its representations.[59] New versions gave rise to bladder Manchu man in **Figure 1.1**, which was copied by the twentieth-century physician Cheng Dan'an (1899–1957).

Hua's meridian men and Yang's stacks of organs contributed to distinct graphic lineages, sometimes appearing together, sometimes appearing apart. They each visualized, referenced, and located taxonomies of Qi. Yang's images circulated through copying by figures like the itinerant monk Kajiwara Shōzen (1266–1337), the political advisor Rashīd al-Dīn Ṭabīb (1247–1318), and the diplomatic physician Muhammad ibn Mahmud al-Amuli (1300–1352). So ubiquitous were Yang's images that by the twelfth century, they appeared in texts that did not refer to its original author. They served as graphic templates for the physicians, translators, and monks who moved along the many Silk Routes. In each iteration, illustrators offered their own flourishes.

Figure 1.9. (oppposite) Yang meridian men, Qing dynasty (1644–1912). Unknown illustrator, woodblock print reproduction. Six of the 12 primary meridian men show paths associated with the Yang organs, including the stomach, small intestine, large intestine, bladder, triple burner, and gallbladder. Meridian men display different versions of Chinese masculinities and ethnicities. Some carry Daoist, Confucian, or Buddhist ritual objects and talismans like flywhisks or feathered fans. From Hua, *Guben jiaozhu, Shisi jing fahui*, 1969

Some added a kidney system, more ventricles to Heart Qi (心氣), removed ventricles from Heart Qi, straightened the lines in the gut, or curled the borders lining the organs to resemble clouds.

Organs, meridians, and related Qi referenced a flexible materiality that did not take for granted corporeal states. Illustrators engaged abundantly with lines yet spared anatomical detail, relying on texts to guide the imagination. Readers contemplated these entities through metaphor, occasionally attentive to Yin/Yang orientations and their associated sounds and colors. But whereas images traveled readily, their cosmologies did not.

Cold and Wet Nerves

Nerves existed within a humoral cosmology of color, sound, and temporality. Embedded in the body, nerves registered fluctuations in the four humors, each relating to a season, time of day, and musical mode.[60] Red blood corresponded to spring, mornings, and lydian/hypolydian modes; black bile to autumn, afternoons, and mixolydian/hypomixolydian; yellow bile to summer, midday, and phrygian/hypophrygian; white phlegm to winter, evenings, and dorian/hypodorian. Humors operated on hot/cold and dry/moist axes instead of Yin/Yang and hands/feet.[61] Blood was hot and moist; black bile cold and dry; yellow bile hot and dry; phlegm cold and moist. The four humors linked to organs reminiscent of Yin orientations: blood to the heart, black bile to the liver, yellow bile to the spleen, phlegm to the brain.[62]

Sitting in the brain, phlegm was the wettest and coldest of the humors. Hippocrates had supposedly felt it himself.[63] "Of its coldness, touch is the only criterion," instructed Galen.[64] In wet climates, phlegm descended from the head.[65] Foods like raw apples, pears, cucumbers, oysters, eels, and goat also induced phlegm, being "difficult to break down."[66] Yet these were broad suggestions because humors mixed in the body rather than existing in isolation.[67] Touch affirmed the elemental nature of phlegm; its clammy viscosity was intimate proof of its existence.

Similarities between meridians and humors were significant yet superficial. Their double axes made ontological claims—Yin/Yang were relational, hands/feet physical; hot/cold and dry/moist were qualitative. These axes pointed to what historians of medicine have already expressed: that transformation was a basic characteristic of the fluid body. Fluids harbored knowledge only made legible through motion. For instance, fourteenth- and fifteenth-century German spiritualism saw blood as an unmediated truth, omniscient in criminal cases.[68] Blood not only demarcated interior and exterior spaces, but also signified relationships among organs. For instance, the heart, made of blood, was defined by a natural spirit; the liver, also made of blood, was defined by a vital spirit.[69] Blood likewise delineated the four temperaments—choleric, melancholic, sanguine, and phlegmatic—and served as the source of emotions like anger, sadness, optimism, and laziness. Its evaporation produced heat, dryness, and anger. Blood actively composed and transformed bodies, with its fluctuating materiality encoding a range of meanings.

Then there was the cold, wet brain. As discussed earlier, Galen distinguished the brain from other organs. Unlike the arteries, it contained no blood. The bloodless brain and nerves were thus cold and wet. As cold, wet organs, the brain and nerves became sites where philosophers situated the movements of the rational soul and animal spirits. In the thirteenth and fourteenth centuries, medieval Galenists like Arnold of Villanova (1240–1311) further articulated this materialist view of the brain.[70] Arnold emphasized that physical objects had tangible effects on the body's subtle elements. External qualities altered internal composition. For instance, humidity in the air could weigh down animal spirits. The tangible world imprinted itself on the brain. These notions continued into the early modern era, as philosophers like René Descartes (1596–1650) insisted that invisible forces registered physical causes. Such forces manifested as physical attributes mediated by fluid and solid things. The more refined the fluid, the more complex the phenomenon.

None of these statements were empirically grounded.[71] The mind, rational soul, and animal spirits were too invisible to touch. Although nerves registered fluctuations in the four humors, which were visible and material, they were animated by unseen animal spirits. Visualizing the fine and invisible fluids that animated the nerves required one to wonder. Empiricism was an endeavor of the imagination. Philosophers could only speculate how the world registered on the animal spirits in the sense organs.

Improvising Nerves

Illustrating these structures further required imagination and improvisation. Improvisation served as a way of knowing. Illustrators invented and labeled corporeal structures as ontologically distinct. For instance, physician and natural philosopher Thomas Willis (1621–1675) famously located the rational soul in the brain, attempting to reconcile its contradictory immaterial yet embodied nature. As seen previously, the rational soul was considered sacred, distinguishing humans from animals as God's unique creations, while also operating independently of brain and nerves. Yet Willis insisted this cogitating soul resided in the brain. Historian Alex Wragge-Morley described how Willis arbitrarily placed it above the cerebrum—a non-empirical decision.[72] The cerebrum offered a physical division separating inferior animal spirits below from the rational soul's superior attributes above. In doing so, Willis conceded the rational soul was fundamentally both material (rather than immaterial) and imperceptible.

Willis expressed this in his images (**Figure 1.10**). The rational soul now occupied the finer extensions of the brain—themselves imperceptible yet important to physically situate the newly corporeal, material soul. Their visualization also necessitated imagination and improvisation. In other words, illustration of the nerves with their manifold branches and numerous labels conjured imperceptible objects through

Figure 1.10. (opposite) Intercostal nerves, seventeenth century. Unknown artist, engraving. Thomas Willis (1621–1675) described this image as showing "the origins and ramifications of the same nerves . . . as some of them are found to differ in brute animals compared to man" (*Eorundem Nervorum qui in priori Tabula describuntur origines & ramificationes exhibet, prout illarum aliquae secus in Brutis animalibus ac in Homine reperiuntur*). From Willis, *Cerebri Anatome*, 1663, 435, table 10

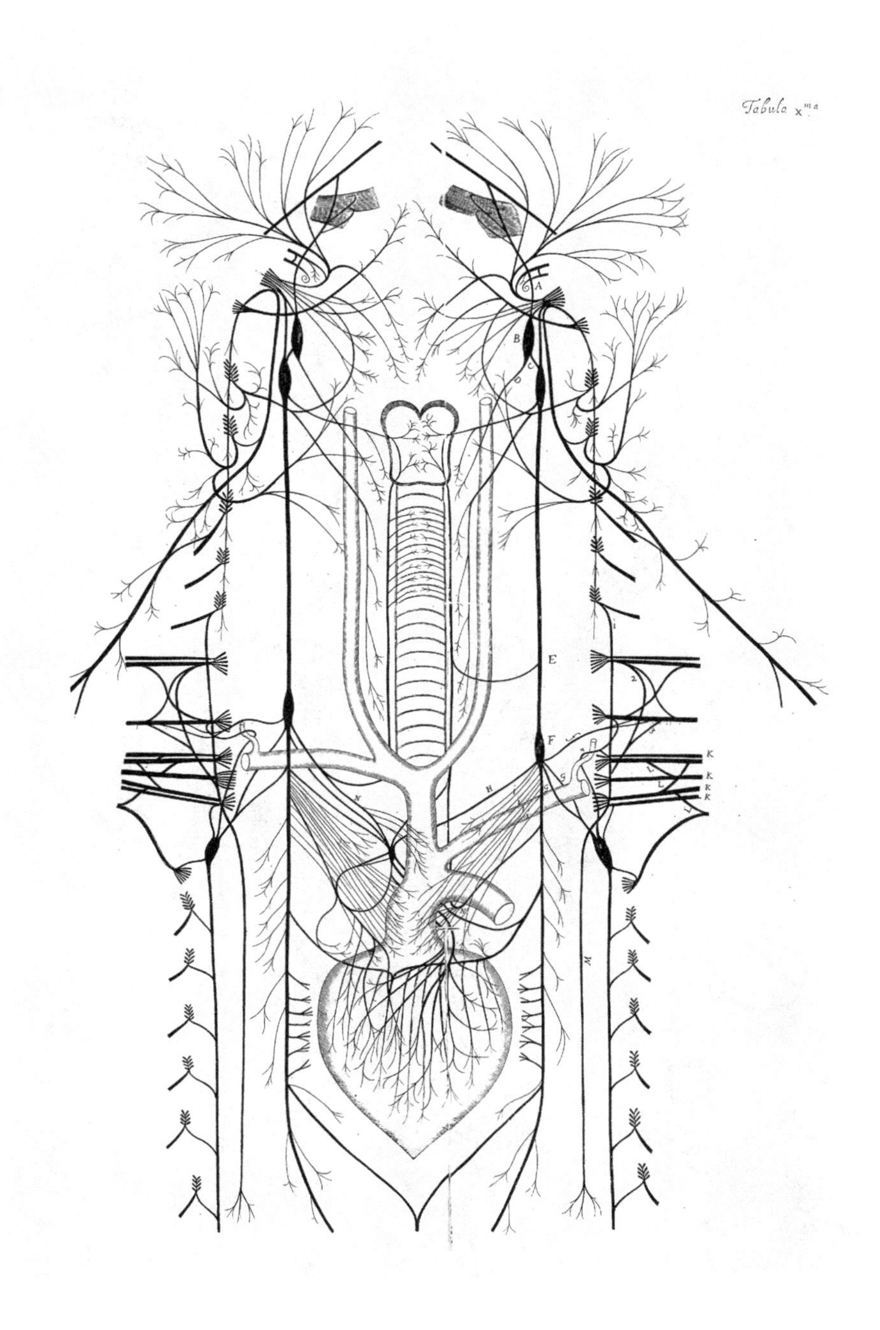
Tabula x^ma
A
B
C
E
F
K
K
KK
L
M
N

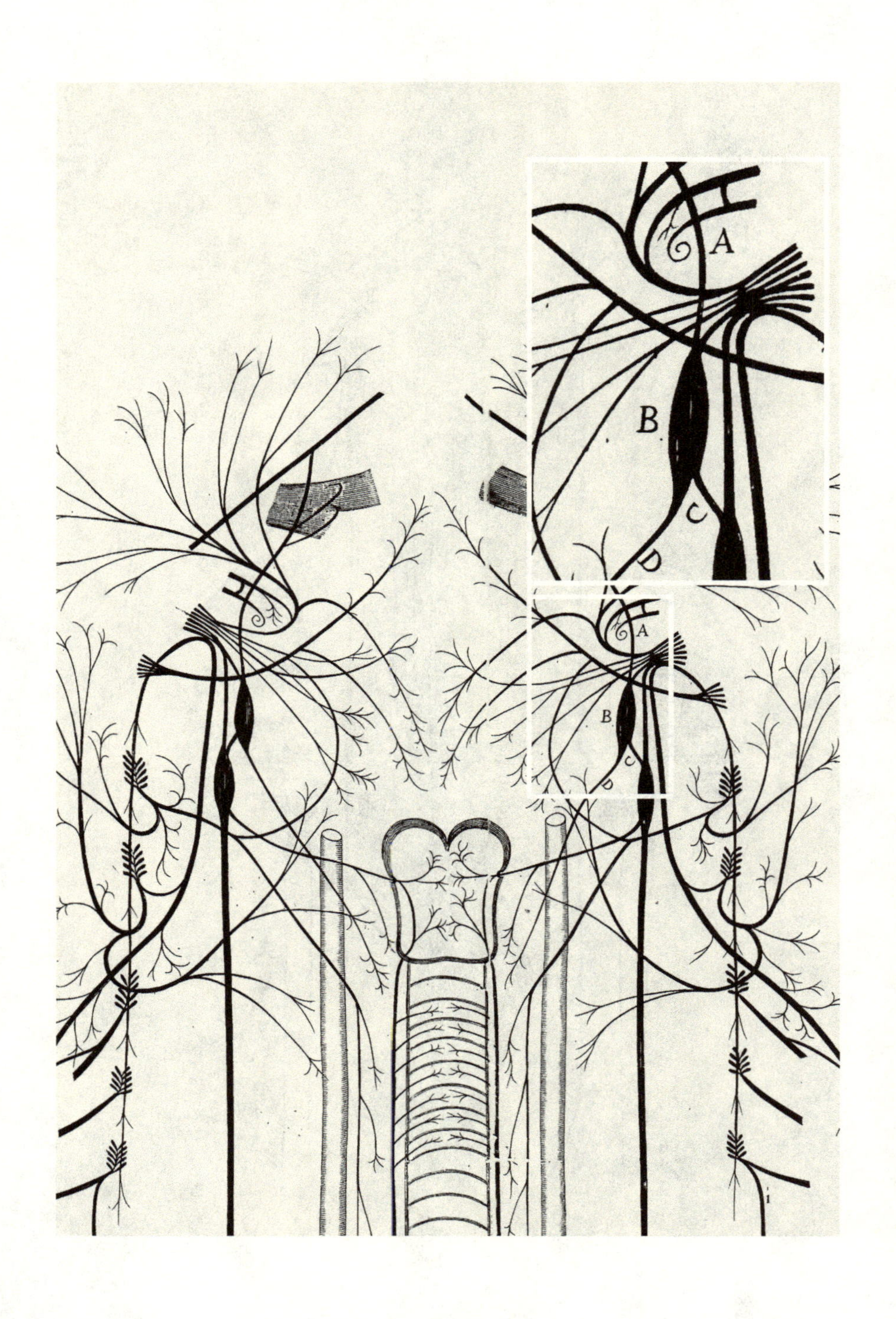
A
B
C
D
A
B
C
D

the simple technology of being made by hand. In Willis's 1663 treatise on nerves, *Anatomy of the Brain* (*Cerebri Anatome*), he offered a vision of the intercostal nerves. This intangible anatomy appeared as tentative fragments.

In the image, twisted lines resembled an algae-like mass. Upon closer inspection, the lines grew more convoluted, abstract, and idealized (**Figure 1.11**). The image seemed to resemble stalks, branches, and twigs. Tiny sticks piled at the center of the page, balanced on a loop that tilted sideways. Branches reached above and below the twig pile, bending to stabilize it. The upper branches impaled a thick bundle of twig, their splinters fanning out further, speared by another chunky stalk. Through organic metaphors, the abstruse anatomy took shape—here was a copse of lines. Willis rendered anatomical uncertainties as arboreal limbs, visualizing imperceptible networks through the metaphor of bark and branch. He transformed the inscrutable nervous system into a botanical fantasy, lines appearing as root and rhizome.

Sparse captions accompanied the image. In one corner, Willis labeled parts A, B, C, and D and defined them thus:

A. The trunk of the intercostal nerve emerging from the cranium.
B. The superior gangliform plexus, excited in the trunk of the intercostal nerve.
C. The intercostal nerve itself emerging from the aforesaid plexus, and poured into the neighboring plexus of the pair of vagus nerves.
D. A branch from the superior plexus into the sphincter of the throat.[73]

Organic descriptions became anatomical objects. Material metaphors like *truncus* for "trunk" and *trunco* for "twig" described the nerves.

Figure 1.11. (opposite) Intercostal nerves (detail), seventeenth century. Unknown artist, engraving. Willis visualized the nerves as living, rooted entities in constant proliferation. *Inset*: improvised rendering of the intercostal nerves labeled A, B, C, D. From Willis, *Cerebri Anatome*, 1663, 435, table 10

Willis depicted nerves as interweaving "twigs" (*surculi*) and "offshoots" (*propagines*), forming gangliform plexuses looking like bulbous tree knots spawning more branches. He traced the origins of each nerve trunk (*truncus*), which split into roots (*radices*) that extended further still. Brachial nerves communicated through a "transverse nervous processes, mutually crossing each other," resembling entangled vines.[74] Later, Willis compared pathways to rivers, where nerves "poured into" (*dimiffus*) neighboring plexuses. It was both water and withe. The image and text each required metaphor to improvise the structures in Willis's mind.

Amé Bourdon (1638–1706), a contemporary of Willis, may have closely studied this image given his own interest in illustrating nerves. Willis's treatise and its engravings may have proved instructive in speculating on the nature of the rational soul and its placement in the brain, which, again, had been arbitrarily designated above the cerebellum. Bourdon even reproduced Willis's image of intercostal nerves in a plate preceding the *écorchée* Venus (**Figure 1.12**). Lines again twisted in an algae-like mass. Tiny twigs again piled at the center of the page, and balanced on a tilted loop. Branches above and below stabilized the pile, bending up and down.

Bourdon's image further elaborated on Willis's. He tethered the intercostal nerves to a cranium. Doing so then suggested that these cold, wet branches potentially attached to a cold, wet brain. He rearranged the letters and added more labels with astrological, musical, and mathematical references (**Figure 1.13**). Upon closer inspection, all the organic parts now had names—smaller branches were marked *1, 2, 3, 4, 5, a, a, b* and larger objects labeled *A, A, A, B*. Below appeared curly feather-like nubs and bands; *C* and *D* curved inward, *ε* and *E* led to *Θ* and *φ*, which multiplied into *ƒ, ƒ, ƒ*, ♂, ♀ marked a threshold, while *A, B, C, D, E, F, H, L, 7*, and *T* cascaded down the plume of twigs and branches.

Figure 1.12. (opposite) Intercostal nerves, seventeenth century. Unknown artist, engraving. Text attributed to Amé Bourdon (1638?–1706) after Thomas Willis. Bourdon copied and modified Willis's depiction of the intercostal nerves. From Bourdon, *Nouvelles Tables Anatomiques*, 1678, figs. 7–8

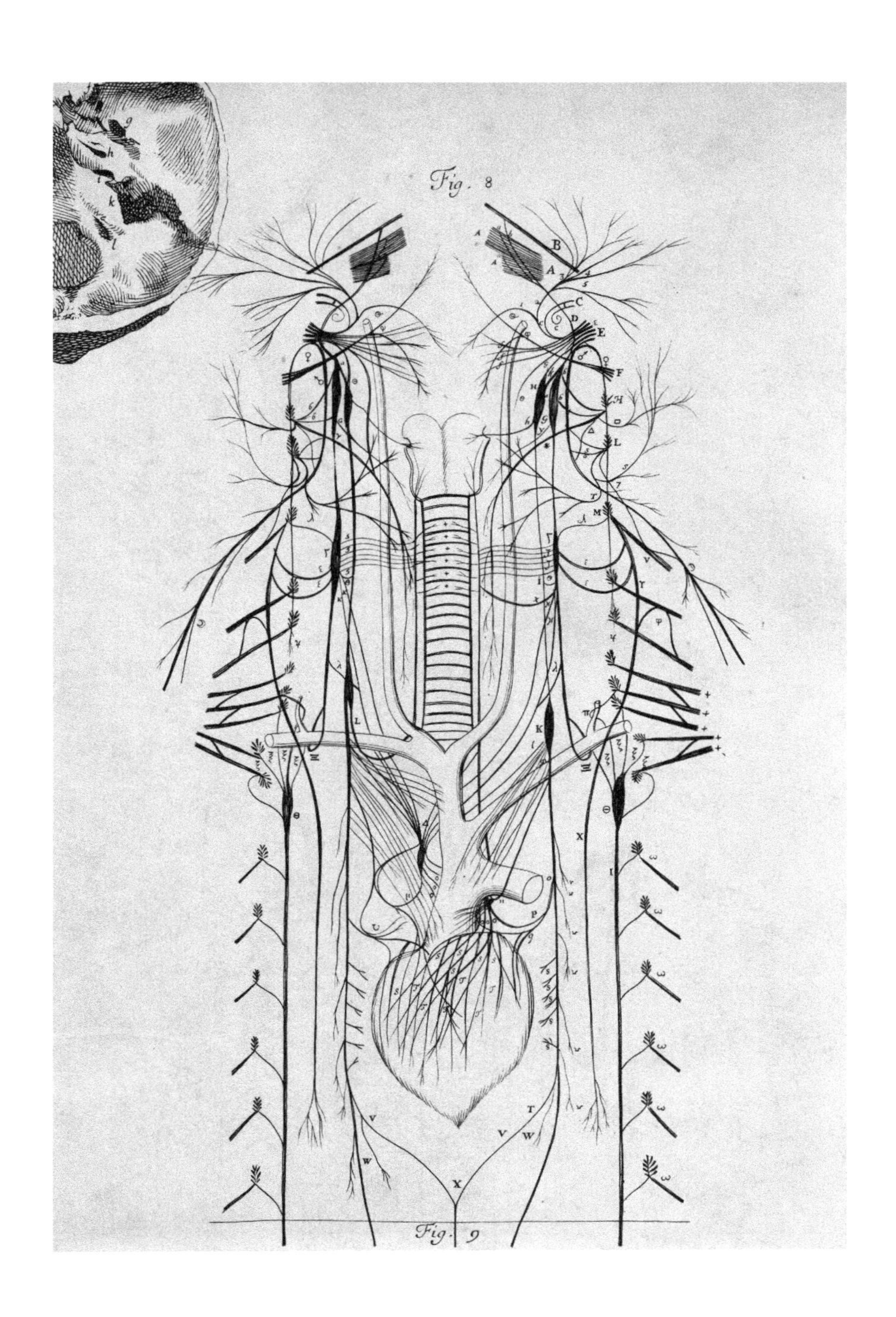
Fig. 8
Fig. 9

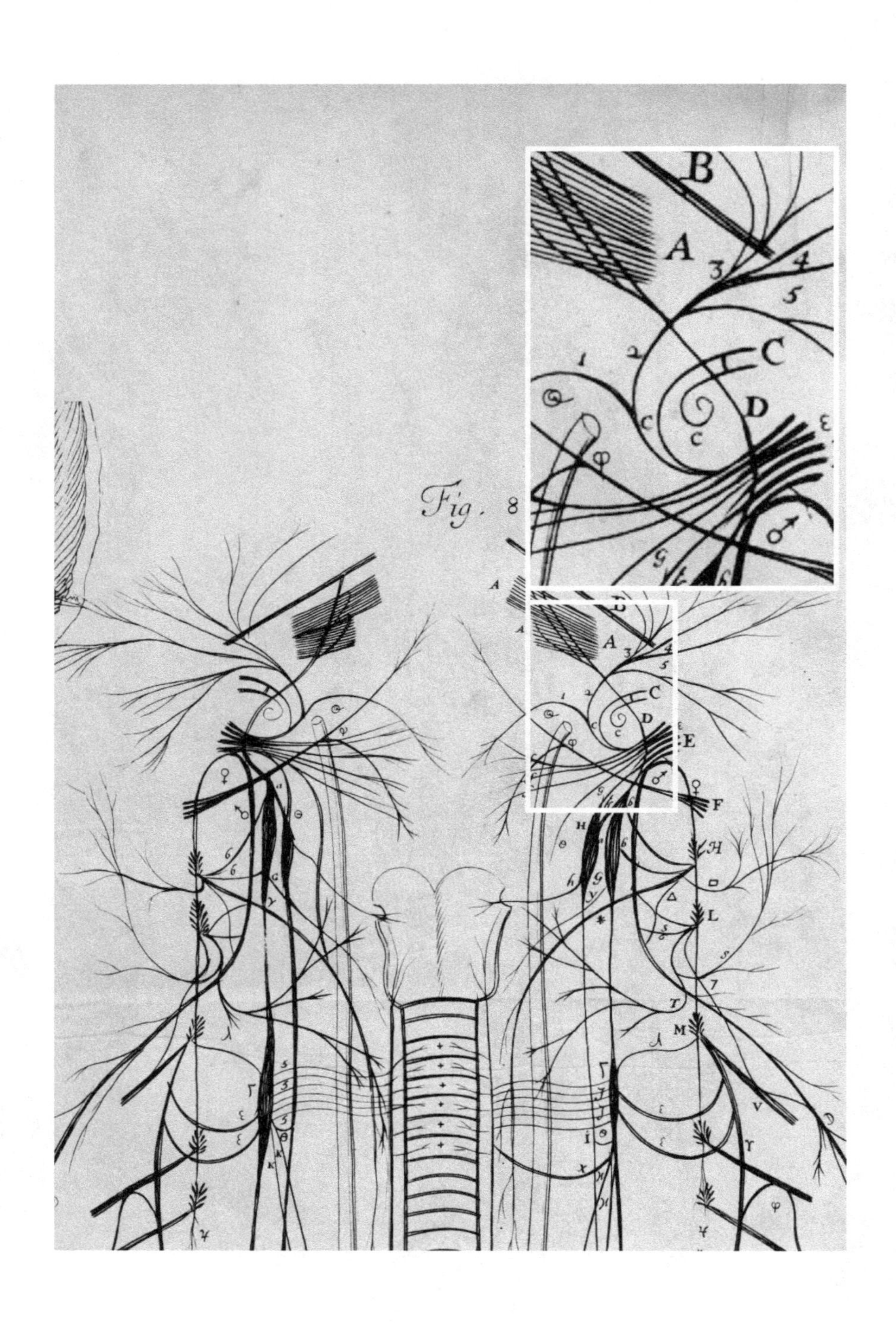
Fig. 8

These labels tracked the non-linear sprawl of nerves and contrasted with other areas that had fewer labels and more tree-like structures. For instance, the spine-like column, divided neatly into segments marked + + + + + +, sat at the center of the page flanked by asymmetrical tubes and branches named 5, 5, 5, Θ, *K*, *K*, and λ on the left and Γ, *J*, *J*, θ, *I*, *X*, and λ on the right (**Figure 1.14**). If these structures had any semantic significance, Bourdon did not say. By adding more labels, Bourdon's graphic improvisation moved beyond Willis's organic metaphors to invoke instead the astrological and mathematical that named the unfathomable.

Images of nerves required embellishment. They were expressed through weeping women, exposed lungs, *écorché* heroes, floating twigs, and proto-batteries. These images sustained the non-empirical vision of a cogitating rational soul that lived above the cerebrum. Other extensions of the brain potentially channeled invisible animal spirits and the imperceptible yet material rational soul. More critically, Willis's illustration and Bourdon's elaboration demonstrated the amount of improvisation necessary for articulating the structural integrity of ephemeral objects seen up close and at a distance. Throughout the early modern period, the changing humoral body relied on images to envision internal fluid movements. At the same time, imagining the more delicate features of the nerves revealed deeper assumptions of what nerves could do in a fluid body that required and rejected the solid parts of the body.

Meridian Landscapes

Meridians also required embellishment. Like Willis's and Bourdon's depiction of nerves, meridian *tu* occasionally eschewed human forms. In some cases, they explicitly resembled landscapes while referencing

Figure 1.13. (opposite) Bourdon's intercostal nerves (detail), seventeenth century. Unknown artist, engraving. Amé Bourdon's elaborations to Willis's original include numerous abstract labels and symbols articulating the nerve branches, with no additional explanation provided. From Bourdon, *Nouvelles Tables Anatomiques*, 1678, figs. 7–8

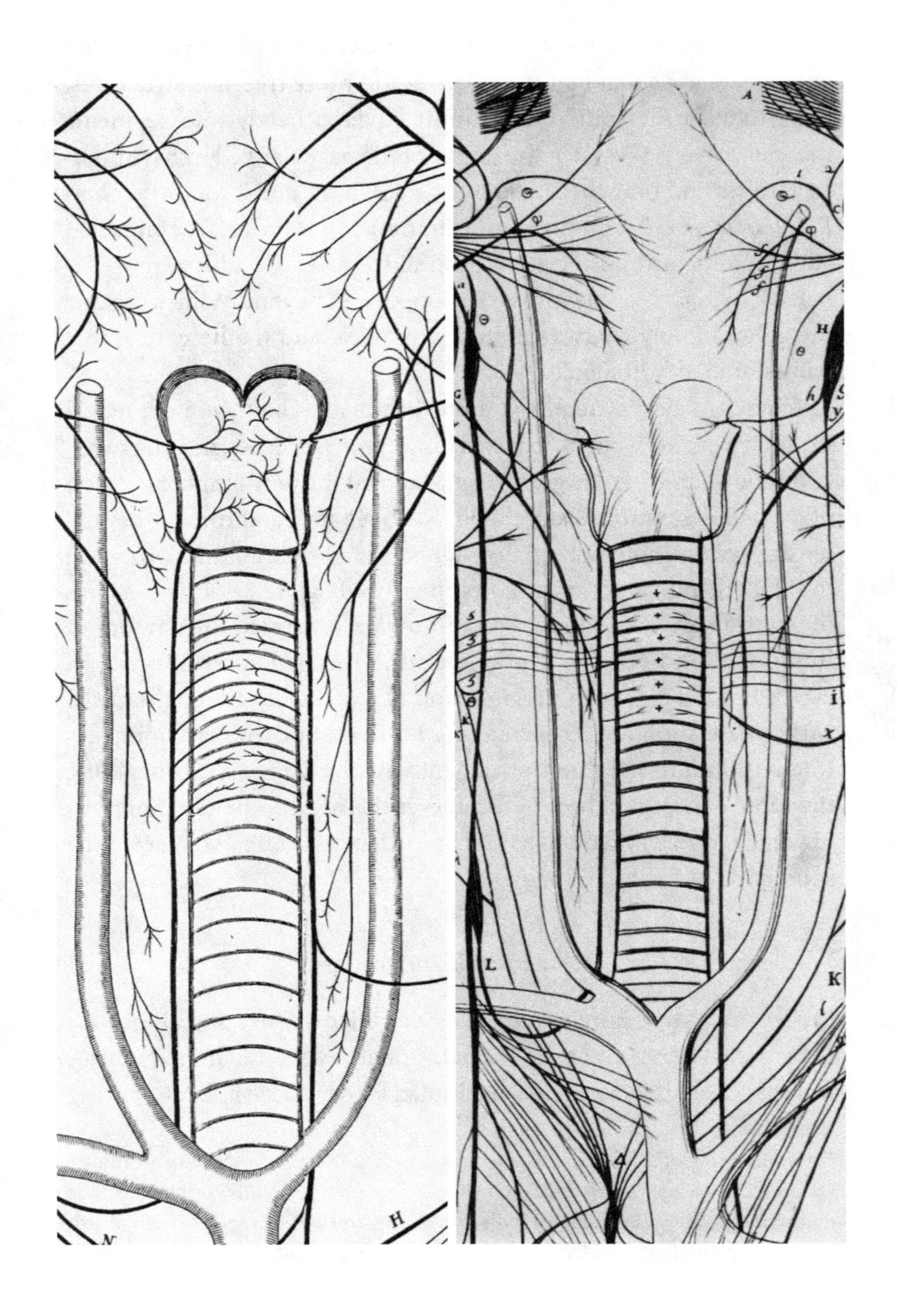

the eyes, ears, organs, hands, and feet. Meridians and their taxonomies of Qi moved within and beyond the physical form. Although abstract, these maps still called on a body, situating eyes and organs within a broader cosmology of things. Early modern scholar-physicians echoed premodern texts and described meridians as having discrete and adaptable measurements. An "inch" or "foot" were flexible units, not fixed lengths. These flexible units meant that as meridians abstracted corporeal features on the page, they did so within a body coherent with itself.

To this end, some images referenced a corporeal human body without looking like one (**Figure 1.15**). Here, meridians appeared as discrete spatial, temporal, and symbolic elements. This was not a famous *tu*; it did not hang in the homes of physicians. It did not flank the doors of entrance halls. Its original illustrator remained anonymous, having contributed this image among many others in a treatise titled *Yi Yin's Grand Method of Decoctions Expanded by Zhongjing*, published in 1234. The book itself—which focused on *materia medica* instead of acupuncture and moxibustion—was authored by a self-taught-scholar-physician named Wang Haogu (1200–1264?).[75] Although historians have offered detailed biographies of Wang Haogu, few have contextualized the ontological implications of this image within the broad scope of meridian *tu*.[76]

Figure 1.14. (opposite) Comparison of nerve branches (detail), seventeenth century. Unknown artist(s), engraving. Bourdon's (*right*) depiction added inscrutable markings, + + + + + + and Greek letters, to Willis's (*left*) original organic forms. From Willis, *Cerebri anatome*, 1663, 435, table 10; Bourdon, *Nouvelles Tables Anatomiques*, 1678, fig. 8

Figure 1.15. (overleaf) Meridian landscape, thirteenth century. Unknown illustrator, woodblock print reproduction. Wang Haogu's image shows six meridians in the hands, with eyes (目) and ears (耳) across the top. The lower left text, *jingluo buju ye* (經絡不拘也), indicates that the paths are not specific meridians. Major organs like the heart (心), stomach (胃), and liver (肝) run down the center of the image, with kidneys (腎) and triple burner (三焦) featured prominently. 會 and 焦 mark intersections. Seven trigrams (☰, ☱, ☲, ☴, ☵, ☶, ☷) are highlighted in boxes. From Wang, *Yi yin tang ye zhong jing guang wei dafa*, 1234, 36

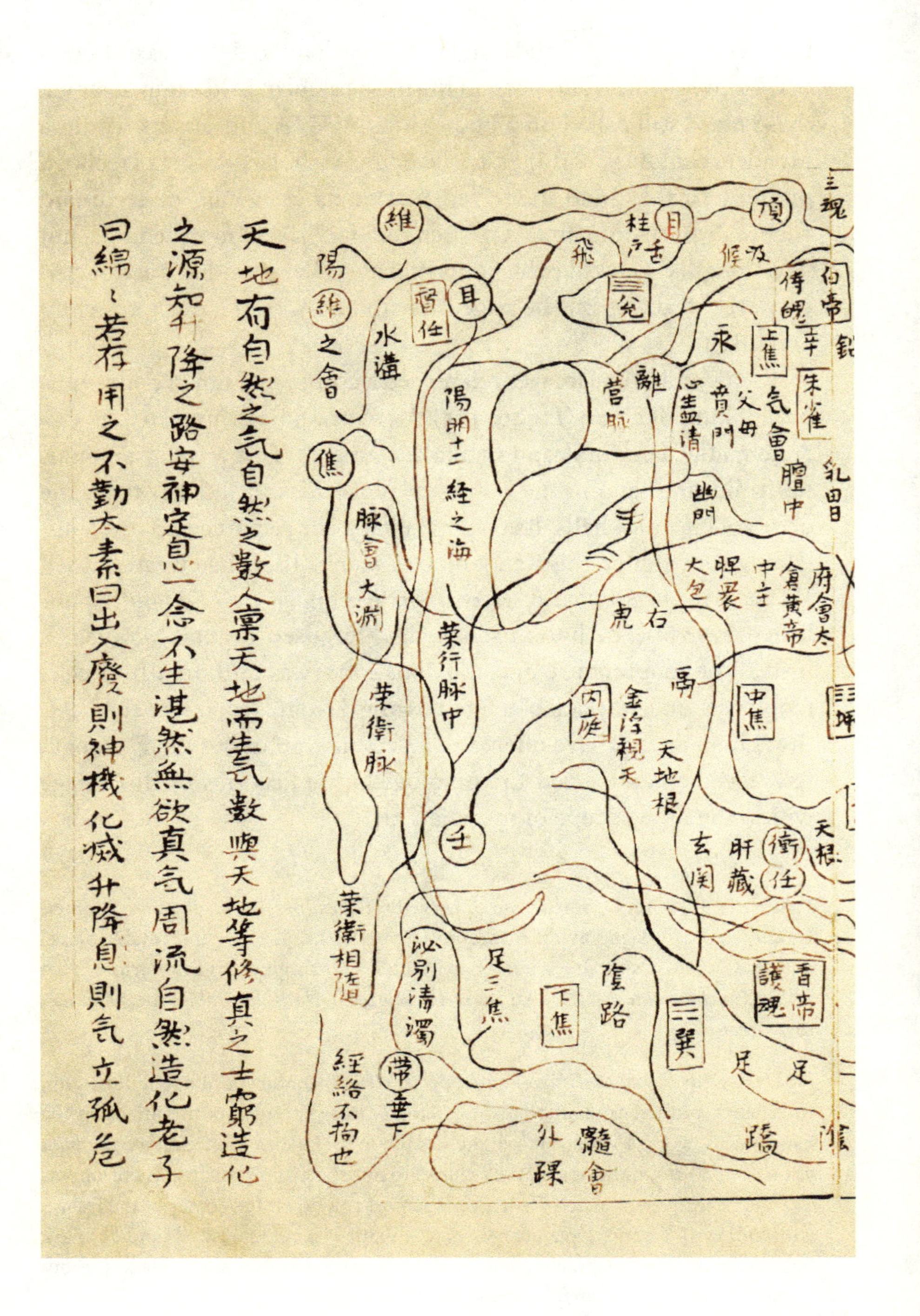

天地有自然之気自然之數人稟天地而生氣數與天地等修真之士窮造化
之源知升降之路安神定息一念不生湛然無欲真気周流自然造化老子
曰綿綿若存用之不勤太素曰出入廢則神機化滅升降息則気立孤危
維
陽維之會
水溝
督任
目
飛
項
三魂
白帝
上焦
朱雀
氣會
膻中
乳田
陽明十二経之海
營脉
離
心盂清
父母
幽門
焦
脉會大淵
榮行脉中
榮衛脉
右
虎
府會太
倉黄帝
中焦
內庭
金液禊天
天地根
玄関
肝藏
衝
任
壬
榮衛相隨
泌別清濁
足三焦
下焦
陰路
三巽
帶
經絡不拘也
外踝
髓會
足
蹻

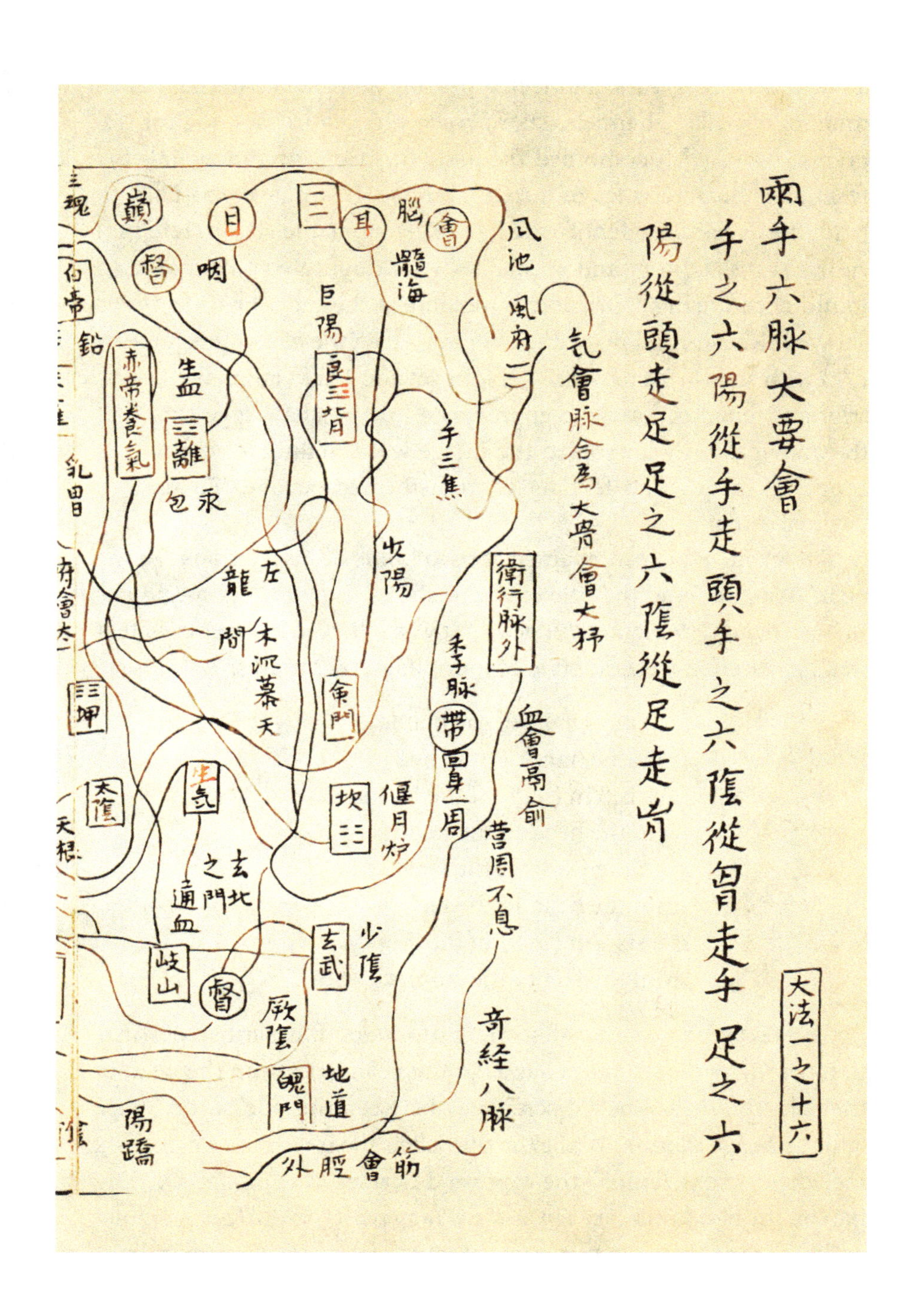

大法一之十六
兩手六脉大要會
手之六陽從手走頭 手之六陰從胷走手 足之六
陽從頭走足 足之六陰從足走胷
氣會脉合爲大骨會大杼
衝行脉外
帶
圍身一周
血會鬲俞
營周不息
奇経八脉
筋會脛外
手三焦
少陽
風池
風府
腦
髓海
巨陽
咽
生血
坎
偃月炉
玄武
少陰
厥陰
地道
陽蹻
岐山
通血

At first glance, the image offered a cartographic representation of a body, with eyes, ears, and feet appearing alongside winding paths, mountain peaks, channels, lakes, reservoirs, and cosmological trigrams (*bagua*).[77] It combined the basic functions of a map with the technical characteristics of a *tu*. Of course, this image was hard to read. It was hard to identify the hands, let alone the six discrete meridian paths. Editors and specialists who duplicated this map also found it confusing.[78] In another version of the map, one illustrator showed two ears and no eyes, having replaced the characters for eyes (目) with sun (日) and moon (月). In yet another version, the Daoist trigrams failed to have the appropriate line breaks, suggesting that the craftsperson who copied the image was unfamiliar with *bagua*. Modified images and text, again, transformed beyond the author's original intent.

The image was accompanied by two kinds of inscriptions. To the right, Wang offered the title of the image as showing "six meridians in both hands" (*liangshou liu mai*).[79] It was followed by a series of text that appeared in a classic style of four-character phrases:

手之六陽	six Yang (*mai*) of the hands
從手走頭	from the hands to the head
手之六陰	the six Yin (*mai*) of the hands
從胃走手	from the head to the hands
足之六陽	six Yang (*mai*) of the feet
從頭走足	from the head to the feet
足之六陽	the six Yin (*mai*) of the feet
從足走胃	from the feet to the stomach[80]

This text seemed somewhat straightforward. It described destinations of the paths from the hands to the head, and from the head to the feet. Yet, there was a discrepancy between the title of the image and the image caption. While the title labels the image as showing "six meridians in the hands," the caption describes six Yin and six Yang *mai* on the hands plus six Yin and six Yang paths on the feet—24 total paths. This inconsistency may seem minor because sets of meridians counted paths on one side of the body. Still, the contrast between title and text challenges a simple reading of the image, complicating assumed relationships between representation and understanding. Rather than clarifying the illustration, the captions expanded on it and assumed more knowledge from the reader.

To the left of the image, Wang included a second caption that described the movement of Qi and techniques for longevity. Rather than appearing in the classic style of four-character phrases, this description appeared in unstructured prose.[81] Wang began stating that Heaven and Earth naturally possessed Qi, following "natural order" (*ziran zhishu*). Humans were rooted in Heaven and Earth; their Qi corresponded to cosmic cycles. "Through cultivation, longevity is attained," Wang mused.[82] Yet people could also exhaust their allotted Qi and succumb to the power of nature. One could bolster their Qi through stillness and emptying the mind. Profound insight could then arise. "Genuine Qi (氣) thus flows unimpeded," Wang explained, "circulating and transforming continuously."[83] Wang then cited the legendary Daoist sage Laozi, who called this existence subtle (*mian mian*) and fleeting. Applying these techniques was limitless and inexhaustible. "Without Qi, danger encroaches, and death is near," Wang warned.[84]

Whereas the right caption presented a straightforward description of the meridians, the left text offered an unstructured and broad description of attaining longevity with allusions to sages and classical texts. The two captions presented different subjects and drew on disparate sources. The text on the right described the location and direction of 24 meridians. The text on the left offered lofty and mundane descriptions of Qi cultivation. Similarly, the *tu* was also straightforward and otherworldly, situating the body within cosmological relationships. The contrasting captions resisted reducing the abstruse landscape to singular meaning and allowed for flexibility. They helped the reader to navigate an interconnected corporeality that joined worldly and cosmic cultivation. Further, meridians facilitated these connections. They linked interior and exterior spaces. They grounded transcendent ideas to an apparent anatomy.

This esoteric image further reflected the social and conceptual transformations taking place at the time. During the tenth and eleventh centuries, a new scholarly culture emerged as individuals outside of government turned to medicine.[85] Once-dominant medical families now worked alongside scholars without claim to an intellectual lineage.[86] An ascendant class of Confucian officials drove reforms in education and textual standardization in medicine and mathematics among other topics.[87] Medical governance transformed with the emergence of new medical institutions, new medical scholars, and new ways of expanding medical relief. As the social landscape of medical practice shifted, so did conversations around medical knowledge.[88] Uninitiated scholars debated on categories of Yin and Yang, the nature of the five phases, and the many types of Qi.

Wang Haogu did not come from a medical family.[89] He instead steeped himself in books, as reflected in the choice of his own professional name, Haogu, or "lover of antiquity."[90] Life had changed in the thirteenth century. Surveys tallied thirty-three epidemics in the twelfth century, compared with just one in the eleventh century.[91] These contagious fevers resonated with Wang's interest in the resuscitated "cold damage," or *shanghan*, tradition for treating external disorders that endangered the organs.[92] Studying classical cold damage texts manifested Wang's antiquarian devotion amidst, or because of, the proliferation of diseases.[93] New medical theories emerged, as did more ways to die. Wang's abstract landscape offered an unconventional integration of meridians and cosmic forces. As epidemics fostered medical reinvention, Wang reengaged past wisdom to move beyond institutional doctrines.

Death encroached in 1232 as Mongolian armies descended on the Jurchen capital at Kaifeng. [94] Generals assassinated ruling family members, igniting riots. This commenced the Mongol domination of eastern Asia, expanding across Eurasia.[95] Alarmed, Wang Haogu fled north.[96]

While escaping the collapsing capital, Wang understood the power of maps as he navigated falling cities and crumbling dynasties—Song remnants in the south, failing Jin in the north, and the consolidating Yuan. Throughout his life, Wang lived in Hebei, Henan, and Beijing, learning from specialists who expanded his knowledge of herbs and *materia medica*.[97] During this pursuit, he remained disappointed in his contemporaries. Their failure to comprehend the body baffled Wang, given the rich legacy of classical texts and collective wisdom.[98] Even with the rich corpus of classical texts and the knowledge of sages and immortals, most physicians still failed to grasp the basics of disease causation. How could doctors be so ignorant?

"Vacant meat sacks!" he cried.[99]

To Wang, healers could not heal because they looked in the wrong place. "You need not seek the magnificent to find it," he insisted.[100] True to his antiquarian name, Wang revisited classics like the *Huangdi neijing*, and the second-century text, *Shanghan zabing lun*.[101] He reproduced images from Yang Jie's illustration of organs. He familiarized himself with the work of the second-century physician Hua Tuo (c. 140–208), who had surgically operated on living patients and conducted his own study of anatomical dissection.[102] Although Wang admitted that he did not always understand what the texts meant, he still referenced them. He reproduced straightforward descriptions of organ function, weight, and structure. He focused on the gut and how eating *materia medica* could counter fevers, indigestion, diarrhea, and other disorders during a pandemic.

Meridian Lengths

Wang was interested in both material transformation and practical medical techniques. He was also interested in precision, particularly in measuring the length meridian paths. These descriptions were straightforward. Specifically, Wang cited the legendary figure Bian Que (407–310 BCE), who had listed the measurements of meridian lengths. These quantities defined the distance from the hands and feet. They could be grouped, added, and totaled. In the style of rote memorization, Wang retorted:

> The three Yang hand meridians are five *chi* each;
> three *zhang* total.

The three Yin hand meridians are three *chi* five *cun* each;
two *zhang* one *chi* total.
The three Yang feet meridians are eight *chi* each;
four *zhang* eight *chi* total.
The three Yin feet meridians are six *chi* five *cun*;
three *zhang* nine *chi* total.
Two Qiao meridians from the feet to the eyes are seven *chi* five *cun* each;
one *zhang* five *chi* total.
The Ren and Du meridians are four *chi* five *cun* each;
nine *chi* total.
These lengths add up to 16 *zhang* 2 *chi* (1,620 cun)
One cycle every two "steps" (*ke*) results in 50 cycles.
13,500 breaths make a living person based on images of water clocks.

Measurement mattered in early modern mathematics. In this case, meridians relied on three units of measurement: *cun*, *chi*, and *zhang*. Similar to a metric system, 10 *cun* measured a *chi*, and 10 *chi* measured a *zhang*.[103] Individual units each had their own history, their own material culture, and their own circumstances for production. These units, although definite and distinct, further relied on approximation as a standard for accuracy. For instance, the *cun* (寸) was an old unit of measurement that referenced different objects in the material world. In one of its earliest forms, the *cun* appeared in the first-century text *Huainanzi* as the size of a millet seed.[104] According to one version of the *Huainanzi*, "a *cun* rises from the millet seed. The millet seed arises from the sun. The sun arises from form. Form arises from shadow. This gives rise to measurement."[105] Although it was possible that the millet seed could have referred to the full floret rather than a single pod, triangulating the height of the sun extended from a single *cun*.

In the seventh century, the famous physician Sun Simiao (581–682) advocated for standardizing the physical dimensions of *cun* based on the human hand. Now known as the "same-body inch," the *cun* used a person's finger as the unit of measurement—specifically the space between the creases inside the middle section of the middle finger (**Figure 1.16**).[106] Whether measuring the total length of 26 meridians or the distance from the ground to the sun, identifying *cun* as the "same-body inch" or as the length of a millet pod quickly revealed the difficulty of relaying absolute units of measurement.[107] The *cun* fluctuated as a relative form of measurement both in history and in practice. It was a useful measurement when put to use.[108] Depending on the material reference of a *cun*, meridian paths expanded and contracted as complete spatiotemporal units, their proportions remaining relatively consistent. Added together, Yang meridians on the hands and feet measured longer than Yin meridians. Yang lengths exceeded Yin lengths. A single hand Yang path could span five *chi* compared to three and a half *chi* for a Yin path. Relative lengths varied, and meridians endured as mutable forms.

In practice, numbers were both symbolic and real.[109] Wang Haogu echoed earlier texts that described the body to contain discrete units of distance, units of time, and units of life as aligning with a "natural order" or "natural calculation" (*ziran zhishu*). This "natural order" manifested in the 26 meridian paths in the body, which included two sets of 12 paths and two additional vertical paths through the torso. When measured out, the total length of all 26 meridian paths was calculated to be 1,620 *cun*. The total length of 1,620 *cun* could also be expressed as 16 *zhang* and 2 *chi*. Meridians in aggregate measured the number of breaths that moved and made human life.[110]

This "natural calculation" further manifested in the movement of corporeal Qi.[111] Qi traveled discrete distances along the meridians. According to earlier texts, each inhalation or exhalation moved Qi by a distance of three *cun*. Thus, one full breath cycle (inhale and exhale) moved Qi by six *cun*. These units of moving Qi were again connected to units of time, which were measured by ancient water clocks that divided the day into 100 *ke* (刻), or steps.[112] By way of analogy, one *ke* approximated 15 minutes, so that one inhale and one exhale that moved the Qi took around 30 minutes for a healthy person. Although, with the multitude of pandemics in the twelfth century, healthy bodies were in short supply.[113]

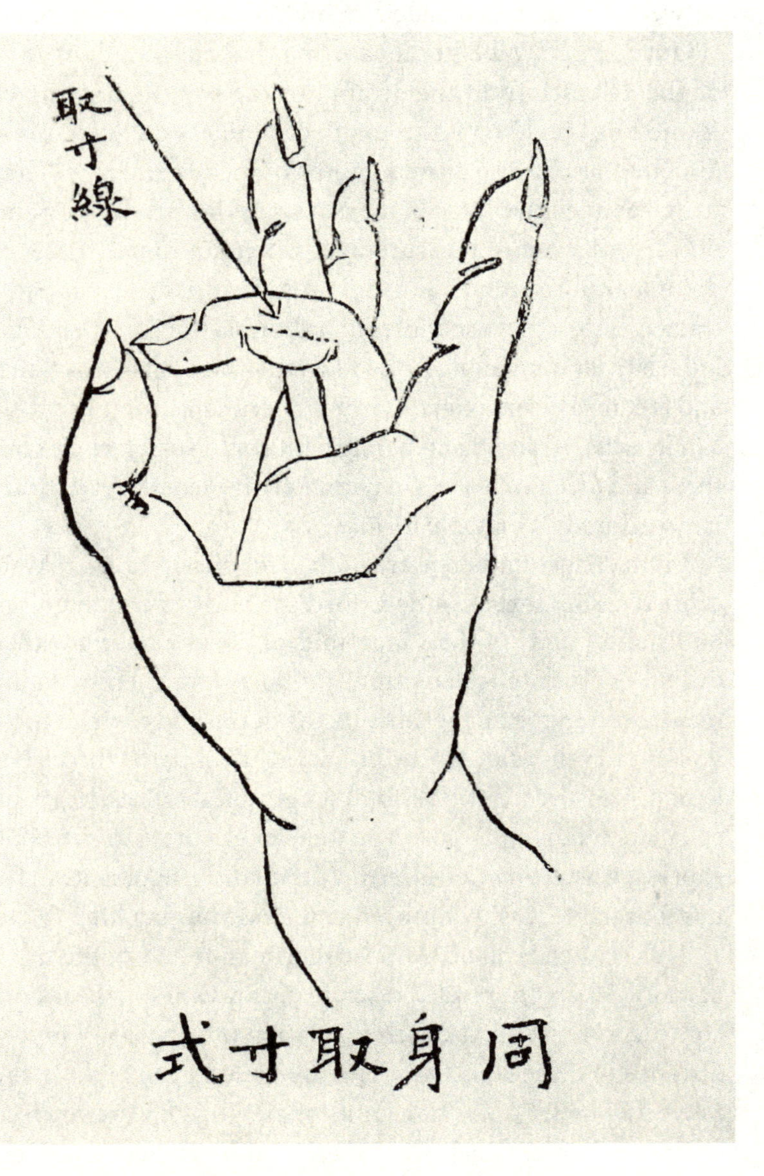

Figure 1.16. Same-body inch, twentieth century. Cheng Dan'an (1899–1957), drawing. The measurement unit called the "same-body inch," or *cun* (寸), was based on the distance between the middle crease of the metatarsal of the middle finger. The location of needling sites were measured in *cun*. From Cheng, *Zengding zhongguo zhenjiu zhiliao xue*, 1931, 55

In other words, human, earthly, and heavenly landscapes shared the same cosmological orientation. Mapping out these orientations unified their order and ontology. Bodies illustrated on paper and moving in the world rested in the open air, lifted by invisible winds, carried by the breeze. These invisible winds placed the body alongside mountains and water, immersed in a stream of "cosmic ethers" that formed a confluence of human vitality.[114] Thus Wang's abstract terrain in **Figure 1.15** was both a *tu* of internal anatomy and the external earth. The cartographic resemblance in this image was not a coincidence. The world and the body belonged to the same vision of being. In this being, meridians moved as a single unit with fixed lengths that extended beyond a corporeal boundary, and measured time, distance, and units of life.

Body Proportions

Like meridian lengths, body proportions were similarly both fixed and adaptable. Illustrations functioned as empirical tools, locking proportions onto paper. Yet these images also referenced flexible bodies, revealing divergent aims of either measuring or expressing form. That is, proportions exposed the need for approximation. Bodies were dynamic, demanding variable models of illustration. This section addresses such differences through two proportion representations: images by scholar Hua Boren (1304–1386) and artist-mathematician Albrecht Dürer (1471–1528). Whereas Hua pursued corporeal averages, Dürer pursued mathematical ideals. What underlay these epistemic incentives was not accuracy for its own sake, but accuracy through approximation.

In Hua Boren's famous fourteenth-century text, *Treatise on the Fourteen Meridians*, proportions appeared on a flat body.[115] The preface showed two men who were "facing up" and "facing down," or *yangren* and *furen* (**Figure 1.17**). They lay sprawled like a flayed specimen.[116] This view from nowhere suggested that proportions provided legible surfaces for comprehending meridians. In the text, Hua dedicated individual chapters to individual meridian paths. He prefaced

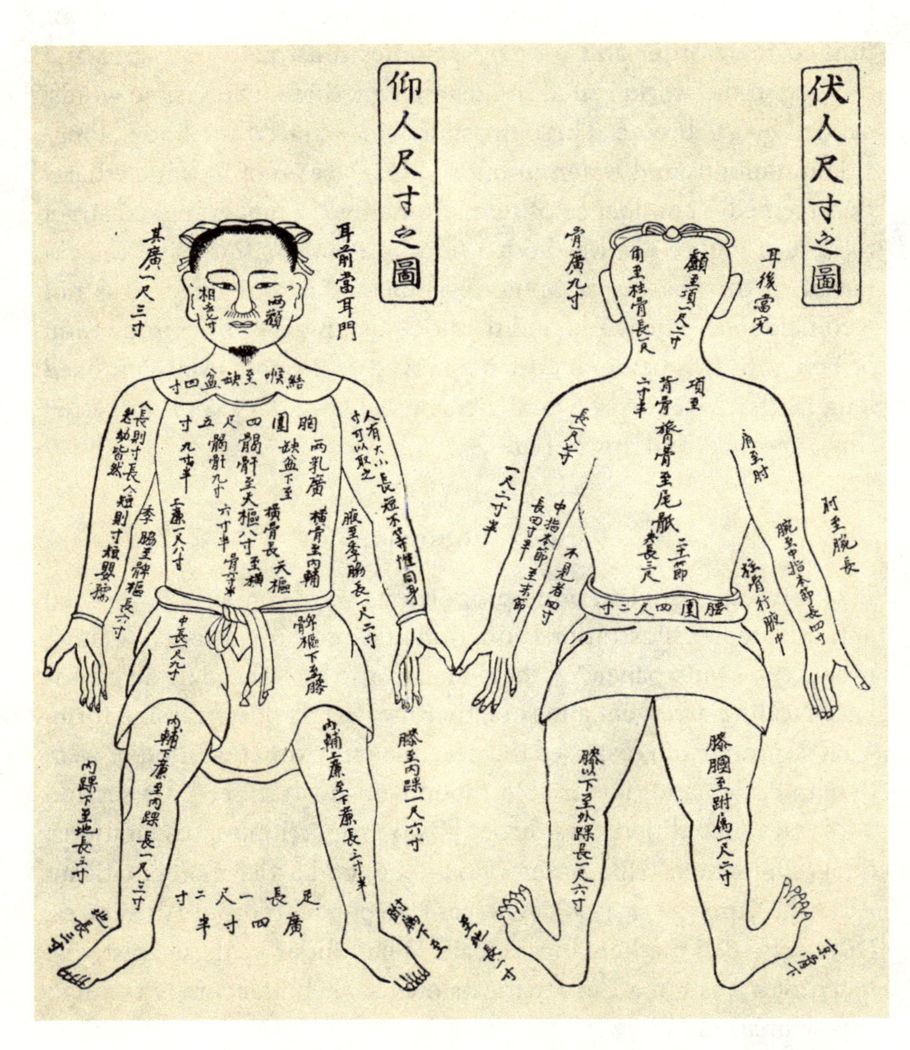

Figure 1.17. Body proportions, fourteenth century. Unknown illustrator, woodblock print reproductions. This image is attributed to Hua Boren (1304–1386), showing men "facing up," *yangren* (仰人), and "facing down," *furen* (伏人). The inscriptions describe distances related to the cheekbones, nipples, underarm, lowest rib, spleen, knees, bottom of the feet, and so on. From Hua, *Guben jiaozhu, Shisi jing fahui*, 1969, 1–2

the book with a general outline of bodily distances, explaining that comprehending the meridians relied on engaging with the body as a legible surface. Inscriptions described the many distances between cheekbones, nipples, underarm to lowest rib, thyroid to jugular notch. Proportions measured visible landmarks, not precise coordinates.

Sites on the body remained ambiguous on the page. Knowing them in practice required tacit and tactile certainty. Like his predecessors, Hua measured distances in *cun*, determining the length between body parts, meridian paths, and meridian sites.[117] Again, the *cun* referred specifically to the middle crease of the middle finger, inconveniently located inside the hand. Physicians thus approximated rather than lining up individual *cun*. The *cun* was not fixed, expanding and contracting based on the size of the body. As the text on the arms explained: "Individuals of different sizes rely on different [units of] measurement. The *cun* grows with the individual."[118] It continued, "If the individual is long, then the *cun* is long. If the individual is short, then the *cun* is short. This applies to the young and old."[119] As functional abstractions, Hua's proportion figures guided approximation.[120] They presented one kind of certainty through the inscriptions on the body and another kind of certainty in the open-endedness of how these inscriptions should be considered. More *tu* than map, Hua's proportions offered technical guidance, and articulated its own kind of didactic accuracy. Proportional *tu* found utility in graphically organizing the body with room for variation. The images accommodated error and difference. Hua's terrain was both speculative and functional.

Hua's proportion figures contrasted with early modern European illustrations, famously associated with German artist Albrecht Dürer (1471–1528). Whereas medical scholars in East Asia remained the masters of anatomy, it was artists in Europe who claimed authority over anatomy—not physicians. For instance, Dürer's later work *On Human Proportions (Von menschlicher Proportion)* prioritized the illustrator's eye over the doctor's touch.[121] These proportions established a vision of the body for artists and mathematicians. For instance, the figures in *Human Proportions* presented recognizable tropes in Renaissance figure drawing. They stood in contrapposto. They gestured to the viewer[122] (**Figure 1.18**). In the accompanying text, Dürer wrote:

> From the crown to the throat let there be one tenth part and one eleventh. To the top of shoulders, two eleventh parts. To the bottom of the chin, 1/7. The cowlick is half-way between the crown and the forehead. From the chin to the base of the hair, 1/10 . . . From the throat to the top of the chest, 1/30. To the armpits, 1/13. To the nipples, 1/10. Bottom of pectorals, 1/8. Loins, 2/11. From the loins to the navel, 1/40. To the curve of the hip, 1/30. To the hip joint, 1/10. To the crotch, 1/8. To the tip of the penis, 1/6. To the lower buttocks, 1/10 plus 1/11. From the lower buttocks to the narrow part of the thigh, halfway down the thigh, 1/18. From the sole to the to the bottom of the ankle, 1/28. From the sole to the top of the foot, 1/20.[123]

In illustrating these images, Dürer fixed bodies using vertical proportions—fractions on a single line that served as a ruler against which to measure.[124] He produced over 30 unique rulers generating 30 unique bodies, some lean and others "stout" and "rustic."[125] Curiously, Dürer's rigid ruler only enabled vertical forms. All 30 different bodies stood upright, none facing up or down. Ultimately, *Human Proportions* represented a mathematical body reducible to a bare, linear frame.

Dürer's idealized types retained a degree of ambiguity. When the philologist Joachim Camerarius (1500–1574) translated the Latin edition of *Human Proportions*, he questioned Dürer's naming of parts, which lacked precise anatomical coordinates.[126] In his preface, Camerarius wrote: "Notice that whenever parts of the body are individually named, their beginning or uppermost point should be understood. This is commonly spelled out in the text, unnecessarily in my opinion. In some places it is downright superfluous—but I felt that in someone else's work it is best not to change too much. (But who

Figure 1.18. (opposite) Varieties of measurement, sixteenth century. Albrecht Dürer (1471–1528), woodcut. Over 30 standing figures by Dürer measure themselves with rulers, demonstrating human proportions for a vertical body. From Dürer, *Hierinn sind begriffen vier Bucher von menschlicher Proportion durch Albrechten Durer von Nurerberg*, 1528

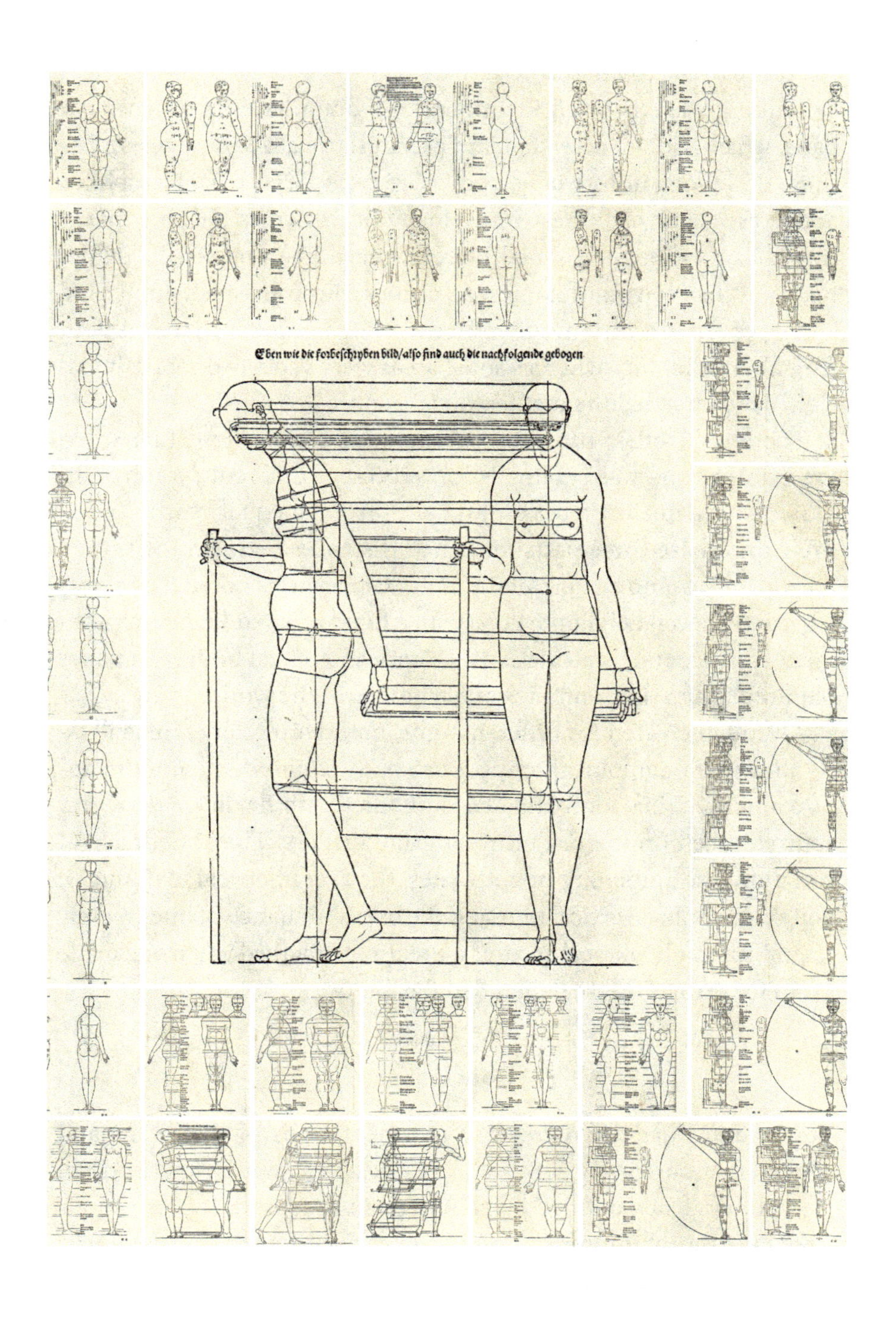
Eben wie die forbeschrybben bild/also sind auch die nachfolgende gebogen

would need the addition of 'top of' when hearing the word 'crown' in reference to the top of the head?)."[127]

Naming sites involved identifying the boundaries between body parts, which had not been standardized and specialized despite—or because of—a long history of anatomy. Greek physicians like Herophilos and Erasistratus had already developed basic names for body parts. But as Camerarius translated Dürer's image, the potentially "same" body part had too many names. As Camerarius translated, duplicated labels proved "downright superfluous." The same leg section was both "thigh" and "femur," whereas some areas were vaguely dubbed "muscles." This impeded the mathematical veneer.

There were other problems. Camerarius also noticed that when the primary line was subdivided to represent different proportions, Dürer gave a final instruction to mark three "unequal" parts[128] (**Figure 1.19**). These three parts measured distances between the throat, the hip, knees, and bottom of the shins. Camerarius noted that marking three uneven height portions with a tilted ruler and made approximate, not accurate, calculations. Dürer's 30 vertical bodies on paper required improvisation to resemble bodies in the world.

Although created for different aims, both Dürer's and Hua's illustrations of topographical proportions both required attention to approximation. Hua approximated on bodies with flexible *cun*; Dürer approximated on the page through tilted rulers. The abundance of variation, the ambiguity of body sites, the asymmetrical mapping of body parts onto a vertical line, and the overabundance of lines merely resembled mathematical accuracy. Precision emerged when one made room for error. Finesse demanded flexibility.

Conclusion

This chapter has explored early representations of meridians and nerves, demonstrating how body maps, body images, and body *tu* (圖) ontologically defined invisible agents. Anatomical illustrations were necessarily imprecise, full of improvised inscriptions as a way of knowing shifting corporeal contours and configurations. Specific

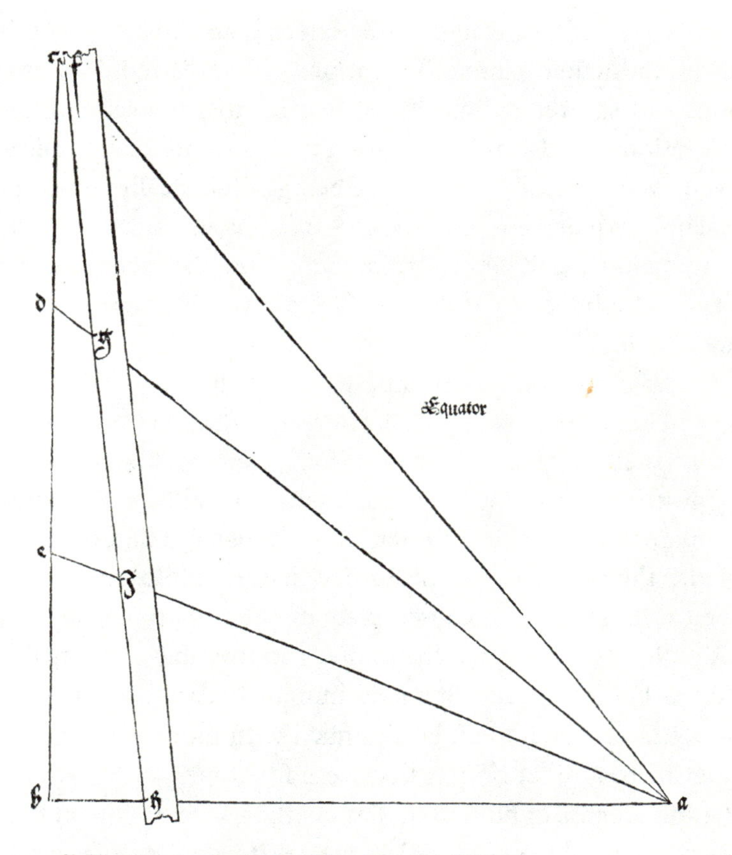

Figure 1.19. Tilted triangle, sixteenth century. Albrecht Dürer (1471–1528), woodcut. Dürer's method for dividing the body into uneven thirds used a tilted triangle, which translator Joachim Camerarius noted could introduce error. "Dürer was probably aware that this proportionality is only approximate," Camerarius wrote (Dürer, Price, and Warnock Library, 2003, B2v). From Dürer, *Hierinn sind begriffen vier Bucher von menschlicher Proportion durch Albrechten Durer von Nurerberg*, 1528, 18

meanings emerged through abundant accompanying text, which both explained images and invited speculation. Meridian *tu* approximated the location of Qi through while offering also discrete lengths measuring individual paths. Illustrations of nerves required embellishment and showed splintering structures that housed animal spirits and rational souls. In both cases, artists invented anatomical detail, posing as empirically useful while being ontologically confusing. Classical images relied on contradiction. Imagination manifested precision. Imagination facilitated approximation and abstraction to articulate corporeal subtleties. Corporeal objects gave tentative shape to unseen movement.

Comparing works attributed to figures like Wang Haogu (1200–1264), Hua Boren (1304–1386), Albrecht Dürer (1471–1528), Thomas Willis (1621–1675), and Amé Bourdon (1638–1706) shows how diverging aesthetics reinforced particular anatomical forms. It allows us to compare the ontological similarities and material differences between Qi and the rational soul. In Europe, natural philosophers used anatomical knowledge to contemplate divine intentions via the human body. Nerves manifested the mind as an invisible capacity that made human bodies special. It made human bodies capable of rational thought; it made human bodies filled with more than animal spirits superior to animal bodies that were filled with *only* animal spirits. The uniqueness of human bodies justified speculating on the special functions of cold, wet, bloodless nerves; it justified situating the mind in the cold, wet, bloodless brain.

In contrast, meridians were not unique to humans. Early modern scholars in East Asia did not place meridians at the center of debates on human supremacy. They were not as special as Qi and Blood. They did not necessitate a brain. Meridians did not express the mind, but they could be affected by the mind. Meridians were stable and idiosyncratic; different sets of meridians manifested in different kinds of bodies in different states of life, health, and distress. They lived in a body that required cultivation. Cultivation through a stillness of the mind allowed for profound insights to arise, and for a form of "genuine" Qi to flow uninhibited. Engaging with this constant transformation kept death at a distance.

As bodies changed across time and place, so did their expressive images. Depictions of meridians and nerves contorted in relation to shifting material and conceptual conditions. Under these conditions, meridian paths fractured and twisted; nerves swelled and shriveled. Both remained important ontological objects despite their abstract appearances on paper. Early modern illustrations exposed the persistence of imagination. Body maps represented something and someone—manifesting cosmological imprints on ordinary things. Meridians and nerves occasionally inhabited divine bodies, and were otherwise ordinary, cliché, unremarkable, and enigmatic. Their unstable forms articulated flexible anatomies that were mutable and multiple.

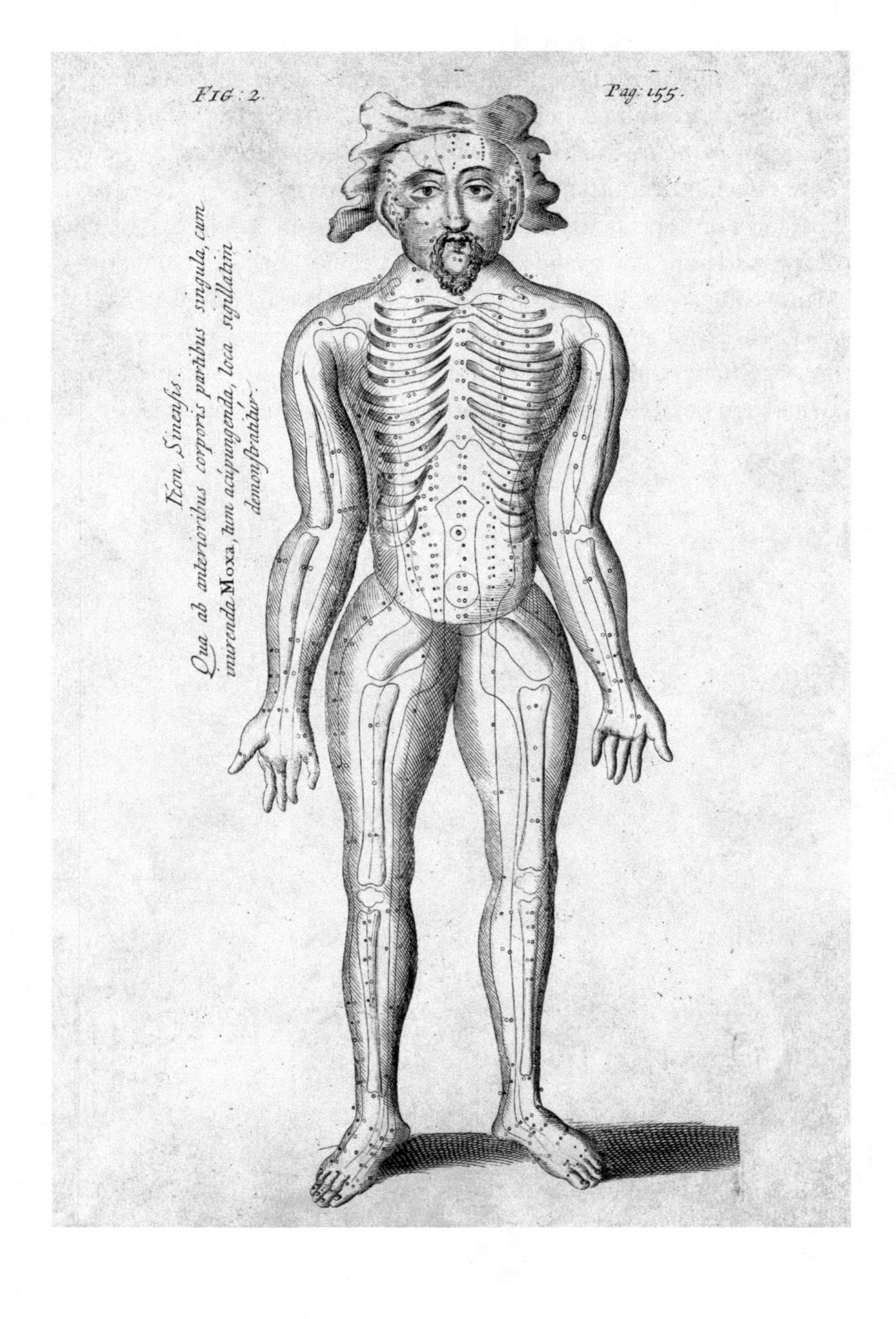
FIG: 2.
Pag: 155.
Icon Sinensis.
Qua ab anterioribus corporis partibus singula, cum
inurenda Moxa, tum acúpungenda, loca sigillatim
demonstrantur.

CHAPTER TWO

Early Modern Metaphors as Translation

Meeting Goatee Man

Two large oval eyes stared ahead (**Figure 2.1**). They flanked a prominent nose that sat above a tidy goatee. The hair on his head—if it was hair—seemed rigid and stiff like a hat. He stood with feet bare, palms exposed, and chest pitted. Finely crosshatched lines shaded the contours of his arms, fingers, and thighs. Even finer lines twisted around his shoulders. They reached his open palms and passed through his ribs before descending down his bulging legs. Bone pressed to his skin like a sleeve. That is, if he had skin.

Was this goateed man Chinese? Historians like Lu Gwei-djen (1904–1991) thought so. Upon seeing the early modern etching, Lu described it as a mere version of a Chinese body. To Lu, this was "some Chinese work" that had been "redrawn in European style."[1] It was only "Euro-

Figure 2.1. (opposite) Man with goatee, seventeenth century. Unknown artist, engraving. An acupuncture moxibustion figure from a volume of Willem ten Rhijne's collected essays. Man with goatee has unconventional proportions, featuring inward-bending arms, broad hips, and bulbous thighs. From Ten Rhijne, *Dissertatio de Arthritide*, 1683

pean" in form and "Chinese" in kind.[2] His face may have appeared European, but his body came from elsewhere.

This was not the entire story.

Goatee man arrived in Amsterdam from Japan via Batavia. He had never been to China. Published in 1683, he traveled across multiple ocean-bound trade routes and was made possible by the activity of merchant vessels, trading posts, medical scholars, local governors, engravers, and interpreters.

Measuring over one foot in length, goatee man was folded into a book by the Dutch physician Willem ten Rhijne (1649–1700), who had famously visited Edo and was credited for authoring one of the first chapters on acupuncture and moxibustion in Latin.[3] Meanwhile, goatee man's original draughtsman and engraver remained anonymous.[4] He was the product of many skilled hands. He was also an imperfect image. The lines and dots did not have names; they did not instruct the reader.[5] The image was less of a guide and more of an impression.

What did he actually represent?

Introduction

This chapter explores the conceptual and material challenges to early modern translations of meridians and nerves. Diplomatic trade policies in the seventeenth and eighteenth centuries limited the terms of commercial and intellectual exchange.[6] Ideas traveled between East Asia and Europe with difficulty. Books on medical practices made their way through official and unofficial channels, moving with the ships of the Dutch East India Company or in the hands of itinerant Japanese monks. Occasionally, information was even coerced through the forced kidnapping of Chinese physicians to Nagasaki.[7] Not everyone was willing to share their expertise; not everyone was equipped to comprehend an unfamiliar language. Translators working under these circumstances invented new words and new ideas for their readers based on limited access to resources.

In this period, Japanese doctors increasingly interacted with warriors, scholars, and philosophers like Itō Jinsai (1627–1705) and Ogyū Sorai (1666–1728), who provided conceptual foundations for botany.[8] In 1754, Yamawaki Tōyō (1705–1762) performed what was celebrated as Japan's first legal human dissection, and in 1773–1774, Sugita Genpaku (1733–1817) produced Japan's first translation of a Dutch medical text with a team of illustrators and translators. To be sure, these conceptual shifts were far from teleological because doctors had been integrating useful techniques like dissection with Chinese concepts of the body over many decades. Through engaging with diverse medical epistemologies, Japanese physicians assimilated European knowledge on their own terms.

As a result, translations were imperfect, misleading, and even deceptive. Goatee man was not merely a neutral object of translation but a struggle to articulate the unknown. He represented imprecise, improvised, and incomplete knowledge that registered and manifested historical contingencies. In this chapter, I examine primarily Japanese sources to illuminate the otherwise naturalized history of *kei* (經) and *myaku* (脈). Like the case of graphic representations of *jingluo*, I will use "meridians" to discuss graphic representations of *kei* and *myaku* and include multiple transliterations to remind readers of the many utterances invoked through Sinographic texts.

Embedded within these imperfect translations was a tension between metaphor and materiality. In the late seventeenth century, physicians redefined *kei* and *myaku*, invented blood vessels and nerves, by speculating on their content and composition. For instance, Japanese scholars selectively appropriated concepts that could be incorporated into their existing intellectual framework. Newly adopted anatomical terms resembling "nerves," such as *seibinseinon* and *zeinun*, served to elaborate meridians inscribed on paper.[9] In the eighteenth century, physicians like Sugita Genpaku (1733–1817) further popularized the term *shinkei/shenjing* (神經) to describe "nerves." *Shinkei/shenjing* was a compound word that paired *shin/shen* (神), representing an otherworldly entity, and *kei/jing*, designating a network of earthly paths.[10] In other words, the ethereal structures of *shinkei/shenjing* became responsible for a nervous body made of flesh, skin, bone, and blood. This new name made a bold claim.

Dutch physicians likewise speculated on the content and composition of meridians to articulate familiar structures like blood vessels and nerves. Ten Rhijne, for instance, never described *kei* and *myaku* as distinct entities. Instead, he explained that *kei* and *myaku* were analogous to blood, nerves, and vessels. He preserved the Japanese transliterations of *kei* and *myaku.* Without assigning them a new name, ten Rhijne did not introduce them as novel objects. Rather, they were already rendered comprehensible as similar to blood, nerves, vessels, and humors. And if *kei* and *myaku* were comparable to blood, nerves, vessels, and humors, then they could simply be folded into the existing humoral framework as a combination of these familiar elements. *Kei* and *myaku* were thus made redundant within the humoral system. They were easy to dismiss.

Then, there were the images. Visual representations offered additional means for assigning new meanings to familiar anatomical structures like blood and nerves, despite the limitations of words—whether newly created or newly Romanized. Images provided novel ways to depict old objects. This mattered because meridians existed graphically in the form of a *tu* (圖). It was in a *tu* that meridians could transgress the boundary of the skin. It was on a *tu* that meridians could fluctuate and flow.

Goatee man was not a *tu*—at least, not entirely. He was an etching of a paper model that followed different rules of representation. As historian Mathias Vigouroux has expertly argued, "goatee man" was a rigid, three-dimensional body.[11] He stood in contrast to two-dimensional *tu* that communicated four-dimensional objects that moved across space and measured time.

Translation invited ontological uncertainty through uneasy comparisons. Reconfiguring image and text applied new natures and new meanings to enduring entities. For Ten Rhijne, meridians were *like* the humors, but they were not the humors. Blood vessels were *like* meridians, but they were not meridians. The "same" object manifested under many names; it was represented by partial images that were created under limiting conditions. The result was not simply a situation of mistaken identity, or ideas that were "lost" through the difficult process of translation. Instead, metaphors *became* the objects that they were meant to represent. Communication bred new entities. Comparisons became binding.

Rigid Flesh

The man with the goatee was a strange object.[12] His body aesthetically deviated from Northern Renaissance conventions of figure drawing[13] (**Figure 2.2**). The inward bending of the arms, the broad hips, and the bulbous thighs did not follow Northern Renaissance genres of anatomy and perspective, which were often influenced by classical motifs. Goatee man did not reference religious scenes, portraits, or mythologies.[14] His thighs pressed against his lower belly as if fixed in place. Even more curiously, goatee man had no genitals, which violated the Northern Renaissance commitment to realism. If male models had genitals, then their nude images would have them too.

In other words, the portrait of goatee man was only a disembodied goatee head. The rest of his body more closely resembled an Edo-era acu-moxa model made of wood and paper (**Figure 2.3**). Stylistically, *tu* based on three-dimensional models had been an established genre of maps, known as "bronze figure maps," although "bronze figure maps" did not always represent figures made of bronze. Two-dimensional imprints of "bronze" sculptures also included models carved from wood or pasted with paper mâché.[15] Larger models featured tiny dents that marked individual needling and heating sites. Small holes led to caverns sealed with wax that were used for exams so that when students punctured the correct site, water would seep from the puncture.[16]

Whether large or small, all three-dimensional acu-moxa models stood upright with rigid arms flanking the sides. Unlike Greek kouros statues that posed with one foot in front of the other, acu-moxa models stood with legs apart and hands parallel to the legs. In some cases, the image faced forward with palms turned up toward the viewer. These models were often crafted as male figures, signified by

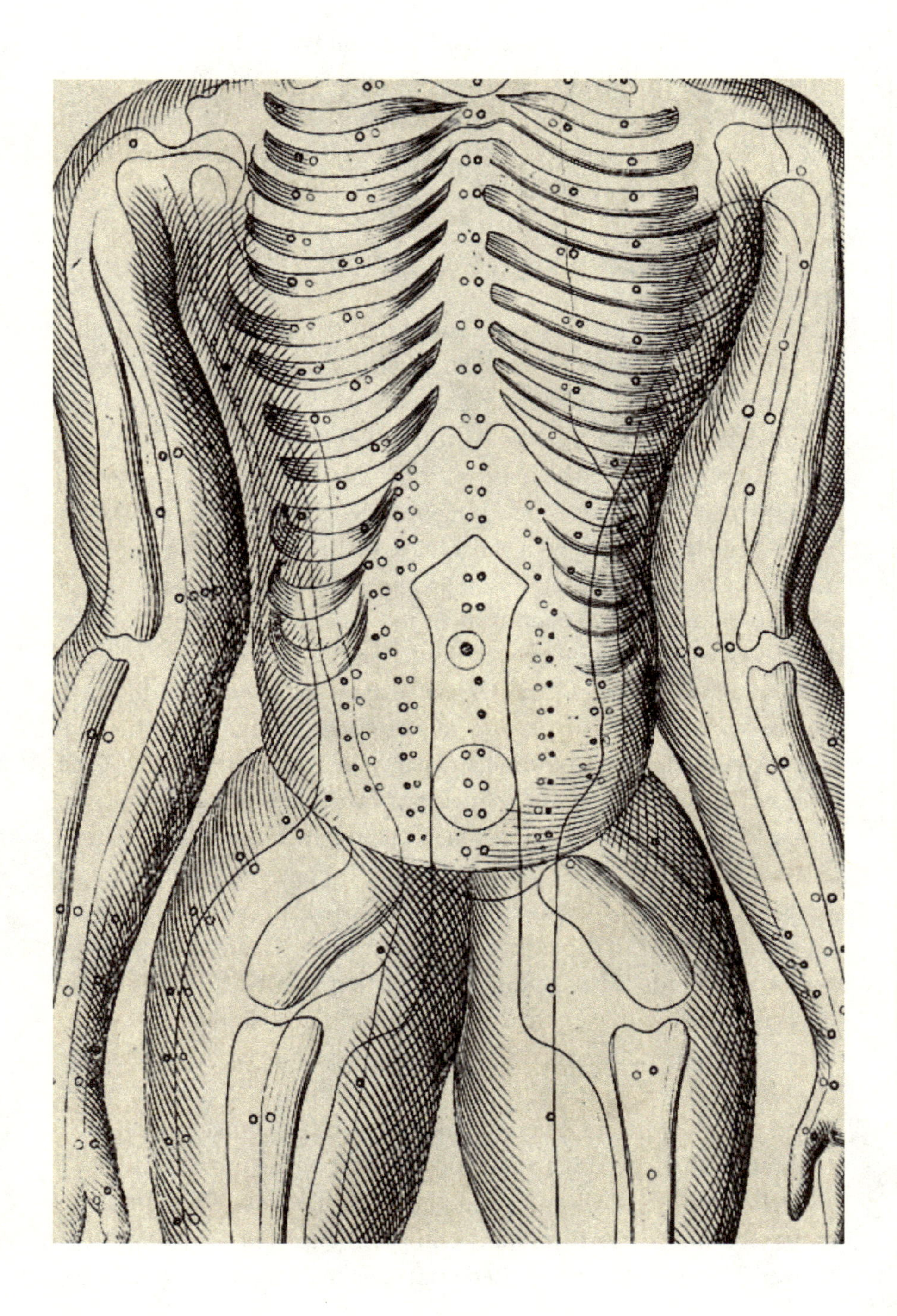

hair bundled into a topknot and a sash draped around the waist. Men without hair or clothing featured a smooth gap between the legs.[17]

Look again at the torso below the goatee (**Figure 2.2**). Here was a body that was both gendered and sexless. The crosshatching signified a bounded body, but not one based on a living human. Instead, it followed a three-dimensional rendering of an acu-moxa model.[18] For these models, flesh would have been made of heavy metals or compressed paper. Their neat contours would have been marked with thin lines and dots. Their hands would have faced outward.

Willem ten Rhijne would have seen many of these sexless and gendered models. They guarded the entrances of expert-practitioner homes. With the help of a guide, he studied them with great interest. "My interpreter indicated holes," Ten Rhijne wrote, "perhaps because those copper machines had been perforated with holes at the locations for needling and burning."[19]

Ten Rhijne's interpreter pointed out that these models themselves were ontologically unique. He gestured to the "holes," or *foramen*, that had been "perforated" onto a solid, static body. These types of models had names, which Ten Rhijne noted, writing that "*[d]onyn* means a figure made from bronze."[20] Whether Ten Rhijne recognized the ontological range of "bronze" models as including all three-dimensional models, he did not say.[21] In any case, here was a model with perforated holes that did not need to be transliterated. Whether in Japanese, Dutch, or Latin, holes were holes.

Or were they?

In a rare anecdote, Ten Rhijne described watching his guide, a garrison soldier, needling himself. The soldier complained of excess heat after a ritual burning. He tried drinking water to cool himself off, but instead felt more nauseated. He then tried drinking wine and eating ginger, which also failed to lift the "trapped wind" or *pertinaciter*

Figure 2.2. (opposite) Man with goatee (detail), seventeenth century. A close-up of the figure shows a sexless body, lacking genitalia. The rendering deviates from Northern Renaissance artistic conventions and commitments to realism. From Ten Rhijne, *Dissertatio de Arthritide*, 1683

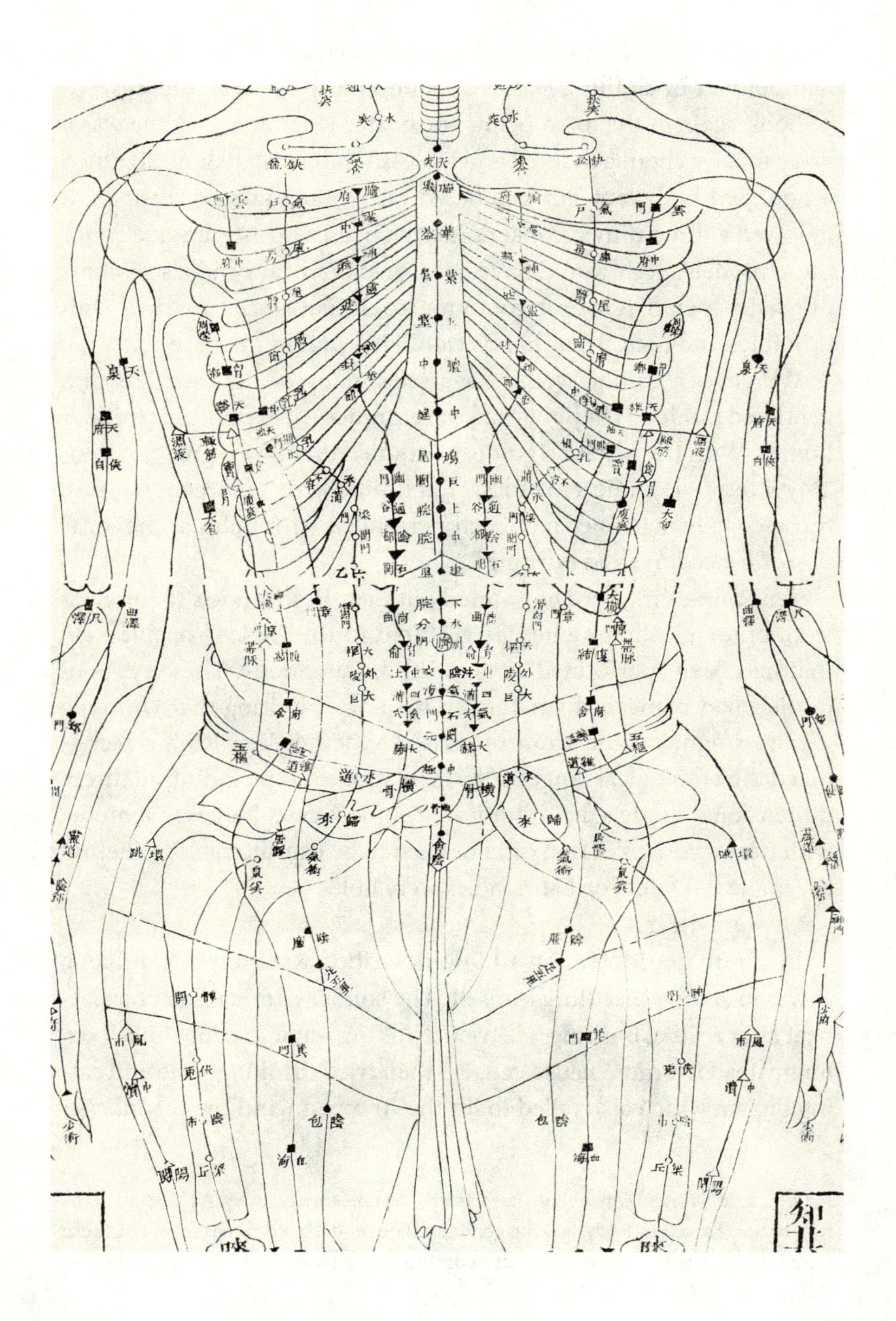

flatum. Finally, he took out a needle and a hammer. Ten Rhijne looked on and wrote:

> Lying on his back, he drove the needle into the left side of his abdomen above the pylorus at four different locations. (For this task, he cautiously held the point of the needle with the tips of his fingers.) While he tapped the needle with a hammer (since his skin was rather tough), he held his breath. When the needle had been driven in about the width of a finger, he rotated its twisting-handle. He pressed the location punctured by the needle with his fingers. No blood, however, appeared after the extraction of the needle; only a very slight puncture mark remained. Relieved of the pain and cured by this procedure, he regained his health.[22]

Willem ten Rhijne watched as the soldier lay sprawled on the floor with his stomach exposed. The man handled himself meticulously, grasping the needle with his fingertips in one hand, and tapping it with a mallet in the other. This was a technique called *uchibari* unique to the Mubun medical lineage.[23] The man held his breath, forcing the needle into his rigid body. His skin was "rather tough," Ten Rhine noticed, so that the soldier took care to "drive," or *adigitur,* the needle into the body. The flesh on the stomach was thick. Like a wooden model, it required needles that had to be hammered into the skin. While doing so, the soldier drew no blood. Where bronze sculptures let out water upon puncturing the correct site, the soldier's body did not bleed. It let out no water and indicated no sign of fluidity. Inside was dry. After the soldier's complaints of trapped wind and excess heat, all that remained following the procedure was a "very slight puncture mark."

Figure 2.3. (opposite) Rigid arms and legs, sixteenth century. Unknown illustrator, woodblock print. This image is attributed to Wang Weiyi (987–1067), a court physician of the Song dynasty who is famous for the corpus *Tong ren shuxue zhenjiu tu jing* (铜人腧穴针灸图经) [Bronze Acu-Moxa Sculpture Point Classic]. Images related to this text were sculptures made of wood, paper, or bamboo. These images also appeared in Japan throughout the fifteenth and sixteenth centuries. From Wang, *Xin kan tong ren zhen jiu jing*, 1515

Willem ten Rhijne was surrounded by rigid flesh. Meridian models were made of bronze, wood, and paper; soldiers hammered into stiff, bloodless stomachs. These textures appeared on goatee man, who was not based on a living or dead human. Instead, lines and dots appeared on a sexless figure that represented fossilized fleshless flesh.[24] All Ten Rhijne could do was observe, as many of his questions were left unanswered by reluctant interpreters who refused to elaborate on the nature of the lines and dots.

"Pig-Headed Interpreters"

When Ten Rhijne arrived in Japan, images had become increasingly important in Japanese medical texts during the Edo period (1603–1868). Earlier works like the *Ishinpō* by Tanba no Yasuyori (912–995) and *Ton'ishō* by Kajiwara Shōzen (1266–1337) contained just a few images.[25] But beginning in the late seventeenth century, illustrations began playing a bigger role in texts for specialists and amateurs.[26] This educational use of images was part of a broader movement in Japanese society to "speak to the eyes."[27] *Nishiki-e* prints informed people about epidemics. Simplified health manuals known as *chōhōki* and vernacular texts spread basic medical knowledge. These images, for instance, depicted meridian paths, pulse positions, abdominal diagnosis, tongue examination, organ placement, skin conditions, and acupuncture needles.

Seventeenth- and eighteenth-century Japan was intellectually dynamic, with growing urbanization and improvements in infrastructure.[28] At the same time, Japanese physicians were suspicious of their European visitors, calling these outsiders "Southern Barbarians" or *nanban-jin*.[29] The Tokugawa government banned the entrance of all Portuguese and Spanish traders following a violent uprising of Catholic peasants.[30] The Dutch, meanwhile, remained allies during the rebellion and became the only foreigners allowed on the artificial island of Dejima, where they established their own communities.[31] Foreign physicians practiced their "red-haired barbarian surgery," a rough reference to *kasuparu-ryū geka*.[32] Foreign physicians, although enterprising, could not be trusted. Some talented surgeons like Engelbert Kaempfer (1651–1716) found an audience with the Shogun.[33] Others, like Philipp Franz von Siebold (1796–1866), who had established his own clinic and medical school in Nagasaki, was deported under suspicions of being a foreign spy.[34]

While Neo-Confucianist schools dominated alongside nativist Shintoism, scholars took interest in Dutch learning.[35] Yet few Japanese physicians spoke Dutch. Medical texts like Andreas Vesalius's *De humani corporis fabrica* (1543), Ambroise Paré's *Chirurgie* (1564), and Adriaan van de Spiegel's *De human corporis fabrica* (1627) entered Japan but fell into illiterate hands. Instead, physicians relied on interpreters to mediate medical knowledge. Known as *oranda tsuji*, interpreters performed specialized labor, translating across Dutch, Chinese, Spanish, Latin, and Portuguese.[36] They occupied an inferior station despite their necessary and formidable skills. In Edo, they acted in teams to facilitate foreign encounters with a single Japanese representative.[37] Interpreters like Narabayashi Chinzan, Kafuku Kichibei, and Motoki Ryōei collected medical texts and instruments before establishing their own schools. They were critical points of contact, facilitating foreign medical encounters while also teaching "barbaric red-haired" medicine themselves.[38]

This period also saw an increase in the publication of annotated medical texts.[39] The sixteenth-century physician Manase Dōsan (1507–1594) had initiated a renewed interest in acupuncture that continued into the seventeenth century.[40] When *Treatise on the Fourteen Meridians (Shisi jing fahui)* first entered Japanese ports in 1638, Dōsan's students eagerly reproduced its images.[41] Soon after, merchant ships began importing texts filled with illustrations.[42] Medical schools in Edo collected acu-moxa treatises and meridian maps even as the same books began disappearing in China.[43] Knowledge exchange happened through translating, editing, and commenting on texts. For instance, at least eight annotated versions of *The Great Compendium of Acupuncture Moxibustion (Zhenjiu dacheng)* appeared in classical Chinese and vernacular Chinese.[44] These texts were so popular that publishers continued releasing new editions of *Fourteen Meridians* and *Great Compendium* into the nineteenth century.

Here was a robust culture of medical diversity. Interpreters facilitated the arrival of foreign physicians. Foreign physicians interacted with Japanese physicians who read and modified imported Chinese texts. Japanese physicians relied on skilled Korean artisans who produced delicate needles by tempering silver mined in Japan.[45] Ten Rhijne noticed the role of expert needle craftsmanship when he examined instruments made of silver and gold.[46] But when Japanese physicians interrogated Korean physicians on their techniques, they were often denied a satisfying response. Korean physicians kept their secrets.[47] They described differences in needling techniques without revealing any useful knowledge like their preferred needle type, insertion depth, or precise needling locations.[48]

Intellectual exchange was robust, but restricted. This might have surprised Willem ten Rhijne when he first joined the Dutch East India Company. Barely 27, Ten Rhijne fled Holland to avoid marriage and Franco-Dutch disputes.[49] He boarded the Ternate as a resident medic and left Amsterdam for Japan by way of Cape Town and Batavia, where he switched ships to later dock at Dejima in the summer of 1674.[50]

On the tiny artificial island of Dejima, Ten Rhijne was trapped. Cut off from urban centers, his interactions were limited to government representatives, interpreters, and prostitutes.[51] He met the royal physician Nishi Gempo Kichibei, who served as personal physician and translator to the Shogun. Nishi Gempo Kichibei apparently gathered a gaggle of physicians along with an entourage of interpreters to interrogate the Dutch doctor. These were not always pleasant encounters.

In his book, Ten Rhijne wrote that Tominaga Ichirobei, the "pigheaded interpreter," had too many demands.[52] Interpreters facilitated and inhibited his visit.

After close vetting, Ten Rhijne was finally invited to tour apothecaries in Nagasaki. He was even personally hosted by the Shogun in Edo.[53] During his travels, he remained under the surveillance of official physicians who continued to evade him. When Ten Rhijne asked about their practice, they replied with "ambiguous comparison." If they did offer an explanation, it sounded to Ten Rhijne like "controversial nonsense" or "verbal globs of honey." It was as if explanation without experience was a trap.[54] So, Ten Rhijne studied their images and models, the "hydraulic machines" and "clear figures" where masters prepared models with pores possibly filled with water.[55]

Within two years, Ten Rhijne finally left Dejima and returned to Amsterdam via Batavia and Cape Town as the Franco-Dutch war drew to a close.[56] In Holland, he compiled a 300-page manuscript from essays written during his travels. The final volume was sent to a Royal Society publisher in London, which included chapters on *materia medica*, arthritis, physiognomy, and monsters.[57] It also featured two short chapters on needling and moxa techniques in Japan. This forty-page section was the same document twentieth-century historians praised as "the Western world's first detailed treatise on acupuncture." It was not a treatise, but a modest translation that featured a series of striking full-body prints, one of which was "man with goatee."

Jumping Meridians

Unable to rely on his Japanese counterparts, Ten Rhijne attempted his own translations and relied heavily on metaphor. In particular, navigation was on his mind. Ten Rhijne had spent over 600 days at sea traveling from Amsterdam to Dejima and back, treating crewmates who suffered from malnutrition and seasickness.[58] He studied the ship's captain, or pilot, who guided the crew through unpredictable weather. Seafaring did not directly apply to the body, but it required a skill Ten Rhijne had not mastered, much like acupuncture and moxibustion. He opened the introduction to his chapter on acu-moxa with a question: "How does a pilot locate the harbor for his ship when he is sailing on the broad expanse of the macrocosm of the ocean? He must know how to establish a course of a degree of latitude, which he charts onto a map with a compass to avoid by forethought sandbanks and rocks, and to calculate the probable progress of his ship, hastened or delayed by favorable or unfavorable or even imperceptible turbulence."[59]

Navigating at sea was overwhelming to Ten Rhijne. It was a "macrocosm" filled with "imperceptible" phenomena. Yet foresight could chart a course, an invisible path made real by safely reaching one's destination. These abstract directions on a map could be realized when the crew completed the year-long journey between Dejima and Amsterdam. For Ten Rhijne, navigation scaled to the body. Like pilots, physicians were guided by "imperceptible" phenomena—turbulent tides, flows, and surges. By comparing phenomena in the body to those at sea, Ten Rhijne rendered the body's lines, its latitudes and degrees, as immaterial objects.

This seemed a reasonable metaphor; seafaring was treacherous where one could veer off course or perish. But the body was not the ocean. At least, not quite.

Meridians on paper behaved differently compared with the relatively consistent grid of longitude and latitude. For instance, consider the illustrated paths in **Figure 2.4**, which I will call "man with hat." Meridians were distributed asymmetrically across the face, such as the single path that curled above the left ear. Meanwhile, other paths broke the boundary of the skin, like the three lines on the forehead that continued into the hat. Two paths on the cheeks broke from the face and wrapped around to the other side, as if flung from the skin. On the shoulders, triangles labeled *jianjing/katai* (肩井) protruded from the top to reference sites behind the back on the shoulder blade.

In other words, the sites and paths parted from the body to inhabit unseen dimensions, marking surfaces that were out of sight. Lines curved around the skin and detached from the cheeks. Sites that would have been on the back floated above the shoulders. Sites that would have been on the scalp hovered above skin and cloth.

Ten Rhijne's illustrator may have noticed these flourishes in "man with hat." If we look closely, we find that Ten Rhijne's illustrator likely copied a version of "man with hat" to create goatee man. For instance, both images had quatrefoil kneecaps, absent from the *mingtang tu*

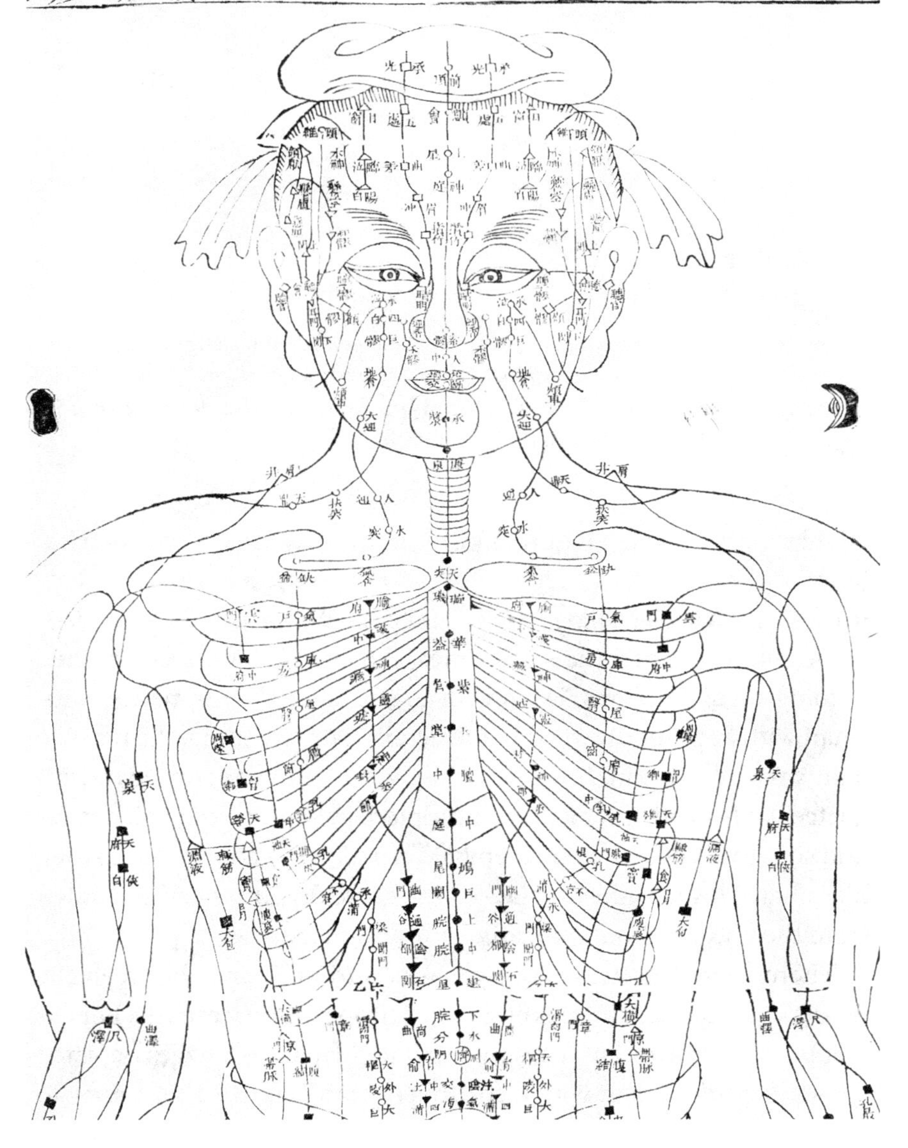

Figure 2.4. Man with hat (detail), sixteenth century. Unknown illustrator, woodblock print. This *mingtang tu* shows symmetrical meridian markings traversing skin and fabric. From Wang Weiyi, *Xin kan tong ren zhen jiu jing*, 1515

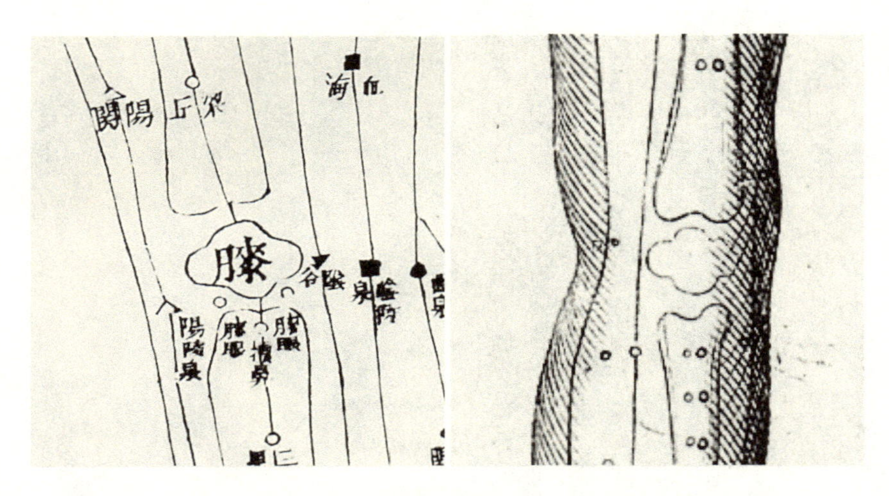

Figure 2.5. Quatrefoil knees (details), sixteenth and seventeenth centuries. Unknown illustrators, engraving and woodblock print. Quatrefoil knees on "man with hat" (*left*) and "man with goatee" (*right*). From Wang Weiyi, *Xin kan tong ren zhen jiu jing*, 1515; Ten Rhijne, *Dissertatio de Arthritide*, 1683

circulating in China (**Figure 2.5**). Both also featured asymmetrical lines on the face, like a rogue path curving on the left forehead (**Figure 2.6**). Both showed *jianjing* (肩 井) sites floating above the shoulders.

Despite these graphic similarities, larger differences established ontological distinctions. "Man with goatee" shared asymmetries with "man with hat" but diverged in the details. For one thing, the hat-like hair was actually skin. If we look closely, the headpiece on "man with goatee" stylistically imitated the headpiece in the *mingtang tu*, but the odd shape was not a hat. It resembled *écorché* skin configured to look like a hat. This hat-like hair lacked the goatee's texture and separated from the body without indicating any acu-moxa sites.

There were other cases of similarity and difference between the Japanese anatomical images and their European counterparts. For example, some European plates that imitated Japanese images depicted a bald, beardless man with a shrunken head (**Figure 2.7**). I refer to this bald, beardless figure as "man with flaps" to acknowledge the *écorché* flaps of skin hanging from his body. This man accompanied goatee man in Ten Rhijne's 1683 publication and resembled another

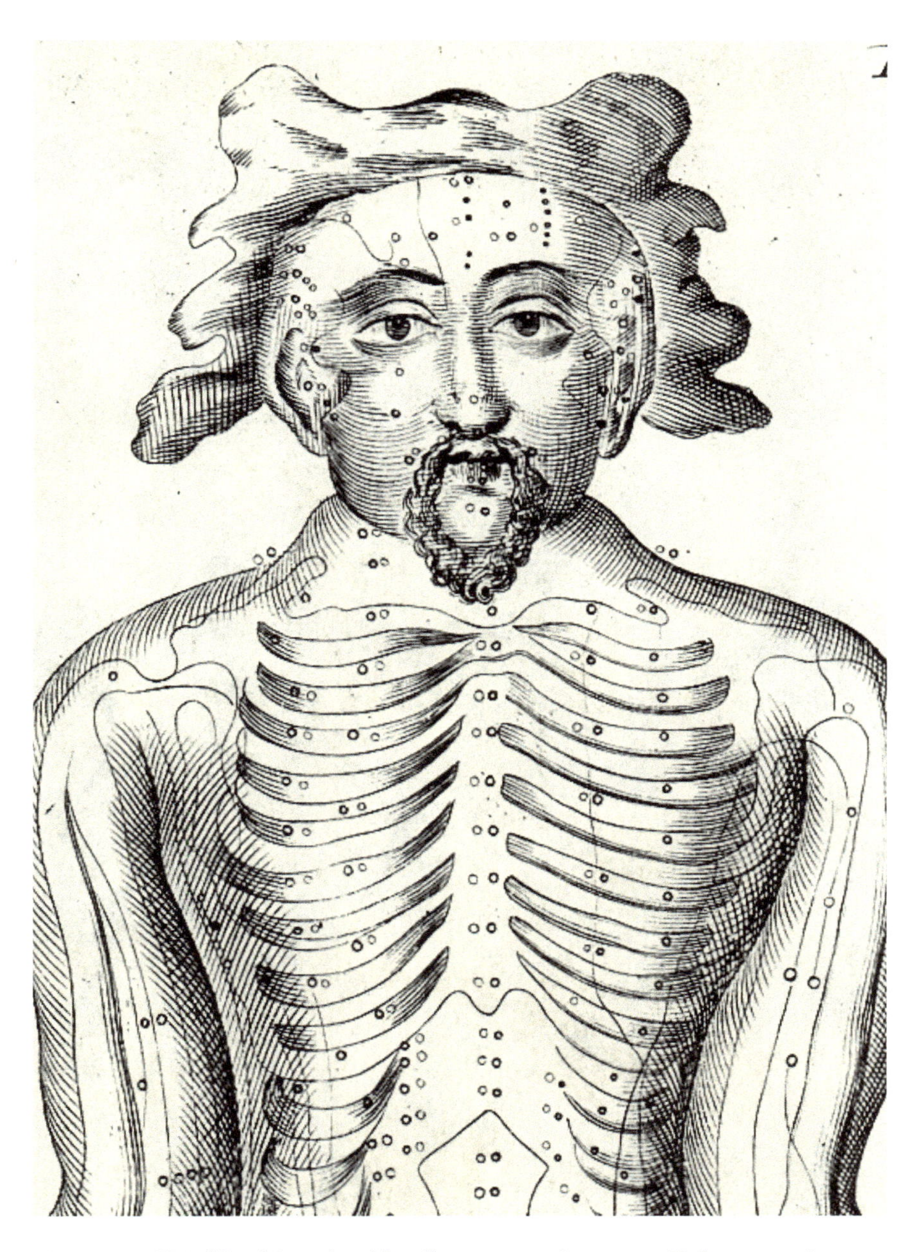

Figure 2.6. Hair-like skin as hat (detail), seventeenth century. Unknown artist, engraving. The figure from Ten Rhijne's manuscript shows an asymmetrical distribution of sites and skin-like hair adoring the head. From Ten Rhijne, *Dissertatio de Arthritide*, 1683

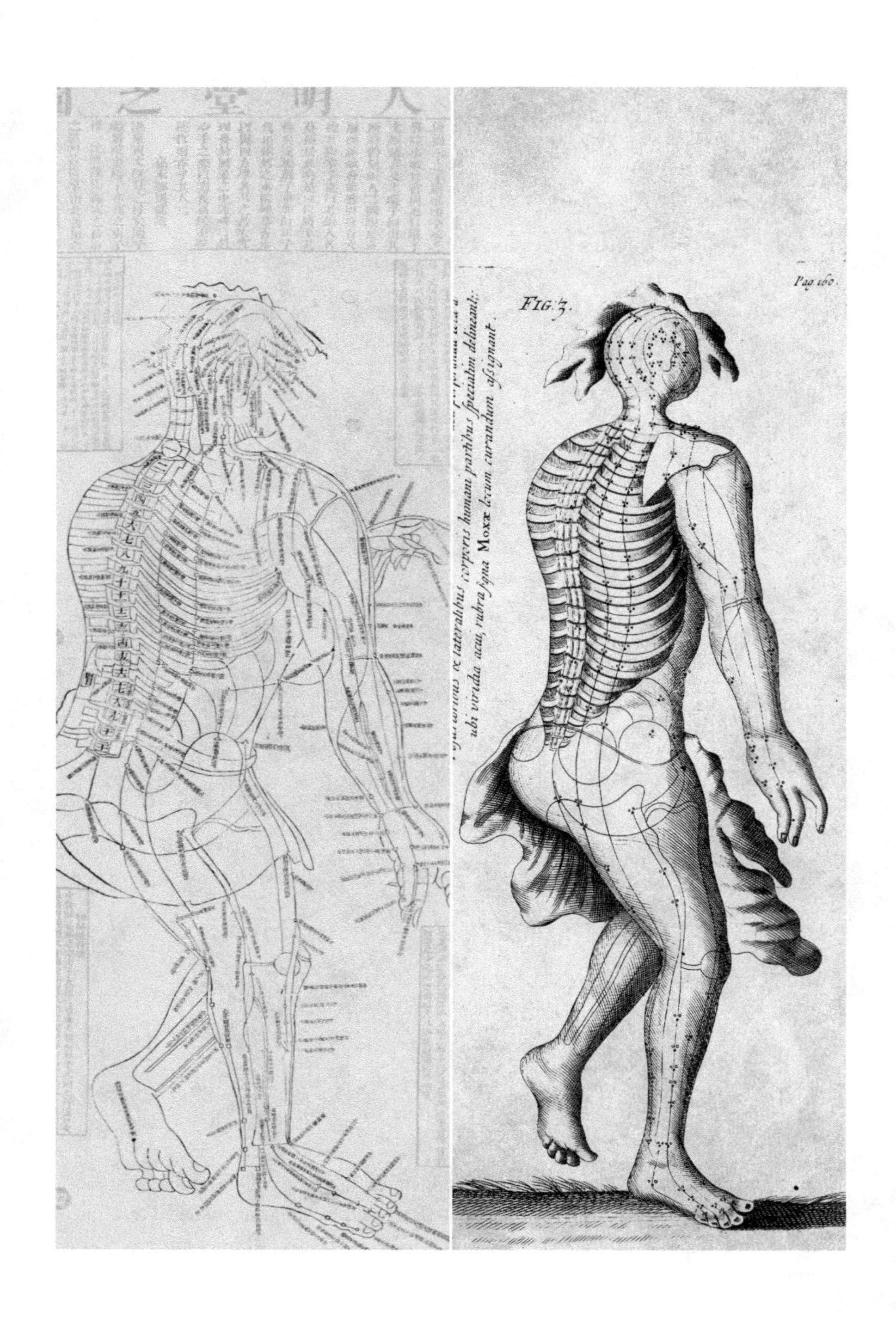
FIG. 3.
Pag 160.
& lateralibus corporis humani partibus specialiter delineant;
ubi viridia acu, rubra signa Moxæ locum curandum assignant.

figure I call "man with robes," which appeared in a 1665 *mingtang tu*. The "robes" here refer to the sash around the figure's waist.

Both the "man with flaps" and the "man with robes" were shown turning away from the viewer, with the right hand extended toward the ground. Each figure displayed markings running from the tip of the toes to the top of the scalp, as well as a prominent row of ribs cascading down the back. Both figures also featured a cloth lining the hips and sets of thin parallel inscriptions marking the joints of the neck, shoulders, elbows, hips, and knees.

Yet there were notable graphic differences between "man with robes" and "man with flaps" that communicated divergent ideas about the nature of meridians. In the "man with robes," the meridian paths were anchored to various surfaces including skin, fabric, bone, and scalp. Other paths easily moved between interior and exterior spaces alongside semi-transparent skin, fabric, and bone. The meridians followed the body's contours but were not confined by them. For example, under the right arm, one path jumped from the torso side and continued up the arm.

In contrast, the skin and bone of "man with flaps" appeared solid and opaque. The illustrator used crosshatching to shade the peeled-back skin to indicate the limits of sight. Here, the meridians only appeared on skin still attached to the body and strictly followed its contours. Unlike "man with robes," the paths in "man with flaps" did not jump out into the open but remained tethered to the outline of bones printed on skin. The skin was not transparent. It had to be peeled back to expose the ribs underneath. The exposed bones were further shaded to show yet another surface that anchored the meridians.

"Man with flaps" may have extended from a lineage of Greek anatomical images meant to conceal the body's interior functions. Known as the *soma*, the physical body was seen as intentionally hiding its inner processes within the flesh, obstructing vision. As classicist Brooke

Figure 2.7. (opposite) Robes and flaps, sixteenth and seventeenth centuries. Woodprint and engraving. A comparison of "man with robes" (*left*) and "man with flaps" (*right*), displaying different kinds of anatomical transparency. From Wang Weiyi, *Xin kan tong ren zhen jiu jing*, 1515; Ten Rhijne, *Dissertatio de Arthritide*, 1683

Holmes has described, navigating these obstacles established a "model for intelligibility."[60] One could not truly comprehend a body, no matter how clearly its parts were shaded, peeled, and bound on paper.

When Ten Rhijne's illustrator translated flesh, bone, cloth, and hands into the image, he actively altered the depths at which meridians flowed and the surfaces in which they were embedded. In "man with robes," meridians leapt freely from the body. Different forms of transparency coexisted. The skin was as see-through as cloth, flesh as solid as bone. By contrast, "man with flaps" disciplined the meridians, pinning them strictly to the skin. On this rigidly delineated figure, meridians were stuck to each surface so that when skin peeled away and lost its form, it also lost its meridians.

Meridians as Humors

Although "man with flaps" explicitly depicted meridians stuck to the solid surfaces of skin and bone, Ten Rhijne struggled to discursively present meridians as fluid objects inside the body. In Japanese, meridians as graphic forms of *kei* (經) and *myaku* (脈) referred to many kinds of structures. *Kei* and *myaku* were polysemous things. They had multiple identities. In response to this multiplicity, Ten Rhijne described meridians based on what he knew—"nerves, veins and arteries."[61]

For Ten Rhijne, the fluid, dynamic nature of *kei* and *myaku*, or *Miak* in his Latin text, more closely resembled bodily humors. He explained that pairs of meridians twisted around each other to produce "radical moisture and innate heat."[62] *Kei* and *myaku* mingled with the hot, moist, cold, and dry qualities of the four humors. However, this did not fully capture their qualities. Ten Rhijne acknowledged his conceptual limitations and echoed classic descriptions that described meridians as "internal vessels" with individual lengths, courses, connections, and names that "continually linked" with each other in the body.

Again, Ten Rhijne tried to make sense of meridians as discrete vessels based on what he knew. He understood that *Miak* as "arteries" (*arterias*) derived from the pulse.[63] They were like pulses, but not quite the expanding and contracting pulse. They could animate blood, though it was not clear what kind of blood. In the fourteenth and fifteenth centuries, blood as a humor was considered as a binding entity that granted access to truth. Early modern German spiritual materialism considered blood as offering a special kind of unmediated reality.[64] Blood joined body and soul. It separated body and soul from the external world.[65] It defined the organs. It was the stuff of heart and liver tissue and the stuff of emotions like anger, sadness, optimism, and laziness.[66] The shape of the humoral body continued to change when blood was let, when one's diet changed, or when one overindulged.[67] Blood moved on its own terms, and when it evaporated, it left its host hot, dry, and angry.

For Ten Rhijne, *Miak* most closely resembled this mutable, volatile version of humoral blood. Still, their precise relationship remained elusive.

In East Asia, *kei* and *myaku* animated Blood, or 血, which was not equivalent to humoral blood. Blood (rendered *xue* in Pinyin and *chi* in Romaji) invoked a broader range of fluids with multiple ontologies—it was polysemous. Blood (*xue*/*chi*) encompassed all fluids nourishing the body, not just the red substance under the skin. Excess sweat could deplete Blood, and overthinking could injure its functions. Reproductive seeds in both men and women were made of Blood. Bad Blood caused irregular flow.[68] Furthermore, the vital functions of Blood relied on the kidneys for water, the liver for storage, and the heart for governance. In women, menstruation and breast milk came from Blood.[69] Blood oriented the deep Yin organs, meridians, and vessels.[70] It could collapse and cause disease.[71]

Blood (*xue*/*chi*) also defined two extra meridians called *chongmai* (衝脈) and *renmai* (任脈). The eighth-century physician Wang Bing (fl. ca. 751–762) described the *chongmai* as the "sea of Blood" vessel and the *renmai* as the "master of the womb." In the eighteenth century, Xu Dachun (1693–1771) further connected these paths to female fertility and menstruation, explaining that the *chongmai* was key to understanding women's diseases.[72] Certain meridians were thus considered specially gendered, particularly in relation to female gestation. Reproductive bodies were distinct entities, differing at times across organs, ages, and degrees of gender difference.

By presenting meridians as "nerves, veins, and arteries" within a humoral body, Ten Rhijne's translation allowed other bodily fluids like humoral blood to overshadow the meridians. In other words, the humoral body did not require meridians to make sense. They could be veins or arteries because "the Chinese . . . use[d] the term loosely."[73] By first equating *myaku* with veins, and then describing veins as arteries, Ten Rhijne applied distinct material qualities to the meridians while blurring the distinctions between the vessels that defined them. Like the volatile humoral blood, *Miak* was a dynamic entity embedded in a fluid, constantly moving body.

In this context, the solidity of veins and arteries brought order to the body. Etymologically, arteries and veins also referenced different physical structures. In sixteenth-century English, "arteries" denoted windpipes whereas "veins" referenced cracks carrying water.[74] Historically, larger arteries carried wind while smaller veins carried water. Both were conceived of as solid structures before becoming blood vessels. Ten Rhijne drew on these connotations to approximate meridians with veins and arteries.

Despite the ontological differences between veins and arteries, Ten Rhijne was more concerned with how their distinctions were defined by their proximity to the heart, as outlined in William Harvey's 1628 publication *De Motu Cordis*.[75] Ten Rhijne focused on the heart as the prime mover of blood. He wrote, "[The practitioner] must understand the motion of the heart (the wind compass of our body), the position, limits, ebb and flow of the rivulets of blood, and avoid injury during the operation. He must be certain of the location which each pain marks with its own signs, and of the accelerated movement of fluid. He must also know that it is safe to burn moxa where whirlpools of higher than normal blood lie concealed in fleshy areas."[76]

Here, the heart animated tiny "rivulets" of blood that ebbed and flowed, only taking form when in motion. The heart was the center of these movements, propelling fluids like an *amusii* (wind compass)—

measuring, regulating, and setting things in motion. If wind and fluid fluctuations were disrupted, if an inherent circulation were disturbed, the practitioner risked unleashing incoherence in the body. More than locating the physical needling site, one had to perceive and anticipate the "accelerated movement of fluid." Movement itself held a certain truth, manifest in the ebbing streams and blood "whirlpools" disturbing the surface.[77] It was within this ecology of spinning mechanical objects that meridians as "nerves, veins, and arteries" enacted discipline; they, too, became solid structures that regulated unruly fluids.

In the case of *kei* and *myaku*, the heart as a vessel or *xin* (心) also guarded the body. The *xin* governed the movement of Blood (*xue/chi*). But Blood relied on the kidneys and liver as well—two Yin organs that lived deep in the body. As a result, Blood was related to the Yin meridians. Ten Rhijne's Japanese counterparts neglected to explain the conceptual and practical implications of a Yin orientation. Equally, Ten Rhijne would have struggled to articulate the constantly mutating humoral body, with its moist red blood, cold phlegm, dry black bile, and hot yellow bile.

Unable to fully translate *kei* and *myaku* beyond Latinizing them as *Miak*, Ten Rhijne resorted to explaining meridians through the familiar humoral model. Here was a body filled with the nerves, veins, and arteries that facilitated radical heat and innate moisture. Perhaps *Miak* simply manifested these humoral qualities. Or perhaps meridians already existed within the humoral body, rendering distinct concepts like *kei* and *myaku* unnecessary to its functioning. Lacking the language to convey their precise ontology, Ten Rhijne merely wedged meridians within the corporeal structures that he understood.

Blood Meridians

In the seventeenth century, Japanese physicians likewise rendered foreign terms redundant. They folded objects like nerves into existing frameworks of *kei* and *myaku* paths that had been used to explain the already elaborated circulation of Qi and Blood (*xue/chi*). For instance, historian Noriaki Matsumura has offered a compelling analysis of Japanese doctors who selectively appropriated concepts into an existing intellectual framework.[78] Structures that resembled nerves did not supersede *kei* and *myaku* but instead further articulated them. Physicians like Arashiyama Hōan (1633–1693) developed new terms like *seibinseinon* and *seinon* to articulate *kei*, *myaku*, and the movements of Qi/Blood (気血).[79] Similarly, Sōtei Nakamura developed the term *zeinun* to also characterize types of Qi/Blood circulation and twelve pairs of primary meridians.[80] Even when Yoshimasu Motoki (1628–1697) identified the term 筋 as muscles that were distinct from meridian paths, this did not appear as a novelty.[81]

New anatomical terms were emerging in a period of pluralism, innovation, and neo-traditionalism in Japanese medicine. In the sixteenth century, Japanese medicine began moving away from secrecy and familial transmission toward a more open educational model.[82] This shift from theoretical to practical learning revolutionized medical education as private schools proliferated alongside the state-run shogunate school in the eighteenth century.[83]

Interest in classical texts prepared the ground for assimilating humoral anatomy. Confucian scholars like Itō Jinsai (1627–1705) and Ogyū Sorai (1666–1728) developed a critical philology of ancient texts that transformed assumptions about textual authority and empirical evidence. As Daniel Trambaiolo has argued, dissenting physicians focusing on the "Ancient Formulas" (古方) criticized speculative orthodoxy and sought to revive classical knowledge, while others like Yoshimasu Tōdō (1702–1773) and Yamawaki Tōyō (1705–1762) took on integrating meticulous study of ancient medical writings alongside direct clinical and anatomical investigation.[84] Members of the Kohoha "Old School" like Gotō Konzan and Kagawa Shuan sought to revive empirical classics like *Cold Damages*.[85] These scholars revived techniques like abdominal palpation diagnosis that inspired new interest in anatomy, spurring Yamawaki Tōyō's landmark 1754 human dissection and 1759 anatomical treatise *Zōshi*. With the banning of human dissection, physicians like Gotō Konzan (1630–1716) actively conducted animal dissections to establish new empirical approaches to anatomical study.[86]

With these shifts underway, Japanese physicians began translating Blood (*xue/chi*) into something like blood as a humor in the eighteenth century. Turning Blood (*xue/chi*) into blood/humor did not depend on translating the substance itself, but instead relied on translating the capillaries that contained it. Japanese physicians initiated this ontological transformation through reading and reproducing images. It was through images that Blood (*xue/chi*) turned into blood as a humor and from blood as a humor into blood as a capillary. Blood became an anatomical object with individually labeled parts. Instead of remaining a fluid object, Blood as capillaries turned rigid.

The transformation of Blood into a humor and then into a network of capillaries appeared most prominently in the famous eighteenth-century medical treatise *Kaitai shinsho* (*New Book on Anatomy*), published in 1774 and attributed to the Japanese physician Sugita Genpaku (1733–1817). Although Sugita had attended dissections, he and his team were also interested in existing anatomical atlases. As a result, *Kaitai shinsho* was a chimera of sorts. Its five volumes represented Sugita's interpretation of many sources, interweaving didactic descriptions, translations, and images from various treatises. *Kaitai shinsho* drew on books by the Spanish engraver Juan Valverde (selections from *Anatomie*) and the German anatomist Johann Adam Kulmus (selections from *Ontleedkundige Tafelen* in Dutch), which were both influenced by Vesalius's 1543 *Fabrica*. Other *Kaitai shinsho* illustrations derived from the French barber-surgeon Ambroise Paré (1510–1590), the Danish physician Thomas Bartholin (1616–1680), and Dutch doctors Steven Blankaart (1650–1702), Volcher Coiter (1534–1576), and Govard Bidloo (1649–1713).

Among the dozens of translated woodcuts was an image of capillaries, or something approximating capillaries. Here stood a skinless, boneless, faceless figure with slightly bent legs and arms (**Figure 2.8**). Sugita's team had translated Kulmus's version of Vesalius's capillaries in *Fabrica*. At the center of the image were two thick trunks, curving around each other and meeting at the chest before splitting into broad shoulders. To the left, a web of trunks and branches vanished at the fingertips. To the right, branches fed into a solid, opaque arm. They

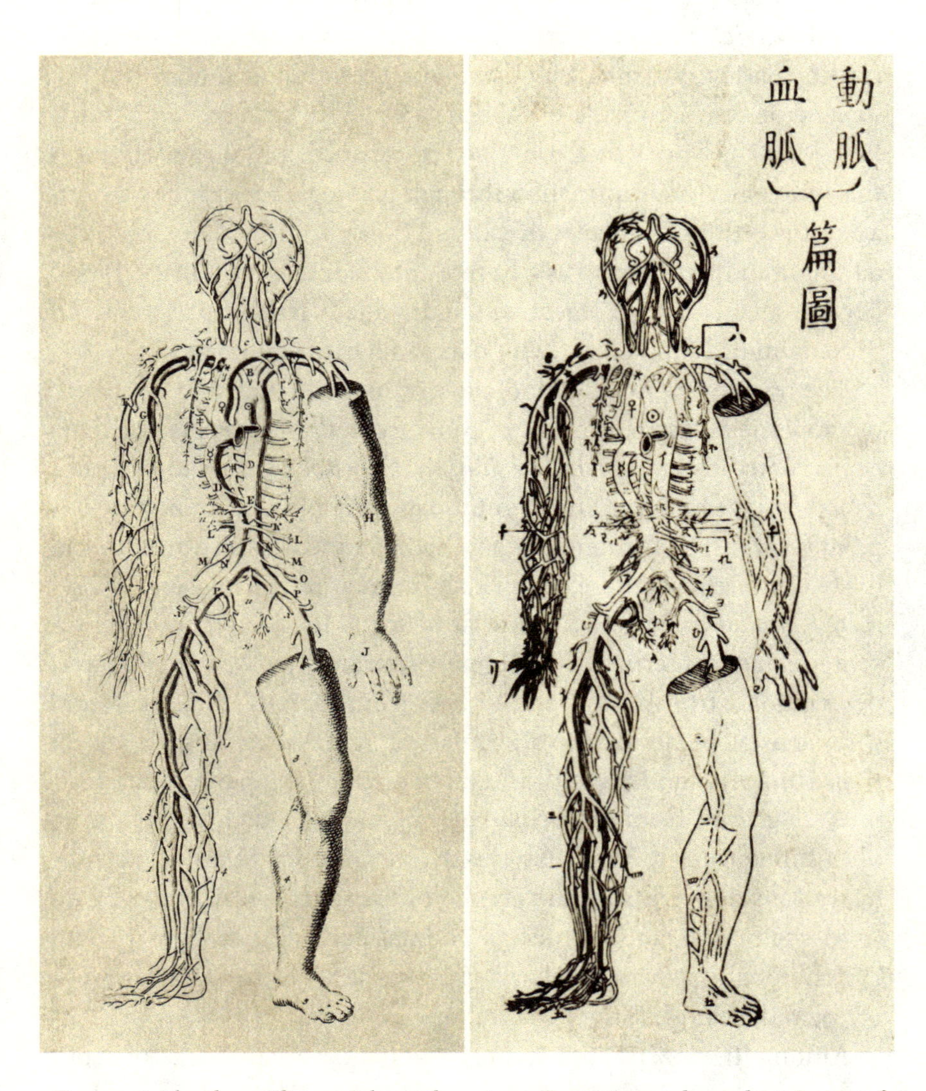

Figure 2.8. Blood meridians, eighteenth century. Engraving and woodcut. Image of capillaries from *Kaitai shinsho* (*right*), likely derived from *Ontleedkundige Tafelen* (*left*). Letters were replaced with katakana characters; arteries and veins were renamed *monmyaku/menmai* (門脈) and *dōmyaku/dongmai* (動脈), invoking direction and movement. From Kulmus, *Ontleedkundige Tafelen*, 1734, plate 16; Sugita et al., *Kaitai shinsho*, vol. 1, 1774, 11

pressed against the surface of the skin and formed light indentations. The engraved lines imprinted interior structures on an exterior flesh, transforming capillaries into a flattened topography.

In the process of translating the image from an etching into a woodcut, the capillary body became something like a *tu* (圖). Unlike the crosshatched original, the new version was almost transparent. Rather than bulging from the skin, the vessels simply appeared. And despite this *tu*-like rendering, the unfamiliar labels did not guide readers. Some characters seemed clear, while others appeared vague or garbled. Greco-Roman signs like Venus ♀ and Mercury ☿ stayed the same, while numbered legs changed from 1, 2, 3, 4, and 5 into 一, 二, 三, 四, and 五. Latin letters like H, J, L, N, and O transformed into imperfect katakana resembling 千, 九, テ, カ, and ヨ. Other characters disassembled into illegible marks.

Sugita invented new names for this image, captioning it as showing *dōmyaku* (動脈) and *ketsumyaku* (血脈), which might have meant something like "moving meridians" and "blood meridians."[87] Together, these terms offered separate identities for blood: one as motion, the other as *xue*/*chi*. While classical texts discussed arteries, they lacked terms for an array of capillaries. As a result, these new compound words *dōmyaku* and *ketsumyaku* had ontological implications. The character *myaku* (脈) contained a flesh radical and pictograph of branches, the same character that imagined meridians as fine networks of 脈 (*myaku* or *mai*), unlike the larger vessels found in *jingluo* (經絡).

In other words, blood as a humor from the Dutch engraving became a kind of fleshy meridian in the Japanese woodcut.

Sugita explained these delicate webs of *myaku* formed an extensive network of very thin, fine things. The specific *dōmyaku* (動脈) did not represent the entire capillary system, but a particular set of smaller capillaries. These finer entanglements were part of a broader taxonomy defined by their relationship to another class of vessels termed *monmyaku* (門脈). *Monmyaku* was a specific kind of *mai*/*myaku* that connected to the internal organs. As Sugita wrote, "*Monmyaku* is one of two blood meridians. Together, they enter the liver and join at its interior [to] form tiny paths of blood [that] wrap around and [form] a network."[88]

Sugita continued to impress that *dōmyaku* blood only became blood when it arrived at the *monmyaku* via the liver. It had a specific destination. It transformed through motion. When the stuff of *dōmyaku* entered the liver, it would turn into the stuff of the gallbladder. *Dōmyaku* thus referred to both a substance and a structure. It was defined by what it looked like, where it was located, and the mutability of its contents. It was both blood as humor and Blood as *xue/chi*. As a relational object, it was smaller than *monmyaku* and situated below it near the stomach, where it branched into finer forms.

These descriptions contrasted with the graphic image of *dōmyaku* and *ketsumyaku* in **Figure 2.8**. "*Ketsumyaku* are delicate," Sugita stated.[89] Yet no finely rendered branches clustered around the liver or stomach in the illustration. Even though the organs defined the *dōmyaku*, no organs appeared on the page. The skinless, boneless body had no heart, no lungs, no gallbladder, no large intestines, no small intestines, no triple burner, no bladder, no pericardium. The only detail from the text that was reflected in the image was the idea of *ketsumyaku* as bulky "waterways" (*suidō*) forming a "fine branched network" (*xizhiluo*). While featuring both bulky and fine structures, the figure matched little else in Sugita's textual account, underscoring the disparities between description and depiction, between text and image.

Rendering Blood (*xue/chi*) as a physical humor was an issue of scale and visualizing branches that did not appear on the page. The engraved capillary skeleton offered a partial translation of invisible networks. Meanwhile, the neologisms *dōmyaku* and *ketsumyaku* had implications for the materiality of Blood (*xue/chi*). By invoking the fleshiness of *myaku*, they linked meridians to vessels and vessels to meridians. Describing vessels as "waterways" further fixed Blood into a concrete apparatus so that it became both the fluid object and the solid vessels containing the fluid.

The asymmetrical, muscular, ungendered capillary figure thus involved material, conceptual, and discursive translations. It translated etching to woodcut, Latin to katakana. Through the solid rendering of conduits, it also translated Blood (*xue*/*chi*) into a humor approximating blood by representing the capillaries. On the page, blood meridians looked as solid as bone.

Useful Images

Kaitai shinsho took three years to produce.[90] Sugita Genpaku's team included translator Maeno Ryōtaku (1723–1803), illustrator Odano Naotake (1749–1780), and physician Katsuragawa Hoshū (1751–1809), who ensured that the shogunate approved the text. Although *Kaitai shinsho* spread rapidly across Japan, introducing new anatomical terms, it remained a curious object.

Ironically, Sugita Genpaku was illiterate in Dutch and unfamiliar with Latin or Roman script.[91] During translation, he relied on his mentor Maeno Ryōtaku, who had spent some time in Nagasaki and was familiar with Dutch syntax. But this was not enough. His team still spent hours staring at the pages of *Ontleedkundige Tafelen*, hoping that its contents would somehow reveal themselves.[92] Sugita wrote, "[I]t was as though we were on a boat with no oar or rudder adrift on the great ocean, a vast expanse, with nothing to indicate our course. We just gazed at each other in blank dismay . . . As far as I was concerned, I knew nothing of the Dutch language, not even the 25 letters of the alphabet . . ."[93] Sugita's team was lost. They were adrift "on a boat with no oar or rudder" on the "great ocean," gazing at each other in dismay. They did not recognize the words for things like "eyebrows," and could not distinguish between description and metaphor. Many expressions "remained a puzzle."[94] Some words dissolved into symbols. For instance, "*zinnen*" made no sense, and the team instead represented it as a divided circle. Even deciphering a few Latin sentences was considered a triumph. After three years of struggle, some sections still remained illegible.

Sugita Genpaku deliberately avoided interpreters, spurning their dictionaries and general expertise. In Nagasaki, interpreters like Inomata Dembei and Katsuragawa Naganari had gained surgical training under Dutch doctors and produced their own texts.[95] They even established surgery schools such as the Kurisaki School, Katsuragawa School, and Casper School.[96] Yet Sugita shunned these resources, considering interpreters too low in status to be true scholars. He claimed their schools only taught the very basic techniques for applying plasters and ointments, doubting the value of the interpreters' annual interviews with Dutch traders.[97] Sugita instead preferred his amateur scholar collective. He relied heavily on artists like Hiraga Gennai (1728–1780), who specialized in Dutch representational art, and master engraver Odano Naotake to collect, translate, and rearrange sections of the book. While interpreters' empirical exposure cultivated practical knowledge, Sugita fixated on the text and images as embodiments of unmediated truths.

Sugita's own practice of dissection derived from Yamawaki Tōyō (1705–1762), a leading figure in "Ancient Practice" (*ko-ihōha*) who performed autopsies. At the time, autopsies were rare and considered taboo. Dissection was further considered a menial task reserved for non-physicians and outcasts known as *eta* or *kawata*. Regardless, Tōyō reconfigured existing techniques so that he could open and peer into bodies. When struggling to translate Latin, Sugita's team attended public dissections and marveled at the accuracy of Dutch images. "We found that they were nothing like those described in the old books," Sugita wrote, "but were exactly as represented in the Dutch book. We were completely amazed."[98] Dutch images, Sugita proposed, rendered Chinese medical texts useless.

However, the 1771 dissection that Sugita and his colleagues attended may not have been as singularly decisive in their embrace of Western medicine. Japanese doctors were already eager to confirm the accuracy of Western anatomical illustrations.[99] Their intentions were already socially determined. They wanted to elevate Dutch anatomy. Weary of these predetermined expectations, Sugita's expert artist collaborators Hiraga Gennai and Odano Naotake remained skeptical of the promises of Dutch image production. As masters, they were self-conscious about the existing circumstances that determined ways of seeing and modes of image reproduction. Flattening the body required rendering some aspects visible while obscuring others. According to art historian Svetlana Alpers, Dutch visual styles that defined Northern Renaissance visual practice extended from seventeenth-century mapmaking. Surveyors, artists, painters, and non-specialists all tried to map the Netherlands, which was quite flat. These kinds of maps generated a class of "realistic landscapes" that focused on surfaces at the expense of volume and solidity.[100] The flat canvas created an illusion of seeing the "truth" of a city.[101]

Without a graphic category of *tu* that distinguished between maps as paintings and as technical artifacts, seventeenth-century Netherlandish viewers wrestled with the discrepancy between the image and its textual description. For mapmakers, maps were not always "templates for action" but representations that joined many perspectives, from the surveyor who scaled towers and looked upon the land to the artist who painted the image into a frame. Looking at the image then seemed "contriving" and full of "deceptive visions."[102] Unlike the Italian tradition of treating the canvas as a window, the Netherlandish style of looking at paintings could slip between many modes of comprehension.

Japanese artists understood these problems of seeing.[103] They, too, were struggling. In the sixteenth and seventeenth centuries, imported Dutch glass introduced objects that facilitated new modes of vision. As art historian Timon Screech has noted, glass simultaneously exposed the inner contents of a vessel and separated them from the viewer.[104] Glass seemed to manufacture otherness. Mirrors and reflective surfaces introduced complications. Even Sugita Genpaku felt disturbed by his refined reflection, describing new self-awareness and "double vision."[105] He existed in the mirror, in the reflection, and in person. When he encountered an external version of himself, he felt compelled to judge it.

Sugita's precarious translation projects further responded to sociopolitical pressures.[106] By the mid-Tokugawa period, Japan was facing an economic crisis in the process of undergoing an agrarian-to-commercial transformation.[107] These economic shifts led to an increasingly indulgent elite class and rising debt. Poverty intensified and infanticide increased. This mobilized artists such as Satake Yoshiatsu (1748–1785), one of the founders of "Western-style" art in Japan, to distinguish useful from useless images. Artists needed to prioritize addressing social needs rather than individual experiences. Classically trained artists who expressed personal perspectives now seemed irresponsible. Self-expression was self-aggrandizing. Artists instead needed to distance themselves from their subjects; affective relationships were wasteful.

Under these circumstances, Sugita felt compelled to produce *Kaitai shinsho* as a socially useful work. Although not an expert illustrator or translator, he urged others to see bodies as he did, or rather as his images did. He was afraid of blind ignorance. His ideal of self-cultivation, influenced by Neo-Confucian empiricism, held that sages and commoners differed only in perspective. The difference was in *how* they saw the world.[108] *Kaitai shinsho* was as much a moral undertaking as a medical one, promoting specific ways of seeing that responded to social and economic pressures that informed Sugita's own positionality.

"Spirit Nets"

It was amid ongoing economic crises and social pressures for useful imagery that Sugita Genpaku encountered the transliterated term for "nerves" as *seinun/zenuw* in *The Book on Wounds* (*Kinsō no sho*).[109] Perplexed by this unfamiliar object, he referred to an etching in Juan Valverde de Amusco's 1560 *Anatomia del corpo humano* (**Figure 2.9**). In both Valverde's version and its Vesalian source, nerves tangled with blood vessels and bone rather than existing as an independent entity.[110]

The image appeared to be a view from the back. One thick vessel in the shoulder gave rise to numerous tendrils that hung from the body in the shape of a dangling arm. The shoulder vessel led to a spinal core (rather than the spinal cord) where individual vertebrae sat neatly in a vertical stack. A slender set of tubes threaded through each vertebra. The spinal core seemed to be knitted together by arte-

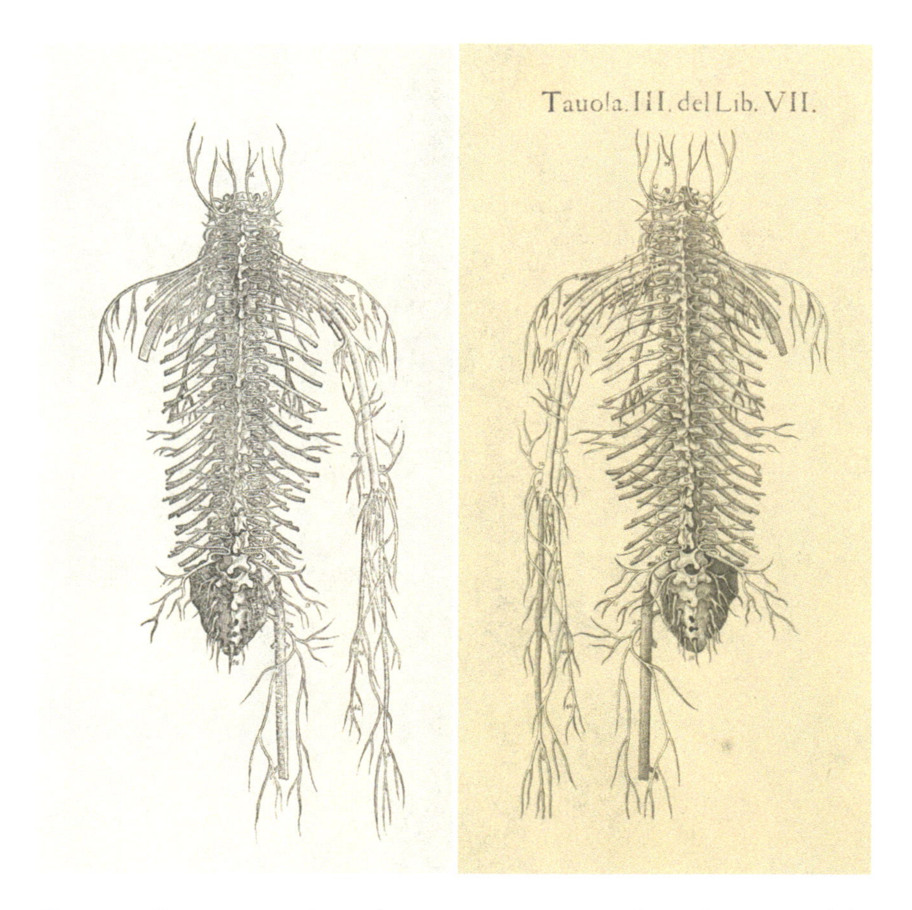

Figure 2.9. Bony nerves, sixteenth century. Engraving and woodcut. One of the images in Andreas Vesalius's 1543 *Fabrica* (*left*) displayed vertebra bones, blood-supplying ventricles, and cranial nerve pairs. The image was roughly copied, not traced, in Juan Valverde's 1560 publication *Anatomia* (*right*). This second engraving most likely made its way to Sugita Genpaku. From Vesalius, *De humani corporis fabrica*, 1543), 333, plate 52; Valverde, *Anatomia del corpo humano*, 1560, 152

rial tubes. From the spine, what looked like a set of ribs made of bone turned into bone-like blood vessels. In other words, nerves were hard to see. The structure folded nerves in a composite copse, where ribs of bone became bone-like blood vessels tethered in the spine.

The *Kaitai shinsho* version of bone/blood/nerves lost an arm (**Figure 2.10**). Two sets of branches extended from the neck and reached

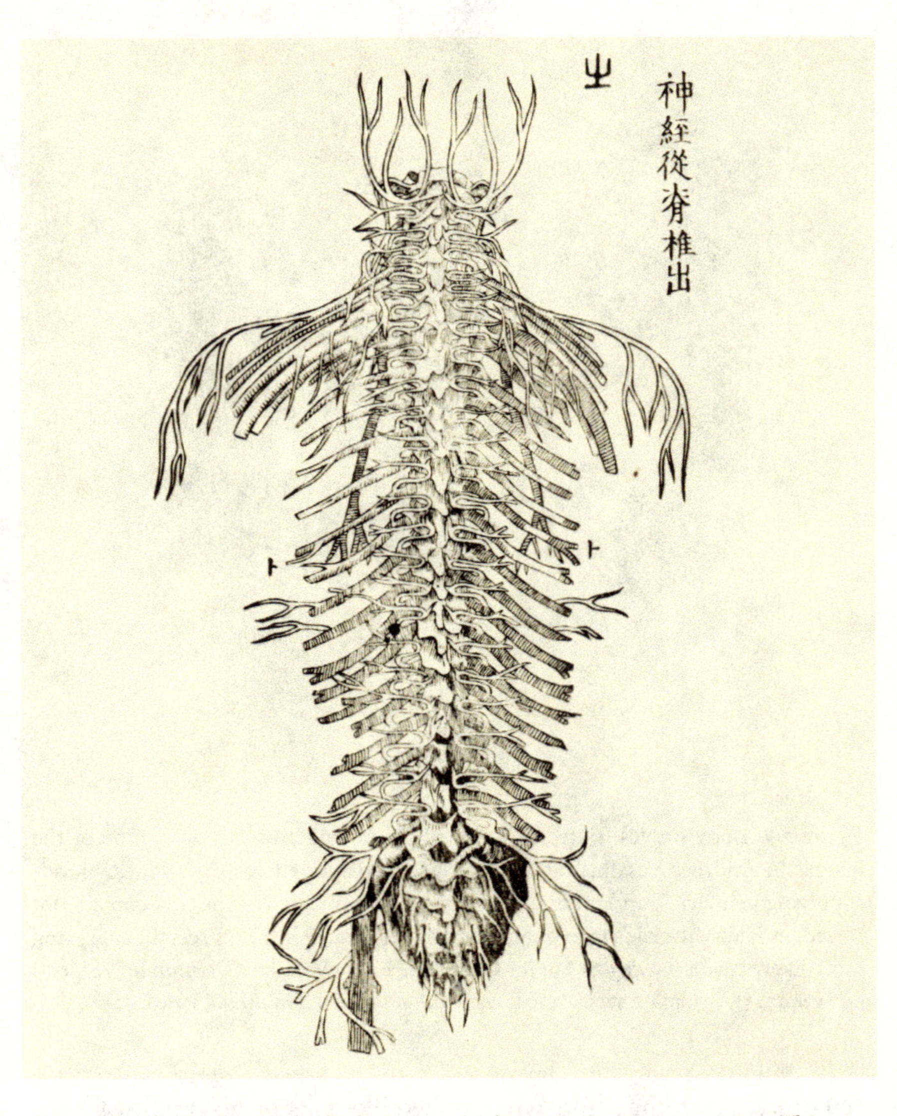

Figure 2.10. Spirit nets, eighteenth century. Unknown engraver, engraving. This unusual image in *Kaitai shinsho* was likely copied from Vesalius or Valverde. The name of bone/blood/nerves appears as *shinkei* (神經), or "spirit nets." From Sugita, *Kaitai shinsho*, vol. 1, 1774, 10

upward to suggest the presence of a head. A vertical stack of vertebrae again lined the back. A lump of thick whiskers at its base protruded from a cavernous mass. A twisting row of tubes again threaded through vertebrae. Blunt branches mimicking bones flanked the spine and contributed to an illusion of symmetry. These branches did not come in perfect pairs. Sixteen stumps lined the right of the spine, and 15 stumps lined the left, depending on where you started counting.

The hand that carved the faceless capillaries in Figure 2.8 also carved this image. It was again hard to discern where the bones ended and the soft vessels began. Nerves were again difficult to see. The headless body had thick structures in the shoulder that resembled *ketsumyaku* (血脈), which looked heavy, solid, and fleshy. Sharp branches that resembled blood meridians but were not blood meridians curved to form the rest of the shoulders. To describe these branches, Sugita Genpaku's team offered two new words: *shinkei* (神經) and *sekitsui* (脊椎). This caption further suggested that the image showed *shinkei*, quite literally "spirit nets" or "spirit meridians," that sprouted from *sekitsui*, or the bones in the back.

Naming the branches "*shinkei*" was an odd choice. Whereas *ketsumyaku* suggested fleshy meridians (or, more accurately, the capillaries that disciplined blood), *shinkei* suggested something more ethereal. Etymologically, 神 (*shin*, or *shen* in early Chinese literature) described rarefied things. It invoked divine power and transcendence.[111] The old character of *shin* (申) also referred to lightning, which further linked 神 to the physical forces of nature.[112] Thus, 神 made reference to things within and without the physical world. As Mencius put it, 神 was a state beyond the human sages.[113] It was always out of reach.

According to sinologist Justin Winslett, 神 represented the "extrahuman."[114] It was everywhere. It intervened in the human world alongside other extrahuman things, including ghosts [*gui* (鬼)], earth spirits [*qi* (祇)], generic gods (*guishen*), and rare gods (*shenqi*). Meanwhile, 神 also referred to human characteristics. Compound words like *shenming* and *jingshen* described one's uncanny intelligence. Other words like *shensheng* described particularly sensitive individuals.[115] So common was this ecology of human and extrahuman 神 that Judeo-Christian missionaries in China and Korea also took notice.[116] Missionaries in particular avoided using 神 in their translation of "God," and instead use words like *shangdi* (sovereign on high) and *tian* (heaven) that were explicit in establishing social hierarchies.[117] These were distinct from the descriptive ontology of 神, which would come to further characterize the nature of nerves.

Whereas *myaku* assisted in the translation of fleshy blood vessels, *kei* (經) assisted in translating the ephemerality of nerves. *Myaku* turned blood into flesh; *shinkei* turned nerves into something divine. Sugita occasionally invoked the divine, vaguely making references to God when his team was frustrated with the words that failed to translate: "[W]hen we had vainly struggled with an unintelligible word, we would exclaim in despair, 'Let's give it the bit with a cross.' "[118] In other words, relegating unknown concepts into "the bit with a cross" was almost done in jest. "We used to say, 'Man proposes, and God disposes,' " Sugita added. The team of ragtag translators made their frustrations known. Anything that did not pass through translation was not assigned a new name but turned into a sign that commemorated the moment when Sugita and his team had given up.

Sugita's seemingly frivolous treatment of text reveals the effects of translation on the material quality of the body. For instance, creating the word *shinkei* for nerves made them distinct from *ketsumyaku* for fleshy meridian blood capillaries. The choice of words made material distinctions. Nerves were extrahuman; capillaries were not. Nerves were almost like meridians—being of the body and beyond the body—whereas fleshy blood meridian capillaries were tied to the flesh. Nerves, through the creation of *shinkei,* almost replaced meridians as meridians. Meanwhile, Blood as a humor almost anchored meridians as vessels.

This distinction is curious because early modern Dutch physicians had used blood to conceptualize nerves. Blood was foundational to understanding nerves. The blood in the veins manifested as blood in the nerves. Physicians presumed that something like blood also flowed in nerve vessels, which they called "nerve juice." This idea was famously credited to the medical instructor Herman Boerhaave (1668–1738) who fixated on developing an explanatory framework for nerve juices based on existing theories of blood and circulation.[119] According to Boerhaave, life depended on the "condition of several fluid and solid parts of the body."[120] Blood in the nerves had to move for the body to access the world.

Early modern natural philosophers like Boerhaave understood that sensing relied not on nerves alone, but on the collective forces of all the juices flowing through the vessels. Nerve juices perhaps more closely resembled the inclusive object of Blood (*xue/chi*). Physicians described how all fluids derived from blood and gathered in the "Sensorium Commune." The Sensorium was not the brain itself, but the center of sight and sound, distinct from the heart. Within this center, nerve juices were a subtler type of blood. As a devout Calvinist materialist, Boerhaave aimed to describe how even the smallest vessels animated the body.[121] Diseases emerged from the Sensorium when its circulation was interrupted.[122]

Despite Boerhaave's materialism, the Sensorium Commune was an imaginary construct. It could not be revealed through anatomical illustration or dissection. One could not open the body, peel back its layers, and encounter the center of sight and sound. Despite the visual articulation of branching vessels, its contents—blood, nerve juice, and other finer fluids—only came to life through speculation.

It is hard to say if Sugita Genpaku took the Sensorium as a material object, a metaphor, or both. Regardless, *shinkei* already referenced a known material object and functioned metaphorically. The divine *shin* (神) provided form to unfamiliar nervous branches. The ontological familiarity of minor spirits (*shin*) and the physiological familiarity of *kei* (經) served as material metaphors for the unknown object of nerves, and the further unknown object of the Sensorium. Even as Sugita prized "useful" vision, these unseen objects gave external structures meaning. Nervous branches on paper were meant to be as self-evident as spirits and meridians. Together, they joined otherworldly objects and empty vessels. Grafting foreign concepts onto existing frameworks remade the body.[123]

Conclusion

Translation was less about knowledge "loss" than a space where social conditions, commercial restrictions, economic anxieties, material culture, and graphic conventions shaped how physicians came to see, articulate, re-present, and misrepresent unfamiliar versions (and visions) of the body. When itinerant physicians moved images from Japanese to Latin and from Latin to Japanese, they also moved the images through discursive and visual transformations that multiplied the ontology of meridians and nerves. On paper, meridians could have been animated by Blood (*xue/chi*) or blood as a humor. Nerves could have been engaged in the Sensorium or been animated through the existence of extrahuman spirits.

Illustrators of images like man with goatee underestimated the ontological range of *kei* (經) and *myaku* (脈). The authors of woodblock prints showing spirit nets overestimated the explanatory power of nerves. These transitory objects each demonstrated how partial perspectives relied on metaphor to make sense. Metaphors were more familiar, more persuasive. They named unfamiliar structures and gave form to unseen fluids.

For instance, Willem ten Rhijne's transliteration of meridians as *Miak* in Latin described it as a combination of blood, nerves, vessels, and humor. *Maik* became redundant in a humoral body that already facilitated "radical heat" and "radical moisture." The humoral body did not need meridians.[124] This was perhaps, in part, why images like man with goatee failed to impress.[125] Upon reading Ten Rhijne's meridian section, British theologian William Wotton (1666–1727) called the entire enterprise mere "banter," dismissing the image as "whimsical" and "tedious."[126] Meridians were irrelevant.

This, of course, was only part of the story. The process of illustrating, tracing, copying, and naming images created the possibility for inventing new anatomical objects. Despite the poor responses to man with goatee among Ten Rhijne's contemporaries, meridians served as a medium for translating objects like capillaries and nerves in Japan. Blood, flesh, bone, and nerves intermingled with meridians. In the case of Sugita Genpaku, who had struggled to translate Dutch texts, meridians came to serve as a basis for translating blood as heavy, solid, and fleshy *ketsumyaku*. Blood meridians took references to flesh in the word *myaku* and disciplined Blood (*xue/chi*) so that it became something like humoral blood captured and contained by the capillaries.

Meanwhile, the wiry *shinkei* that emerged from bony *sekitsui* transformed nerves into something both self-evident and other-worldly. In this process of translation, meridians as *myaku* assisted in the translation of fleshy blood vessels, whereas meridians as *kei* assisted in the translation of ephemeral nerves. Where *myaku* (脈) turned blood into flesh, *shinkei* (神經) delivered nerves to the divine.

By the mid-nineteenth century, Japanese physicians continued to translate Dutch medical texts. Medical schools dedicated to German and Dutch styles of anatomy and physiology expanded as Japanese physicians hired German surgeons to perform procedures in internal medicine, surgery, dermatology, and obstetrics, which fell into a broader category of "Dutch-style surgery." At the same time, medical plurality persisted with physicians studying classical texts like the tenth-century treatise *Ishimpō*, or *The Heart of Medicine*. They continued to offer new words for sensation;[127] and they still invoked the extra-human attributes of *shin* (神).[128] These practices facilitated a multiplicity of images and words that implicated different versions of meridians, transforming ethereal blood into everyday flesh and mundane nerve juices into extra-human spirits.

Images of meridians—whether pinned to the skin or flying off the face—were created by multiple hands and multiple interests. Among these images, meridians and nerves took on multiple identities. Early modern images made anatomy seem self-evident—as if vessels and capillaries existed through uninhibited observation, rather than the through the affective force of metaphor.

To twentieth-century scholars, decontextualized images like "man with goatee" often appeared as unproblematic Chinese maps.[129] Even modern translations of Ten Rhijne's engravings described them as "copies of an ancient Chinese chart" as well as "a Japanese imitation."[130] It seemed to suggest that regardless of its type or style, the engravings traced back to China. Despite the imperial physicians who worked closely with Ten Rhijne, such as Motogi Shodayu, Iwanaga Soko, and Yokoyama Yozaemon, the sites and paths on the re-represented body were essentially Chinese.[131] Although the image was marked by a particular visual style, for many, it originated from a single primary source. These narratives manifested political imperatives that reinforced legacies of a modern Chinese medicine with traditional, unchanging, unchallenged roots. Of course, the story is never so simple.

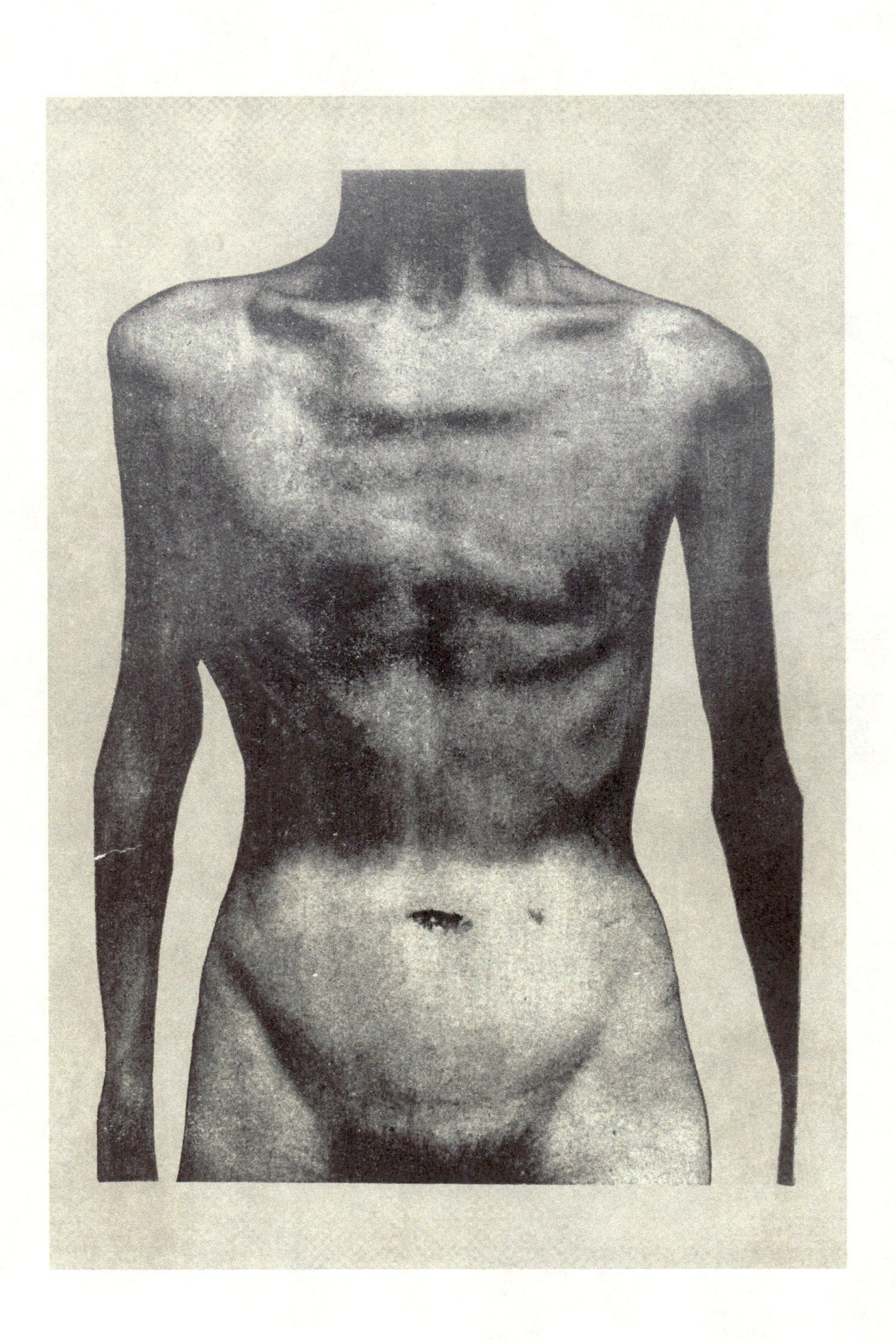

CHAPTER THREE

The Limits of Anatomy through *Tu* (圖)

Meeting the Model

Cheng Dan'an (1899–1957) took out a pocket camera.[1] In front of him stood a naked man who looked frail and gaunt. His rail-thin arms loosely dangled from his shoulders. Cheng positioned the man under a light that raked down his chest. It highlighted his rigid collar bones, pronounced rib cage, and emaciated waist. Cheng extended the folded lens of his camera and lifted the shutter. The man turned around. Cheng took a picture. He captured the man's front, back, and sides. Cheng then closed in on details in the arms, hands, legs, and feet (**Figure 3.1**).

Cheng used these images to create new meridian maps. He developed the photos, cut out the man, removed his head, and started drawing. He inscribed a series of white and black dots and connected them with white and black lines. This made one layer of the map (**Figure 3.2**). Cheng then glued a second piece of rice paper on top of the original image and traced the body for a second time. He outlined

Figure 3.1. (opposite) Headless man, twentieth century. Cheng Dan'an (1899–1957), photograph. From Hunan Museum of Acupuncture-Moxibustion Archives. Used with permission

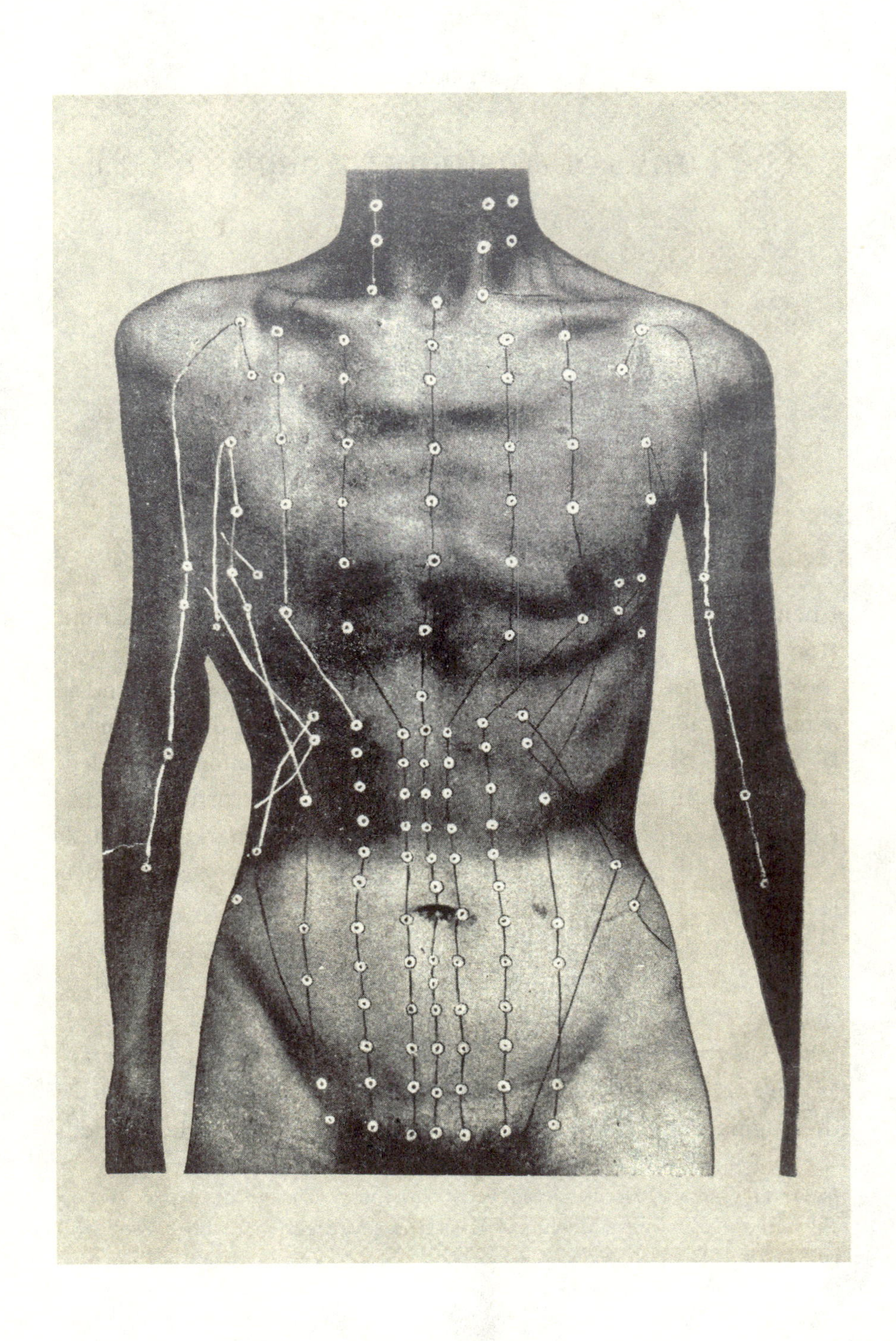

the bony arms, rigid collar bones, and dented belly before adding the dozens of lines, dots, and individual meridian site names (**Figure 3.3**).

Cheng had layered a *tu* (圖) on top of a photograph. This introduced a conceptual challenge: the *tu* and the photograph offered competing ontological expectations. A photograph functioned as a *xiang*, or a likeness, which limited the imagination. The inscribed lines functioned as a *tu*, which welcomed the imagination. Whereas the photograph elaborated on classical genres of portrait painting, the *tu* referenced a longstanding style of images associated with woodcut prints. Cheng incorporated both styles in his modernist vision of a body. Yet, the graphic contradiction between the *xiang* and the *tu* meant that Cheng could not claim absolute mechanical objectivity. The hand-drawn lines on one layer interrupted the camera's mechanical eye on another.

Cheng Dan'an never published these images. Instead, he kept them neatly tucked away, folded in a 1929 medical manual. Years later, Cheng would eventually abandon photography altogether.

Introduction

This chapter interrogates the limits of representing anatomy through technologies of visualization in the nineteenth and early twentieth centuries. I argue that these discursive and material technologies introduced a contradiction between the visual appearance of anatomical structures and their functional capacities. In the first half of the chapter, I examine early modern natural philosophers in Europe who fixated on the presumed uniqueness of the rational human mind, believed to be animated by an invisible soul. To represent this rational soul, they employed metaphors that stabilized the visual depiction of nerves while also widening the gap between the appearance and function nerves.

Figure 3.2. (opposite) Headless meridian man, twentieth century. Cheng Dan'an (1899–1957), photograph with hand-drawn additions, marked with black and white meridian paths and points. From Hunan Museum of Acupuncture-Moxibustion Archives. Used with permission

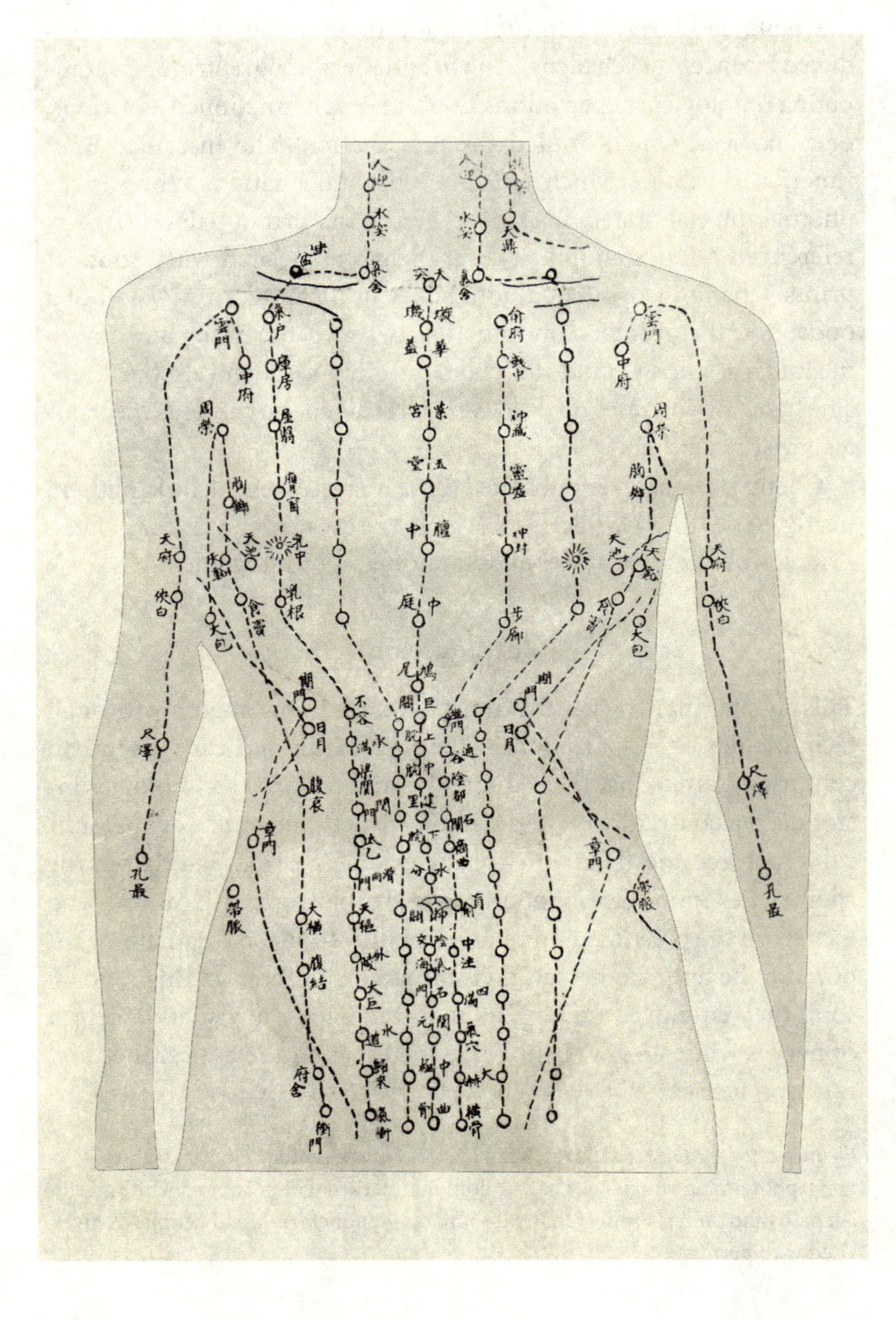

人迎
水突
缺盆
氣舍
天突
天鼎
璇璣
華蓋
紫宮
玉堂
膻中
中庭
鳩尾
巨闕
雲門
中府
氣戶
庫房
屋翳
膺窗
乳中
乳根
俞府
彧中
神藏
靈墟
神封
步廊
周榮
胸鄉
天谿
食竇
天池
天府
俠白
尺澤
孔最
大包
期門
日月
章門
帶脈
腹哀
大橫
腹結
府舍
衝門
不容
承滿
梁門
關門
太乙
滑肉門
天樞
外陵
大巨
水道
歸來
氣衝
上脘
中脘
建里
下脘
水分
神闕
陰交
氣海
石門
關元
中極
曲骨
幽門
通谷
陰都
石關
商曲
肓俞
中注
四滿
氣穴
大赫
橫骨

Meanwhile, the rational soul and brain did not exist in traditional East Asian medical cosmology as essential organs. Historically, the term *xin* (心) referenced the heart, mind, and emotions. Representations of the *xin* expanded from an organ of the mind to also encompass the capillaries in the nineteenth century. In contrast, the term *nao* (腦) shifted from denoting part of the skull to becoming an organ associated with cognition. Whereas the *xin* thus expanded in anatomical location and function, the *nao* closed around a new meaning of being the physical brain. Despite their divergent histories, both the *xin* and the *nao* came to share invisible capacities of human thought and cognition.

These visual, discursive, and ontological shifts had further repercussions. The second half of the chapter explores these consequences through two case studies that grappled with objectifying nerves and meridians. The first case study begins with the attempt to map nervous sensations *onto* anatomy by the British physician Henry Head (1861–1940), who tried and failed to use photography to capture sensations. The second case study examines the attempt to transform meridian paths *into* anatomy through nerves with the case of the camera enthusiast Cheng Dan'an (1899–1957). In both instances, Cheng and Head critiqued assumptions about the nature of anatomy. Dreams of anatomical nerves as fixed, solid structures did not align with embodied sensations that were flexible, volatile, and ever-changing.

By juxtaposing Henry Head and Cheng Dan'an, this chapter examines the challenges of describing nerves as anatomical objects, the risks of using nerves as an ontological framework to anatomize meridians, and the limitations of photography as an epistemic tool. For both men, nerves held great promise. Head turned to photographing sensation maps because standard nervous anatomy failed to explain patterns of physiological expression. Cheng Dan'an used photography

Figure 3.3. (opposite) Headless meridian *tu*, twentieth century. Cheng Dan'an (1899–1957), photograph with hand-drawn additions. Cheng's outline of the photo showing meridian points as circles connected by dotted lines, with abbreviations for each site. From Hunan Museum of Acupuncture-Moxibustion Archives. Used with permission

in hopes of justifying meridians as anatomical structures through nerves. Both tried containing subjective sensations by cutting, burning, and inscribing the skin. They used cameras to capture illustrations of invisible corporeal structures, only to realize that photography presented an illusion of accuracy. The basic expectations of photography as an epistemic tool conflicted with the ontological ambiguity of nerves. This threw into relief the failures of anatomy and its unfulfilled explanatory powers. Nerves would illuminate everything and nothing.

Conflicting Cognitive Hierarchies

Toward the end of the early modern period, discourses on sensation took a new turn.[2] The mind had been relatively more embodied until the strict mind–body dualism that emerged in the seventeenth century. Yet, with this strict dualism, investigations into the physiology of the brain and nervous system presented ontological problems. Rather than clarifying the separation of the mind and the body, philosophical enquiries instead raised new questions about how they precisely related to one another. Clearly, the mind could be disturbed by material sensations, and nerves apparently facilitated this disruption. Nerves tethered aspects of the soul to the body. In other words, nerves sat ambiguously between being facilitators of material sensations and the immaterial mind.

With sensations encroaching upon the mind, natural philosophers in Europe further wrestled with its unsettling implications. How was it that lowly forms of sensation could shape cognition, evade cognition, and potentially present as a *form* of cognition?

Cognitive hierarchies threatened to collapse. Human cognition had defined human bodies, whereas nervous sensations defined animal bodies. But could sensations actually resemble higher cognition? Natural philosophers resisted this notion.[3] They had already created arbitrary divisions between human intelligence and animal ignorance, generating similarly arbitrary explanatory objects. For instance, "animal spirits" supposedly mediated sensation, while the "rational soul" governed the mind. Both were *like* fine fluids in that both *were* fine fluids. However, animal spirits and the rational soul were fundamentally different. Animal spirits were malleable, impressionable things that received sensory signals but did not contemplate them. The "rational soul," meanwhile, existed only in humans. Unlike the material animal soul, this "rational soul" generated thoughts untethered to the outside world.

Even with this framework, "animal spirits" and "rational souls" remained theoretical conjectures. They were an exercise of the imagination. Their existence depended entirely on the theological cosmology that delineated the special qualities of humans. According to René Descartes (1596–1650), humans alone were conscious of their own consciousness. Humans alone could comprehend God's profundity. Humans alone had rational souls. Animals and plants, even if sentient, apparently lacked awareness of their sentience. To articulate this difference, Descartes famously relied on material metaphors.[4] He claimed that the body operated as a hydraulic machine, with animal spirits flowing through nerves like water through pipes. The body was a mechanical device manifesting physical causation. Nerves and blood vessels were solid, straightforward conduits for unremarkable animal spirits. In contrast, the rational soul existed beyond the physical body, untouched by its machinery. As such, it housed the immaterial mind, separating it from the body. This was the Cartesian split.[5]

Yet, Descartes could not empirically verify the difference between animal spirits and rational souls. He could not physically locate the uniqueness of human cognition. He had no evidence of these invisible objects.

Enter the brain. In the seventeenth century, the brain became the material home for the rational soul in early modern neurophysiology. The brain had long been an essential organ in humoral medicine, but now took on new importance. This was credited to British physician and natural philosopher Thomas Willis (1621–1675), who asserted that functional differences between animals and humans could be attributed to the brain. Even if large animals had the capacity for memory, humans were still special. Willis identified the cerebral cortex as the site of human cognition, grounding the rational soul and giving the human mind a physical home.[6]

But like Descartes, Willis offered no empirical evidence. As we have seen earlier in this book, Willis relied on imagination rather than observation to assert his claims about the workings of the animal spirits and rational soul.[7] Despite his dissection skills, he could not demonstrate how the human cerebral cortex was unique. When carving into the brain, anatomy alone could not corroborate his ideas. The rational soul was nowhere to be found. Faced with these epistemic limits, Willis also turned to metaphor. Rather than describe what the brain was, he characterized how it functioned. Mental impressions came in the form of "rough seas" that roared through the body.[8] They pressed upon the many chambers of the brain that worked like baking apparatuses that cooked up ideas. Animal spirits supposedly flowed in currents along nerves toward the cerebrum. More powerful currents moved beyond the cerebrum through the corpus callosum, colliding with thoughts before looping back to the cerebrum as memories. All the while, the rational soul remained safely housed in the cerebral cortex, shielded from the disruptive currents of animal spirits.

So effective were Willis's material metaphors that physicians took these hierarchies for granted. Animal spirits moved like waves through the nerves, unable to pass beyond the cerebrum and corpus callosum, while the rational soul percolated in the cerebral cortex. The fine fluids of animal spirits—which no one could actually see—were anchored in the body. The rational soul, meanwhile, was anchored in the rarefied parts of the brain. This divide between human and animal led to the divide between brain and body, and eventually the divide between the autonomic and sympathetic nervous systems.

In the eighteenth and nineteenth centuries, the physical structure of nerves gained greater visibility. The hydraulic metaphors of Descartes and Willis conceptually anticipated these efforts. Natural philosophers increasingly suggested that disease entities emerged from structural problems. Anxieties of modernity stemmed from weak nerves. In other words, nerves became capable of being nervous. They vibrated. They broke. They were brittle. Swollen blood vessels could interrupt the flow of spirits through nerves. Nervous problems were a problem of structure. Structure impacted the movement of fluids, a narrative that physicians standardized. For instance, Robert Whytt (1714–1766) prioritized nervous tubes over their internal contents. A century later, Moritz Romberg's *Manual of the Nervous Diseases of Man* (1840–1846) also framed nervous illnesses as caused by physical distortions.

The language used to characterize nerves further characterized the modern subject.[9] Form determined expression. Individuals could be "tightly wound" or "high-strung." Nervous structures as social objects defined social actors. Edward Philips's 1658 dictionary described nerves as "vigorous" and "lusty."[10] "Nervous" connoted strength, desire, weakness, and debilitation. Material metaphors facilitated what nerves looked like and suggested what they did and did not have the capacity for doing. Nerves became hungry, plant-like abominations, causing weakness in poor English citizens with deficient diets. Nerves defined the ailing body.

Nerves took on increasing agency, as if to define human anxieties, desires, and emotions. Although Willis used the brain to hold human cognition, the fluids pushing through nervous vessels could apparently alter one's ability to think clearly. Animal spirits thus encroached upon the brain. Far from immaterial, the mind now physically manifested in the body. The Cartesian divide between mind and body seemed to narrow. The mind was making itself known through the body. Physiologists like Franz Josef Gall (1758–1828) tried to empirically identify the imprints of the mind onto the skull through the infamous field of phrenology.[11]

Other anatomists went further. For instance, Albrecht von Haller (1708–1777) relegated the soul to just one of many things that rattled nerves.[12] Nerves were autonomous entities. Perhaps the brain and spinal nerves were as ordinary as the rest. Perhaps the brain, spine, and body were equally autonomous. Peering through a microscope, British physiologist Augustus Waller (1856–1922) described peripheral nerves as possessing a battery-like ganglion that enabled self-sufficiency. In contrast, the brain and spinal nerves lacked such batteries and relied on external sources of power.

As bodily nerves gained autonomy and brain nerves lost autonomy, cognitive hierarchies, again, threatened collapse. Depending on the metaphor, physiologists might claim either the prominence of the brain or the autonomy of the body. Some metaphors reinforced the importance of the brain, but others undermined it. In their hungry, active, plant-like, sentient, battery-powered state, nerves could resemble vegetative trees, even if individual branches did not function as such. With increasing articulation of nerves, active bodies did not always need heads. As Eduard Pflüger (1829–1910) observed in 1853, a decapitated frog could use its left leg to scratch an irritation on its right leg.[13] Such "reflexes" operated independently of the brain, bypassing it and the spine to render them obsolete for basic functions. Even as neuroanatomists assumed superiority of the human brain, bodily sensations still shaped self-perception through affecting mental distress. Clearly, feelings in the body interrupted the finer qualities of the mind.

Decentering the *Xin* (心), Defining the Brain

In East Asia, the organ for thinking and feeling—the *xin* (心)—historically did not connect to the nerves. Nerves in the Chinese medical context did not exist; they did not need to. Early modern physicians and scholars, despite the influence of Jesuit missionaries, appreciated higher forms of cognition without the anxieties of Descartes and his Cartesian split.

The *xin* was a dynamic object. It appeared as the heart, the mind, and engaged in forms of cognition. The *xin* did not need to justify the presence of a brain. Like nerves, the brain did not need to exist. Instead, *xin* contained the impulses for thinking and feeling and facilitated the processes of thinking and feeling.[14] As a cultural object, *xin* was the stuff of despair in novels, the stuff of life in anatomical illustrations, and the stuff of focus in body-cultivation techniques. It varied across cultures, presented slight variations within Daoist, Confucian, neo-Confucian, and Buddhist texts. The *xin* formed attention and attitudes, channeled care and emotions, reflected intellect and ideas, demonstrated sincerity and sentiment, and reinforced meaning and motivation.[15] It was both invisible and material. Unlike animal spirits, it was a point of concentration that one could actively build and refine. *Xin* was an object that required practice.

By the modern period, nerves entered East Asian medical discourses with the help of Sugita Genpaku's didactic efforts in the late eighteenth century, seen in the previous chapter. Still, nerves did not fully connect to the many cognitive and proto-cognitive qualities of *xin*. Sugita had invented the word for nerves in the form of *shinkei*, or "spirit nets," that were embedded in the body and existed beyond the body. And even with the introduction of nerves as "spirit nets," it was *xin* that facilitated human emotions, human anxieties, and human cognition.

In the nineteenth century, the cognitive functions of *xin* had gradually transferred to the brain.[16] This transference was by no means deliberate or absolute. It was also carried out by a group of translators, writers, illustrators, and editors. These men included Chen Xiutang, Guan Maocai, Leung a-Fat, and Pan Shicheng, who famously worked with the Scottish missionary-physician Benjamin Hobson (1816–1873) to create a series of new anatomical texts and images.[17] The team strategically used classical objects like *xin* and Qi (氣) to introduce new concepts in neurophysiology. As they worked, they gradually managed to articulate a new vision of anatomy that involved decentering the *xin* in the body, assigning a new term for the brain (*nao*, meaning shell), and reinforcing meridians as capillaries that connected the heart and the brain.

Chen, Guan, Leung, Wong, and Pan perceived no stark divides between foreign and native anatomy. "Western" knowledge was not incommensurable or mutually exclusive to "Eastern" knowledge as long as they both vaguely fit within existing cosmologies of the body. This kind of intellectual plasticity had been facilitated by neo-Confucian impulses from the tenth century that encouraged a sense of intellectual universalism.[18] Any form of wisdom was shared knowledge.

This preexisting neo-Confucian universalism was particularly useful for Benjamin Hobson as a foreigner. It allowed him to communicate with collaborators. For one thing, Hobson barely spoke Cantonese. When he landed in Hong Kong in 1843, he relied heavily on Cheng, Guan, Leung, Wong, and Pan to translate his lectures. Hobson scheduled regular meetings with Chen Xiutang and another teacher named Wong Ping to discuss physiology, anatomy, *materia medica*, surgery, and clinical medicine.[19] Cheng and Wong were both recent converts and spent up to two hours daily educating Hobson.[20] As their conversations gradually turned to printing presses, Hobson, Cheng, and Wong teamed up to publish several books. Hobson, per his proselytizing duties, co-authored translations of biblical texts, a book on midwifery, and a dictionary of medical terms.[21]

The group's combined efforts also led to a medical treatise titled *Treatise on Physiology (Quanti xinlun),* which was based on Hobson's own training as a physician and presented his introduction to physiology.[22] Published in haste, the first 1851 edition of *Treatise on Physiology* had no images. For the second 1853 edition, Chen and Hobson added many lithographic prints. By the 1857 third edition, they further refined the woodblock prints.[23] *Treatise on Physiology* was not an original work but proved popular, with the first edition being well-received by Shanghai and Ningbo elites, even reaching Emperor Xianfeng.[24] But like Sugita Genpaku's *Kaitai shinsho* published a century earlier, *Treatise on Physiology* was a pastiche. It summarized and drew from existing works, including those by anatomist-physicians William Paley (1743–1805), William Benjamin Carpenter (1813–1885), and Richard Quain (1800–1887). As with *Kaitai shinsho*, copying, printing, and reprinting involved translation, transformation, and interpretation.

Chen Xiutang cared about images. He worked with the scholar-official Pan Shicheng (1804–1873) to study images from William Benjamin Carpenter's 1843 *Animal Physiology*. Fittingly, the book contained numerous animal illustrations, such as an ostrich, weasel, crayfish, slug, starfish, camel, and dozens more. As he copied the images of ostriches and newts, Chen paid particular attention to the image of a human-like form (**Figure 3.4**). It looked like a tangle of capillaries with a face-less head, boneless arms, and boneless legs. Nearly every part of the image was tacked with hundreds of individual labels that indicated major and minor arteries, and arteries that broke in half.

Chen ignored these labels and created his own version of the image (**Figure 3.5**). He reconfigured the *xin* in relation to the meridians and the brain. Instead of naming and subdividing arteries, Chen simply registered blood vessels as *guan* (tubes or pipes) and *mai* (flesh meridians). *Mai* (脈) pointed to numerous flesh meridians while *guan* (管) structurally defined the capillaries.

The caption then rendered the image as representing "meridian tubes" (*maiguan*) rather than "blood vessels" (*xueguan*). Historically, the word *guan* only ever appeared in Ming dynasty acu-moxa texts as references to straw tubes. For instance, the seventeenth-century *Great Compendium of Acupuncture-Moxibustion* never used the words *maiguan* or *xueguan*.[25] Instead it deployed the word *guan* for naming bamboo pipes, feather quills, diminutives, and guarding against hubris.[26] In other words, using *guan* as an anatomical reference introduced the metaphor of a bamboo pipe to further articulate a new corporeal materiality for blood, flesh, and meridians.

In his redrawing of this image, Chen rendered the *xin* as common as meridians, which were as common as blood-flesh-tubes.[27] These flesh-meridians no longer told time; they were no longer a unit of space, seen in chapter 1. Abandoning the hundreds of labels crowding Carpenter's capillary figure, Chen instead labeled the brain, eyebrows, shoulders, fingers, and *xin*. This was a strange decision given that the labels indicated a brain without showing a brain, the eyebrows without showing the eyebrows, the shoulders without showing the shoulders, the *xin* without showing the *xin*.

Chen then altered another image from Carpenter's *Animal Physiology* (**Figure 3.6**). In the original image titled "Nervous System of Man," the figure stood in the *adlocutio*, or the orator's pose, facing away from the viewer. His left finger pointed downward, and his right

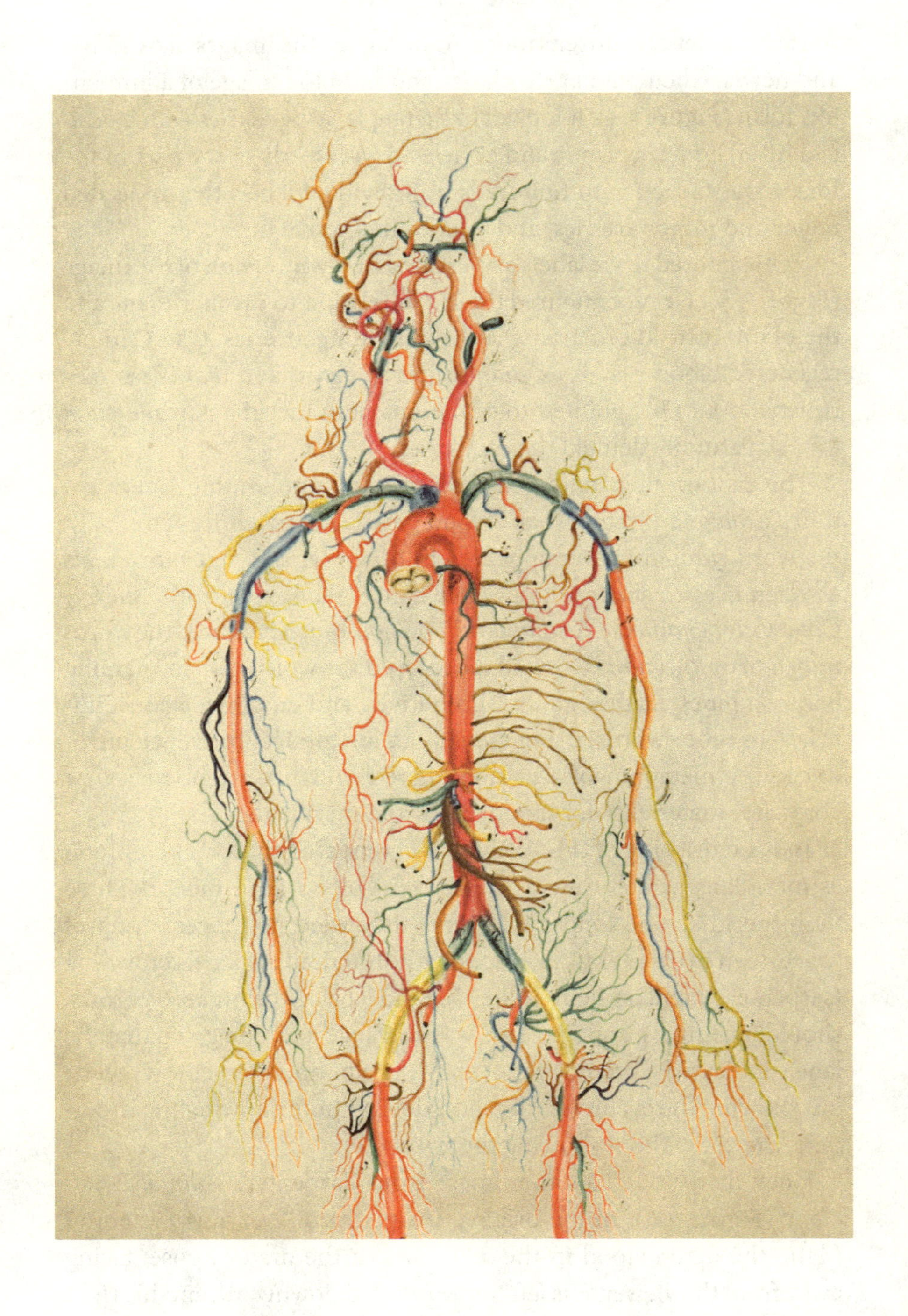

hand was open. Carpenter had captioned it with the cerebrum, cerebellum, spinal cord, brachial plexus, median nerve, and more.

Chen again ignored these captions. He abandoned Carpenter's display of nervous anatomy in a new version of the image (**Figure 3.7**). The man again stood in the *adlocutio,* or orator's pose, with his left finger pointing downward and right hand open. But he came with a new set of instructions. Chen now described 9 cranial nerve pairs and 31 spinal pairs, potentially referencing the 12 primary meridians. Chen's man now had a brain, or *nao* (腦), which was a word that often referred to a shell. In this image, that shell was filled. Rather than holding the sea of bone marrow, the head now became a bony encasement for a brain. This was a radical shift. Whereas the *nao* had referred to a skull that could house the brain, this image allowed *nao* to become the brain.

Most importantly, Chen added a title suggesting the cranial and spinal nerve pairs were animated by *nao qi* (腦氣), or brain Qi. This radically altered Carpenter's Nervous Man. *Nao* (skull) and Qi were two historically unrelated objects. Like *maiguan*, *nao qi* had not appeared in seventeenth-century medical texts.[28] Joining them in the neologism *nao qi* offered something new. The term *nao qi* facilitated the concept of a "brain" that was accompanied by the nerves. It is possible that these were intentional constructions, especially because Chen had read classical medical texts.[29] He understood the multiplicity of objects like the *xin*. He understood the materiality of Qi. He

Figure 3.4. (opposite) Articulated branches, nineteenth century. Thomas Medland (1765–1833), engraving. This image depicts more than 100 labeled branches of the arterial system, including the aorta, coronary, carotid, vertebral, intercostal, celiac, iliac, and tibial arteries. From Bell, *Engravings of the Arteries*, 1801. Image courtesy of the US National Library of Medicine. Public domain

Figure 3.5. (overleaf) De-centering the *xin*, nineteenth century. Chen Xiutang (nineteenth century), woodcut. This image shows a faceless head with boneless arms and legs. The text describes capillaries with references to other body parts, such as brain, eyebrows, shoulders, and fingers. It elides any naming of small arteries, only indicating major ones and an invisible *xin*. From Hobson and Chen, *Zentai shinron*, vol. 2, 1857, 46

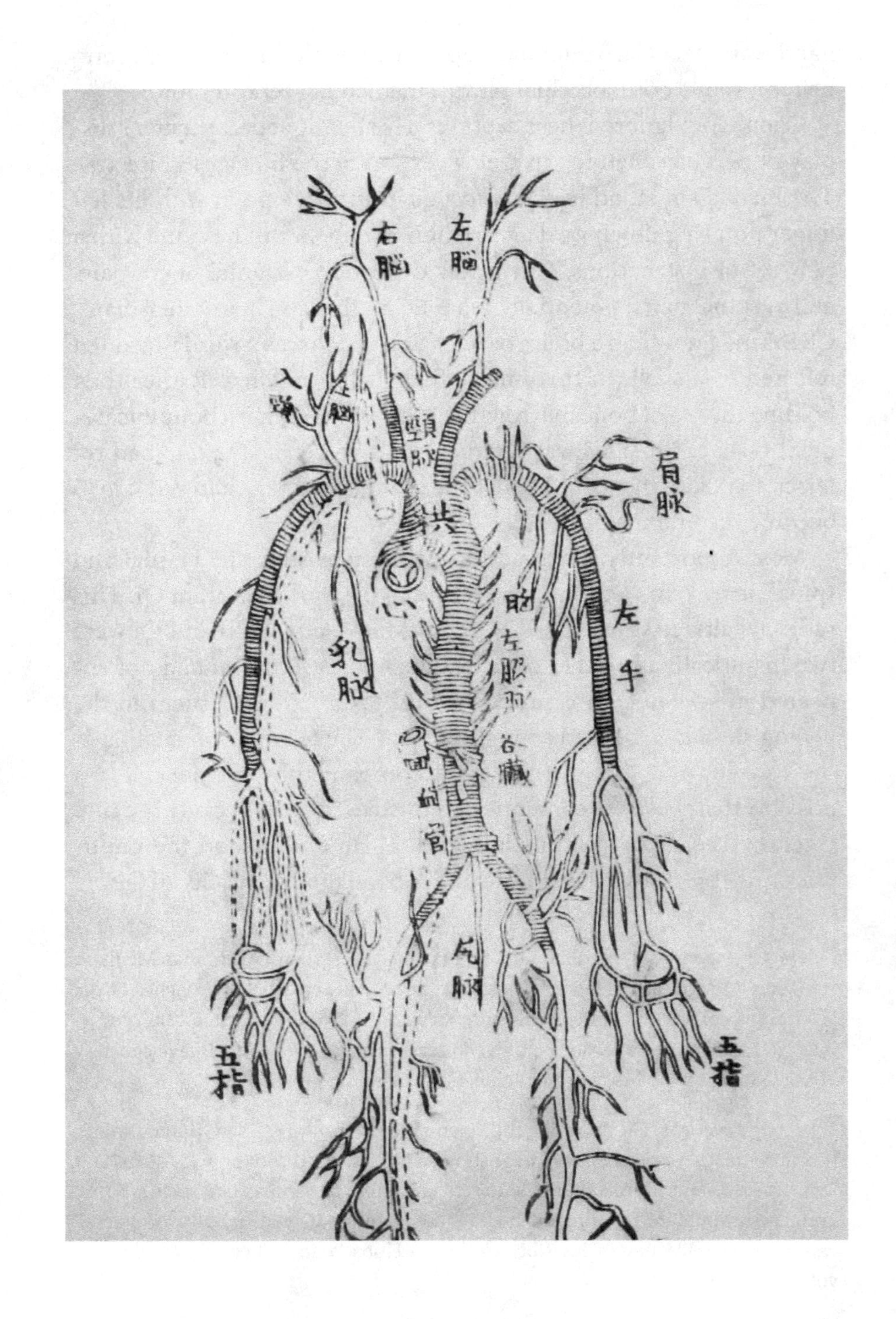
右脳
左脳
頸脉
肩脉
心
乳脉
左手
五指
五指

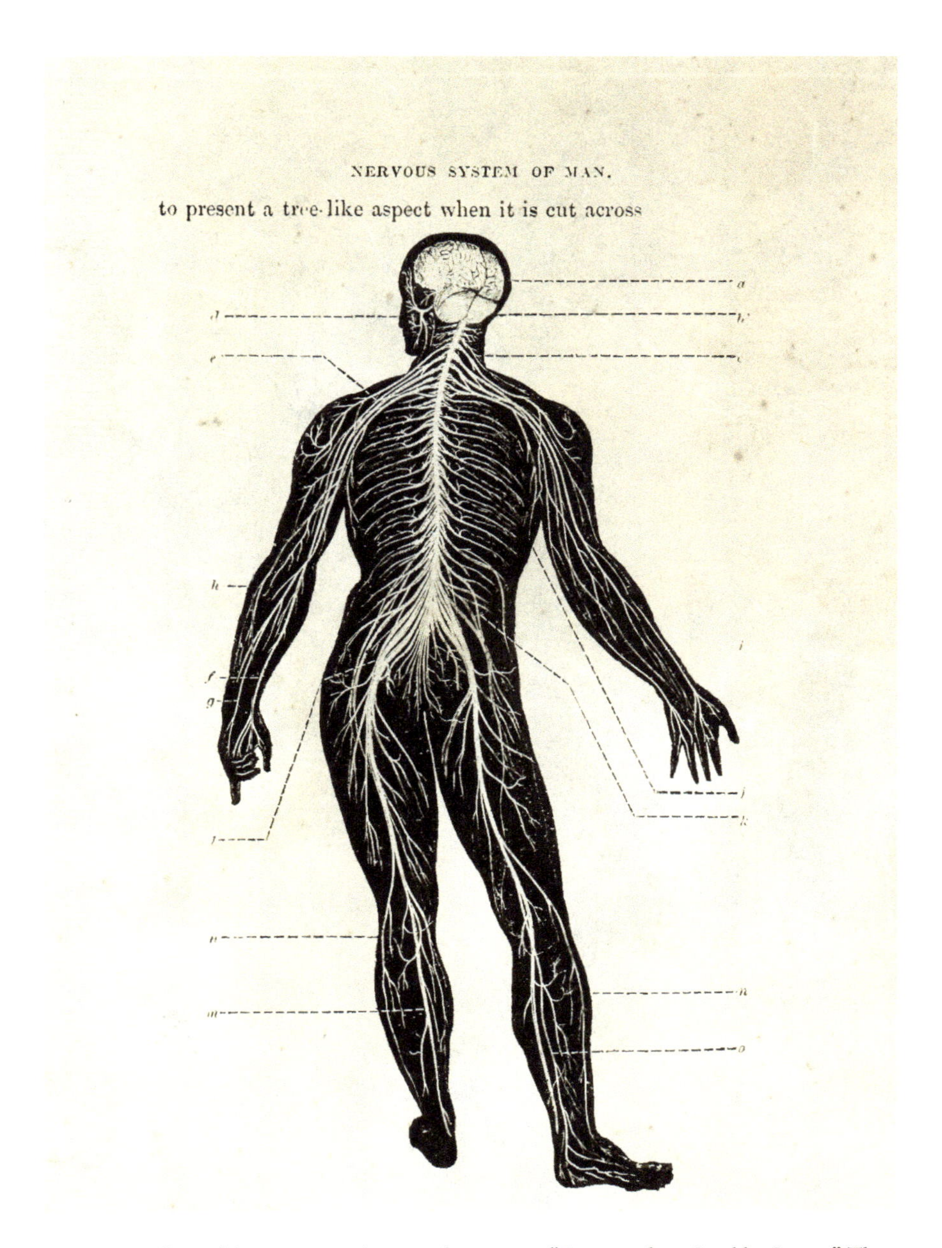

Figure 3.6. Branching nerves, nineteenth century. “Engraved on Steel by Lucas.” The male figure in the orator’s pose now faces away. Letters label an array of paths. Like capillaries, nerves are likened to tree “trunks.” They connect the cerebrum, cerebellum, and spinal cord to nerves in the face, the brachial plexus, the median nerve, ulnar nerve, and more. From Carpenter, *Animal Physiology*, 1843, 349, fig. 184

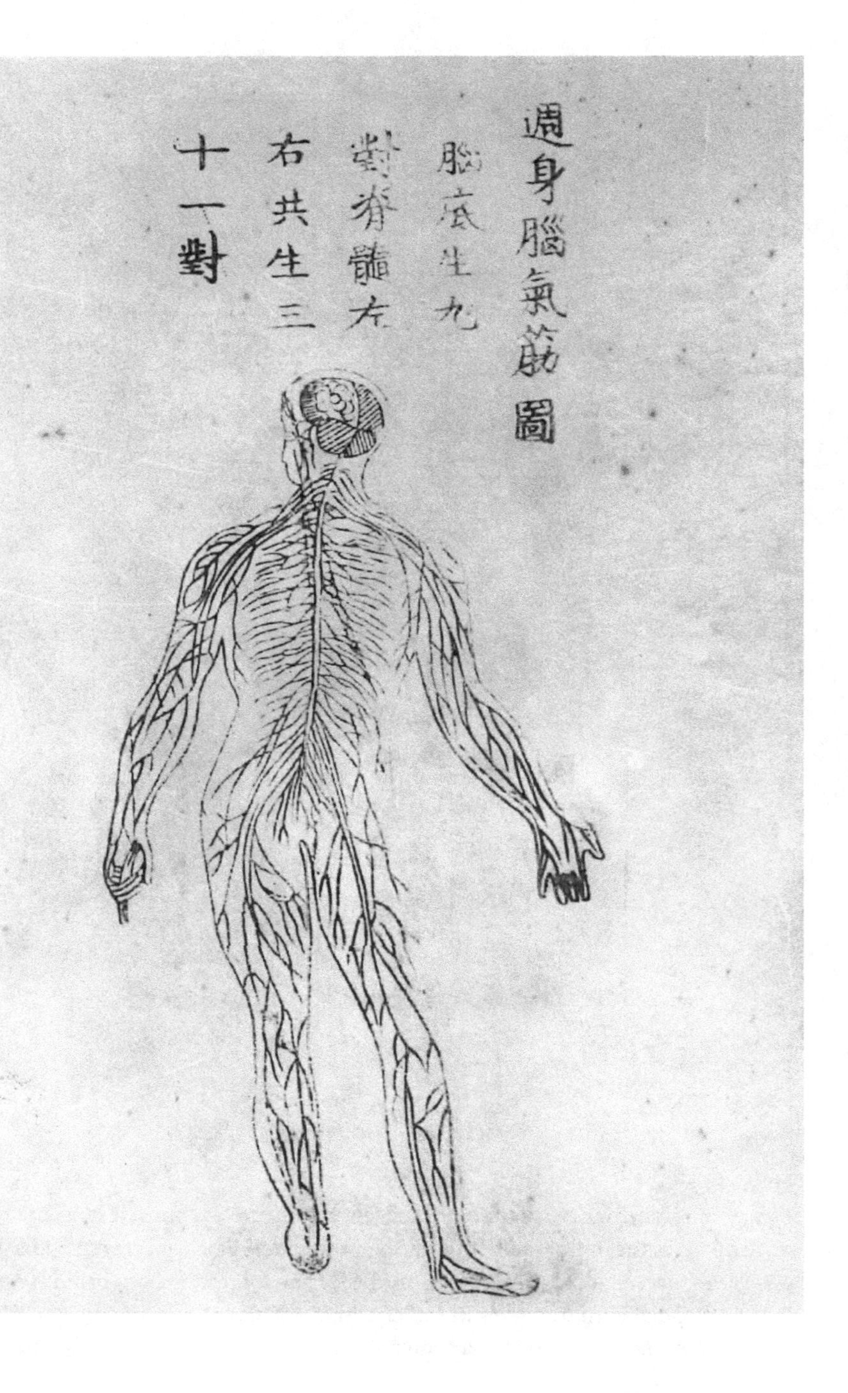
遍身腦氣筋圖
腦底生九對脊髓左右共生三十一對

understood the ontology of meridians. With this knowledge, Chen gradually shifted his emphasis on *xin* as the organ of cognition to an organ that shared the cognitive responsibilities of a brain. He had used these ideas to reconfigure the *nao* as the brain that was animated by Qi.

Benjamin Hobson objected to these modifications. He found Chen's images misleading and underwhelming.[30] He also objected to Pan Shicheng's heavy-handed censorship. One of Pan's new prefaces failed to mention Hobson's qualifications as a physician. Even worse, Pan eliminated all of Hobson's theological references, including any indication of the soul.[31] Annoyed, Hobson complained: "He did not however ask my permission and has thought proper to expunge all that referred to Jesus and God. He has also condensed several of the plates to such the size of the book and consequently has made frightful things of them."[32] For Hobson, the new set of images were "frightful things." They were abridged, condensed, and monstrous. When introducing the brain, Hobson tried to explain that it was the central controller of the body. Nerves were like electric wires, solid and stable, so that brain responded to electrically induced diseases.

Pan left most of Hobson's text unchanged but had his own agenda. Working from classical texts, Chen and Pan displaced the *xin* to the realm of arteries. The *xin* became the heart. It still represented the mind, but it no longer cogitated on its own. It alone was no longer an organ for evaluating, thinking, and feeling. It shared some of this work with an organ that filled the head. And even then, the newness of the *nao* as the brain did not fully replace the *xin*. The Chinese anatomical body remained robust, sustaining new words for both brains and nerves. It became a composite space where writers projected new structures and ideas.

Figure 3.7. (opposite) Branching nerve pairs, nineteenth century. Chen Xiutang (nineteenth century), woodcut. Based on *Animal Physiology*, this nerve figure precedes capillary man of **Figure 3.5**. The caption describes the image showing 9 cranial and 31 spinal nerve pairs. From Hobson and Chen, *Zentai shinron*, vol. 1, 1857, 15

Objectifying Sensations

Structures promised a great deal. They promised ontological certainty, even though nerves themselves were not self-evident. In some ways, this contradiction made sense. The expression of nervous vessels as vessels was dictated by the movements of its internal fluids. Internal fluids manifested the volatility of animal spirits. Nineteenth-century biologists in Europe who followed the work of Descartes and Willis continued to speculate on the remnants of the animal spirits. Like the other natural sciences, neuroanatomy in this period underwent numerous changes.

The impact of imperialism, industrialism, and war initiated a new kind of practice that historian Philip J. Pauly has called "biological modernity."[33] New technologies characterized biological modernity by offering new ways for repackaging old concepts. Under these circumstances, the animal spirits persisted. They lived not just in the nerves but in all cells. Invisible fluids became actual fluids. In the nineteenth century, biologists considered the stuff of cells as the ultimate stuff of life, known as the "protoplasm."[34] Famous physiologists, from Thomas Henry Huxley to Santiago Ramón y Cajal, correlated structure with function, working to understand the ephemeral qualities of sensations that vibrated through fluids and disturbed organisms.

At the same time, sensations were difficult to contain with empirical certainty. They were not localized but displaced. They were not specific but diffused. The increasing distance between what nervous structures looked *like* and *how* they behaved baffled the physiologist-physician Henry Head (1861–1940). His patients reported overlapping sites sensitive to pain. Stimulating one area produced pain elsewhere, which Head termed "referred." Perhaps one area of the body referenced another area in the same way that one cluster of nerves sympathetically resonated with another similar-looking cluster of nerves. Patients with gallstones felt tenderness by the gallbladder along with sensitivity around the ribs and spine.[35] Those with throat and esophageal discomfort reported "definite referred pain" in the chest.[36] Patients who felt pain in their gut described it as diffuse, suggesting that it reached many parts of the body.

Head collected these cases and illustrated how affected areas mapped onto the surface of the body (**Figure 3.8**). Some of the images showed discrete areas that occupied the shoulders and arms, calves and thighs, chest and back, elbows and neck, shoulders and waist, feet and the waist, and anal area and buttocks. They were both localized and dispersed. From collecting these case studies, Head became increasingly perplexed by the relationship between the interior and exterior sites on the body.

Head had trained with the most distinguished histologists and physiologists in Wittenberg, Cambridge, and Prague. He understood the open-ended debates from the early modern period. Like their predecessors, nineteenth-century neurologists still wrestled with empirically accounting for the animal spirits. Anatomists who fixated on structure were limited by their visualization techniques.[37] Prominent researchers in Germany, France, and Italy proposed conflicting principles on the volatile qualities of peripheral sensation, which grew ever-more volatile under disease states. No single explanation accounted for how nerves could overlap at the body's surface while sensations shifted among different these overlapping clusters. Moreover, patients' idiosyncratic pathological sensation patterns undermined the neat structure of the reflex model.[38]

"No view yet is sufficient to explain obvious facts," Head lamented.[39]

Determined to track sensations shifting over time, Head collaborated with surgeon James Sherren (1872–1945) to collect more patient records. The two men met at the London Hospital where Head had helped to establish a new Pathological Institute.[40] Modeled after the Pasteur Institute in Paris, Head's new institute aimed to study pathology "for its own sake" without immediate clinical application.[41] Head aimed to develop a theoretical understanding of pathological processes.

But patients complicated the research goals. They were unreliable subjects with spotty paper trails and often arrived at the hospital without records of their treatment history. When they felt that their

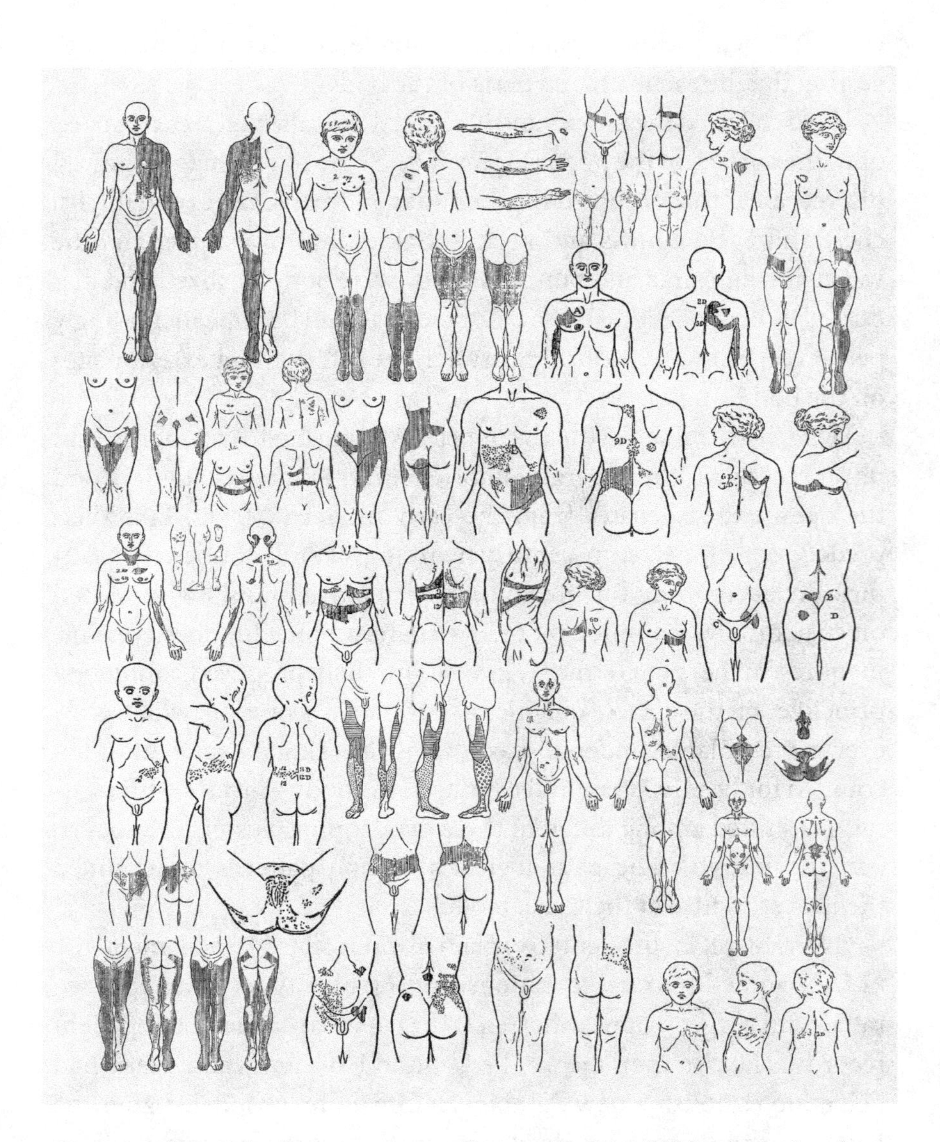

Figure 3.8. External references, nineteenth century. Compiled by the author from the work of Henry Head (1861–1940). Head found that changing sensations of touch and pain correlated significantly with visceral diseases among his patients. Tender areas with distinct borders occupied whole tracts of skin. They also corresponded with other tender areas linked to visceral disturbances and diseases like anemia, malaria, and fevers. From Head, "On Disturbances of Sensation with Especial Reference to the Pain of Visceral Disease," 1893

symptoms had improved, they often did not return. Patients exercised autonomy over their own care, much to the frustration of Head, who complained that they offered poetic accounts of their symptoms that seemed excessively florid, embellished, or vague.

Fed up with clinical research, Head took to experimenting on himself (**Figure 3.9**). In 1903, at 42 years old, Head stated that he was "in perfect health" after two years of abstaining from smoking and alcohol. With his collaborator James Sherren, Head prepared to sever his own nerves in his own perfectly healthy body to understand how induced pathological sensations changed over time. Sherren slipped a blade into Head's arm and guided it over the fold of his elbow. He peeled apart the flaps of skin, found two large nerves, and cut them in half.[42]

For nearly four years, Head tracked how severing nerves affected the sensation in his arm and hand. He created drawings and photographs with the famous poet and physician W. H. R. Rivers (1864–1922). Together, Rivers and Head sat quietly in a large oak room at Trinity College, Cambridge, and evaluated how Head responded to heat, cold, pain, and touch.[43] Rivers meticulously poked at the areas that had lost sensation. He pulled at Head's arm hair, burned his skin with hot water, froze it with ethyl chloride, impaled it with sharp objects, and stroked it gently with cotton.[44]

As Rivers poked and prodded, Head developed strong visual responses.[45] He confessed to his wife, Ruth Head, that the entire process was "excessively unpleasant."[46] Head felt like he was "chained to an oar" for years.[47] When he closed his eyes, he tried to focus his attention away from being both the experimenter and the experimental subject, reporting what felt like a "flash of pain," "wave of cold," or "flicker of heat."[48] But the more Head tried to describe individual moments, the more he doubted his ability to report anything useful: "[T]hough I am not conscious of any difference in myself," he admitted, "my answers are less certain, and it is obvious that the feeble sensibility of the affected part is giving out."[49]

Like his patients, Head was an unreliable subject. His attention, memory, and body generated inconsistent responses to different stim-

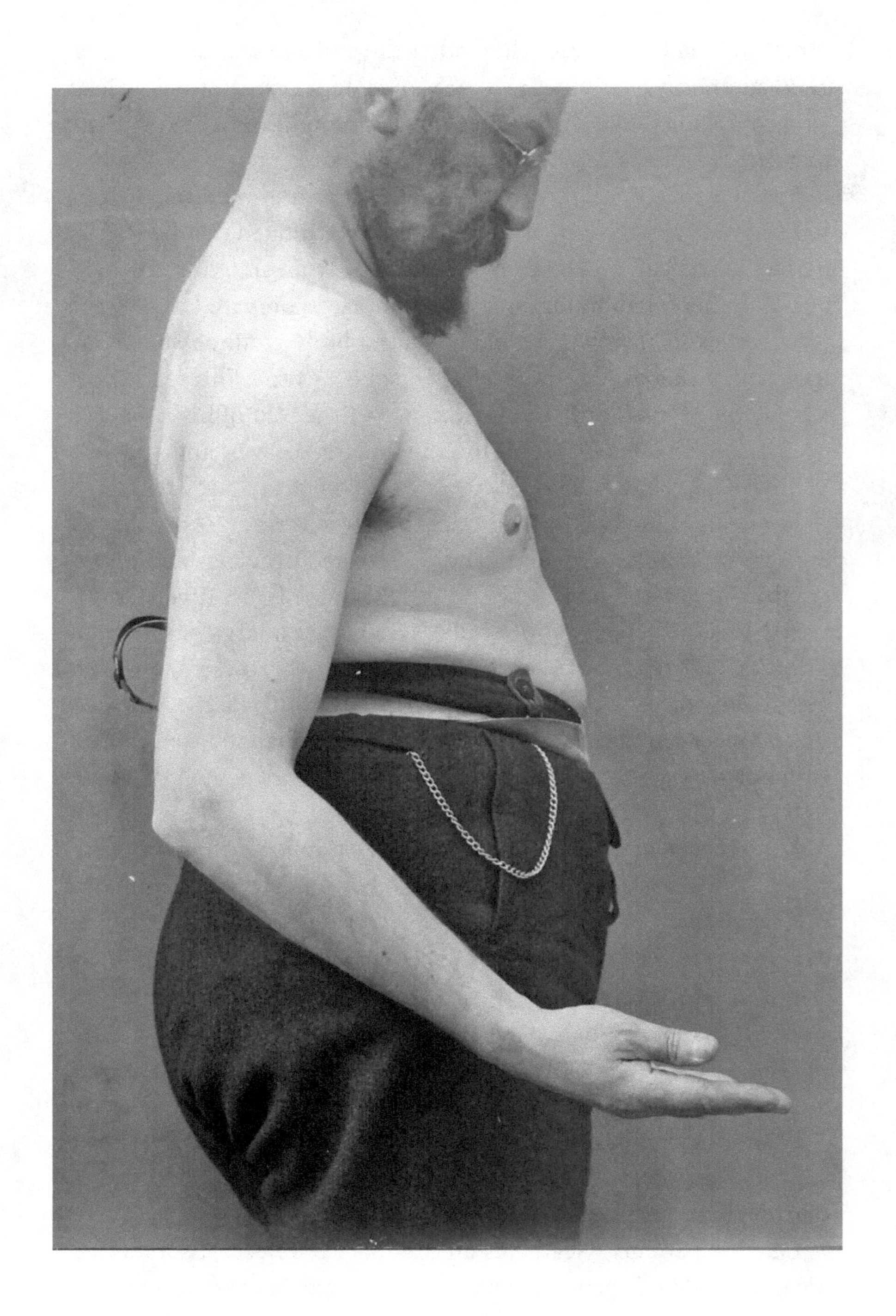

uli. Some areas would register touch but not pain, or pain but not touch. The same area that responded to a pinprick, but not cotton, would on another occasion respond to cotton, but not a pinprick. Sometimes, vivid sensations vanished once the stimuli lifted. Other times, vivid feelings spontaneously returned with the stimuli absent. And although Head expected that discrete sensations would correspond with discrete objects, such as being prodded by a pin or a glass tube, he found that sensations were dispersed. Stimulating one site generated "constellations of spots" at distant sites.[50] Stimulating the wrist created cold sensations up the arm; stimulating the upper arm induced coldness in the thumb.

To record these sensations, Head and Rivers traveled from Cambridge University to the London Hospital to photograph Head's arm (**Figure 3.10**). Rivers applied bold, thin, and scored lines to track Head's shifting reports. Thick lines indicated no pinprick response. Dotted lines marked indefinite boundaries between over-sensitive, under-sensitive, and normal areas. Thin lines that showed insensitivity to cotton and sometimes merged with dotted lines. Then, these regions overlapped. Some areas that did not respond to a pin prick were sensitive to cotton. And over time, borders moved, vanished, or returned.

At one point, Head experimented on patients with redirected colons, pressing hot and cold tubes of water into their anuses. With eyes closed, the patients reported feeling something in their abdomen, but not their anus. In other cases, they would "point into the air," convinced that someone was splashing cold water onto their stomach.[51] It became clear to Head that these sensations were "never localized," but traveled depending on location of the stimulation site and temperature of the stimulus.

Head tried making sense of these vast inconsistencies by theorizing a hierarchy of nerve clusters, which he called "deep," "protopathic,"

Figure 3.9. (opposite) Test subject, twentieth century. Henry Head (1861–1940), photograph. Head used his unaffected right arm as a control after experimentally severing nerves in his left arm in 1903. From Henry Head Papers, HH-BEA9-2, Wellcome Library Archives

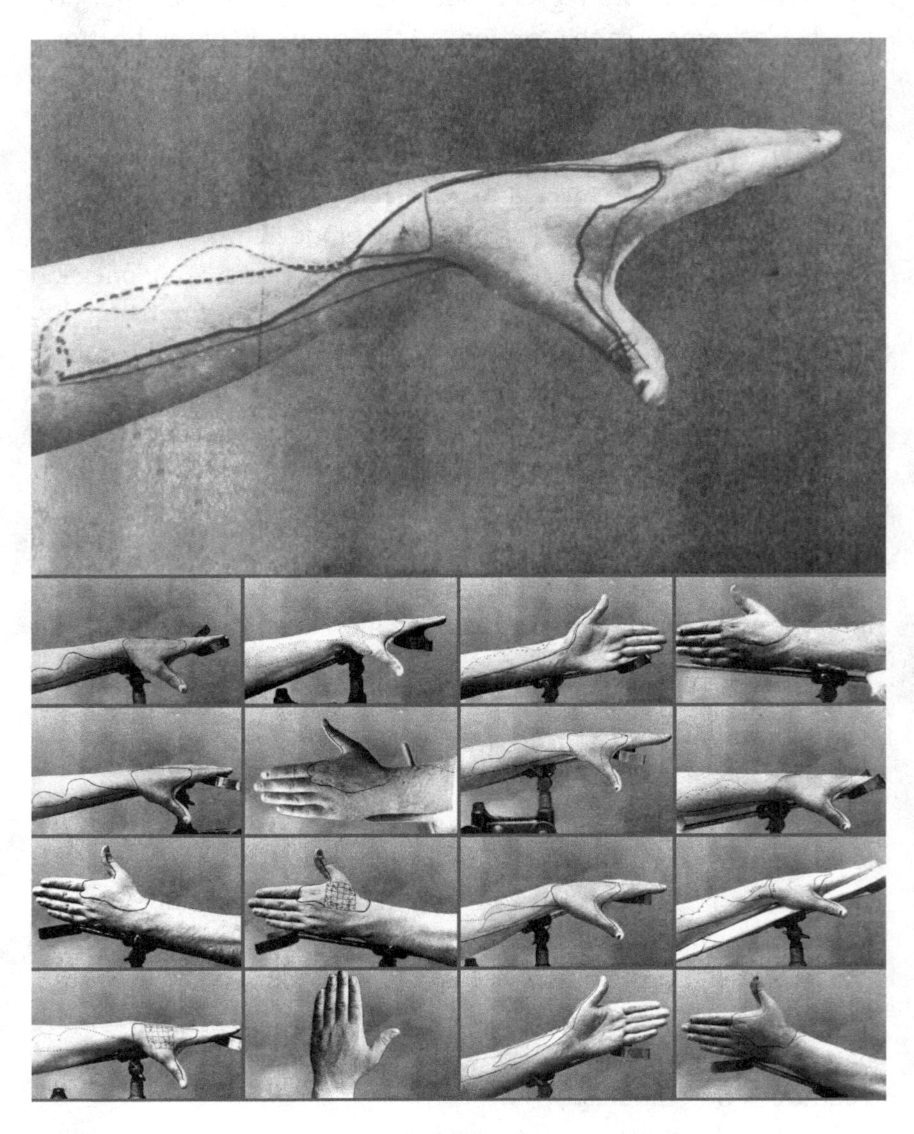

Figure 3.10. Affected sensation, twentieth century. Henry Head and W. H. R. Rivers, photographs. Images from 1903 to 1908 tracking changes in Head's arm after severing his nerves. Thick lines mark insensitive areas; dots mark areas with uncertain sensation; thin lines mark areas sensitive to cotton-wool. These borders shifted over time. From Head and Rivers, "A Human Experiment in Nerve Division," 1908

and "epicritic." This derived from Walter H. Gaskell's (1847–1914) model of vertebrate nervous system evolution, which had shaped Head's earlier work in spinal and cranial nerves.[52] Hierarchies ordered the body. They naturally emerged over time beginning with primitive protopathic sensations, before producing more evolved epicritic sensations.[53] Head reasoned that these functional hierarchies were not just pathological, but that they already existed in the body.

To be clear, Head did not develop this theory through structural observations. Instead, he relied on his own unreliable impressions, visualizations, and imaginations. He used years of poking, burning, and stroking his own overlapping sensations to articulate these ideas. To demonstrate the primitive nature of protopathic sensations, Rivers irritated Head's penis glans, calling it the most "rudimentary" organ with only protopathic and deep sensibility.[54] They experimented again using cotton swabs, algesimeter, and tubes of hot and cold water. Head replied that the tip of his penis only produced vague sensations, unlike the finer epicritic sensations on the arm that gradually recovered to distinguish among finer sensations.[55]

Perhaps they had proven their point—nervous sensations registered differently across different parts of the body. But whether on the arm, in the anus, or the tip of the penis, the sensations in and on the body did not correspond to nervous anatomy. Sensations changed; they were idiosyncratic and temporally contingent. They could not be objectified with cotton-wool brushes or tubes of warm water. The "obvious" facts of sensation were not so obvious. As a result, Head's theories would eventually be lost to history—that is, until they were picked up by acupuncture-moxibustion researchers decades later.

Feeding the Brain

Whereas Henry Head's experiments contradicted the utility of nervous anatomy, Cheng Dan'an (1899–1957) aimed to reinforce the utility of nervous structures. He hoped to appropriate nerves as a means of transforming meridians into anatomical objects. Like Head and

Rivers, Cheng would poke and prod at the skin. Like Head and Rivers, Cheng would photograph the paths that he inscribed on the body. But unlike Head and Rivers, Cheng fixated on a different task. From clinical experience, he knew that acupuncture and moxibustion did more than simply tickle the skin. These techniques directly altered disease states. Perhaps nerves could explain the diagnostic and therapeutic effects of acupuncture and moxibustion. Perhaps nerves could reveal the mechanics of *jingluo* (經絡).

Much was at stake. By the early twentieth century, the Qing dynasty had ungracefully collapsed. Scholars, physicians, and political activists scrambled to justify native knowledge, accept foreign influence, recognize social incoherence, and imagine a better future. Following the famous 1919 May Fourth Movement, two opposing visions emerged: either break from the past or preserve "national essence."[56] One vision celebrated "Western" knowledge as a means for saving China; the other urged for a China that could save itself.

These divergent desires stirred urban life. They were particularly familiar to Cheng Dan'an when he began working as a physician in Shanghai during the 1920s. The city had been partitioned into foreign concessions governed by US, British, and French interests. Amidst the wealth, poverty, tension, and commercial activity, Shanghai harbored radical activists, conservatives, and the new Communist Party.

The Shanghai medical market adapted to these many sentiments. Chinese businesses embraced hybridity when it suited them. Marketing slogans introduced products that were "Chinese-Western," or *zhongxi*; "Chinese-French," or *zhongfa*; and "Chinese-English," or *zhongying* and *huaying*. These labels sold tonics that could treat excess phlegm, coughs, stab wounds, hemorrhages, contagion, or diarrhea.[57] New drugs promised to strengthen bodies weakened under a fallen empire. In the nineteenth century, many mainstream Japanese products claimed to be brain-enhancing, blood-enriching, and kidney-refining.

Brain tonics were especially popular, including entrepreneur Huang Chujiu's (1872–1931) famous product, *bu naozhi*, which promised to reinvigorate the body. Having read Hobson's *Treatise on Physiology*,

Huang incorporated the brain into Chinese cosmology, even using Chen's Nervous Man as a mascot for his brain juice. Like his peers, Huang knew that commercial products combining different visions of the body sold well. Hybridity turned a profit, working within the existing cosmology so that European pathologies coexisted with meridians.[58] These two visions of the body—a nervous body tied to a brain and a meridian body based in the *xin*—were not at odds. Elements of East Asian medicine could even explain the brain. In a 1904 advertisement, Huang's company claimed that the brain simply served as another source of Blood essence (*xue*).[59] This explained why his tonic treated not only neurasthenia, but also other diseases caused by Qi depletion, such as cataracts, fainting, dizziness, headaches, swollen gums, toothaches, troubled muscles, and impotence.[60]

Although Huang himself did not come from a family of physicians, he knew that the brain was not one of the essential viscera, or the five *zang* and six *fu*.[61] It was nothing like the *xin* but was becoming something like the *xin*. He also knew that Daoist practices involved renewing kidney essence, or the "sea of bone marrow" that filled the skull—the space of the brain.[62] His brain tonic thus aided in this process. It nourished the sea of bone marrow. It fed the kidneys. It fed the brain.

But despite Huang's certainty on the material relevance of the brain, literary figures, like the Wu Jianren (1866–1910) still puzzled over this superfluous organ. They wondered whether the brain was the seat of the soul if the soul was not the same as the *xin*. The soul was loosely connected to the nerves but completely absent from acupuncture-moxibustion practices. In other words, the soul seemed redundant in a Chinese body.

Market practices encouraged ontological plurality. Part of the impulse of plurality was the search for similarity and difference. For instance, physicians turned to classical texts and debated the actual nature of Qi and Blood (*xue*).[63] Perhaps *jingluo* were nothing but crude blood vessels.[64] Perhaps *jingluo* were distinct from neurological and circulatory networks. Perhaps *jingluo* joined all of the above.

During this period, the prominent Confucian scholar Tang Zonghai (1851–1908) famously claimed that *jingluo* were a combination of blood vessels and other biological structures.[65] At the same time, *jingluo* and nerves were similar but not the same; nerves were nerves—whatever that meant—and the Qi in *jingluo* could act on the new organs that appeared in imported anatomy books.[66] And when it came to the issue of Qi, Tang claimed that Qi moved the *jingluo* in the same way that steam operated in a steam engine.[67] Mechanical metaphors addressed the invisibility and the materiality of Qi, which worked for a while until it backfired by displacing Qi altogether.[68]

In the early twentieth century, campaigns to converge elements of "Eastern" and "Western" medicine continued with great enthusiasm and great difficulty.[69] Despite Tang Zonghai's legacy, some reformers resisted convergence campaigns altogether.[70] They deemed the experiential epistemology of Chinese medicine inferior to the epistemic virtues of European experimentation.[71] To these reformers, "native" knowledge could not be a "modern" way of knowing.[72] This difference was not only ideological but also economic. Compared with Chinese-run institutions, foreign-run schools were often better funded and educated more elite medical students.[73] Physicians who trained overseas further blamed Chinese medical practices for the demise of the state. "Revive the nation by first curing the people" they urged, "Cure the people by purifying medical theories."[74] Such aggression culminated in the 1919 May Fourth Movement, which sanctified "Western" knowledge as a heroic regime that could completely break with past.

The violence inherent in distinguishing between "native" and "non-native" knowledge led reformer Yu Yan (1879–1954) to actively critique "Chinese" medicine (*zhongyi*) in a way that, counterintuitively, aimed to preserve it.[75] Yu had studied Chinese medicine in Japan and observed that Japan's power extended from state-centered medical systems modeled after German institutions.[76] This inspired Yu to initiate institutional change. When the new central government formed its Ministry of Health in 1928, Yu proposed to abolish the "old-style medicine" that "did not obey science."[77] In 1929, he rallied enough support in the National Board of Health to vote unanimously for regulating Chinese medicine and set explicit standards for its practice and professionalization.

"We must first recognize that references to Yin, Yang, the five phases, and the twelve *jingluo* are total fabrications," Yu announced.[78]

In response to Yu Yan's regulatory campaigns, practitioners representing different forms of Chinese medicine collectively mobilized as the National Medicine Movement.[79] They proposed a new national medicine that rendered "Chinese medicine scientific" (*zhongyi kexuehua*). Leaders of the National Medicine Movement adapted *zhongyi* into the national medical system, even incorporating some of Yu's reforms by partitioning Chinese medicine into different categories: *jingluo* represented medical theory; pharmacology represented drugs; practice became experiential "native" knowledge.[80] This social mobilization had institutional impact, giving rise to the Institute of National Medicine as the first nationwide association for home-grown medical practice.[81]

As different groups mobilized to eradicate *zhongyi*, others rallied to claim its social, cultural, and political value. Controversy over native knowledge gave new meaning to classical meridian maps and new life to acupuncture-moxibustion practices. Making meridian maps became a political undertaking under the fractured and fraught modernist gaze.[82]

Appropriating and Abandoning Nerves

Cheng Dan'an participated in this fraught arena of rigid political binaries and practical medical pluralism. Official campaigns took an inflexible view of knowledge, whereas practitioners and merchants were more pragmatic. Cheng understood the stakes of these debates—however decentralized, contradictory, and confusing. One needed to work in service of a broader ideology while also demonstrating practical expertise. He had a practical background and came from a family of physicians.[83] A precocious student, Cheng had devoured classical texts by age 19. In his early 20s, he apprenticed in surgery. When he developed severe insomnia and back pain, he made a full recovery after his father successfully treated the chronic symptoms with acupuncture and moxibustion.[84] Stunned by his return to health, Cheng dedicated himself to learning these techniques.[85]

In the next two decades, Cheng would become one of the most influential figures in Chinese medical education. He traveled north to Tokyo in 1933 and spent six months at the Advanced School for Acupuncture-Moxibustion in Shinjuku City.[86] Japan transformed Cheng. He visited bookstores and collected medical classics that no longer circulated in China. At one point, discovering copies of *Treatise on the Fourteen Meridians* (*Shisi jing fahui*) from the fourteenth century reduced Cheng to tears.[87] When he returned to Shanghai, Cheng immediately began translating these texts for his students. He continued his work even as the Sino-Japanese War displaced him from Shanghai in 1937 to Anhui, Jiangxi, Hunan, Hubei, and finally Sichuan. There, he continued his practice, recording his clinical encounters and translating classical and contemporary books.[88]

Before traveling to Japan, Cheng created his own meridian men. He used his camera in his early studies to photograph lines on a male model (**Figure 3.11**). As the man stood still, Cheng needled, heated, and marked every point and path on the arms, legs, and head. Cheng directed; the man responded, posing with arms outstretched, legs bent, chin turned, almost mimicking a *pudica* form. Expressionless, the man cast his eyes up, down, and ahead. He lifted his right arm over his head, splaying his hand to display dots on his elbow, wrist, and pinky finger. He obligingly contorted to expose his chest, ribs, thighs, nose, cheeks, hands, and feet. The man served as specimen and spectacle, but more importantly, as an idealized map.

Cheng published the photos in his 1931 opus *Revised Approaches to Studying Acupuncture and Moxibustion Therapy*. The book featured 14 chapters, each one dealing with one of the 14 meridians. Every chapter was accompanied by a single-page collage of his photographs. He had cut out and reassembled the meridian man's chest, arms, head, and thighs. Reactions were most likely mixed. Members of the National Medicine Movement found the images both grotesque and

Figure 3.11. Naked meridians, twentieth century. Cheng Dan'an (1899–1957), photograph. Cheng marked meridian paths on male models, who contorted their bodies to better display areas where he would trace additional paths after developing the photograph. From Cheng, *Zengding zhongguo zhenjiu zhiliao xue*, 1931

innovative. Local Shanghai officials had banned the use of nude models in 1926.[89] Although studio operators had been taking detailed full-body photographs since the early 1890s, Cheng was interested in the naked body.[90] "We need to establish standards of measurement that correspond to the patient's own body," Cheng explained.[91] His new man offered a new template. The objective mechanical gaze had set Cheng apart. "Previous bodies were [drawn] in two-dimension, this encouraged inaccuracy," Cheng added. "Photographing the body allows us to see with greater precision."[92]

With this, Cheng hoped that the virtues of a *xiang*, a likeness, would correct for the generalized simplicity of a *tu*. He stressed that verifying sites via joints, vertebrae, and jawlines would identify the precise location of the potential therapeutic site. He wrote: "Despite the wealth of literature on acupuncture-moxibustion, the practice remains mysterious and miraculous only to be made clear under the guidance of a learned master. Medical diagrams are often inaccurate and rife with errors with numerous points poorly identified. Students who attempt to learn from these texts will find it difficult and laborious, and many have begun to abandon it altogether . . . the technique may one day be lost"; Cheng felt that premodern texts were incomprehensible without the guidance of a true expert.[93] Ignorance caused patients to suffer, faint, or die. Cheng was not wrong. Compared with other therapeutic practices, acupuncture and moxibustion had been unpopular in China, even banned in 1822 when the emperor deemed giant needles, hooked scalpels, retractor hooks, and drains better suited for assassination.[94] Cheng blamed inadequate maps for the failures of the practitioner. And despite the discursive rigidity of "native" and "non-native" binaries, modernist visions for the new Republican state offered reformers a chance to wield needles, scalpels, and hooks as uniquely indigenous technical artifacts.[95]

Still, discourses of accuracy had their limits. Cheng knew this, and it bothered him. In the same text, he acknowledged futility in standardizing across fundamentally different bodies. The person's disease state and physical size, or the time of day and weather, changed the depth of needling and the effects of the burning moxa. Warmer weather required shallow needling; colder weather required deep needling. Smaller individuals responded to sites activated at shallow depths; larger individuals responded to sites activated at deep depths. Even after finding the right site, the practitioner had to wait for the needle generate the right sensation. A single prick could induce heavy, rough, tight, slippery, slow, gentle, delayed, or rushing feelings as part of the therapeutic effect.[96]

Perhaps these sensations manifested in the nerves. This explanation would anatomically ground the therapeutic sites and mediate their expression. "When the needle enters the body, you must ask the patient if the sensation is *suan* (sour), *zhong* (heavy), or *san* (dispersed)," Cheng Dan'an explained. The practitioner "captured" the right site only when she and patient both reported feeling heavy, sour, or tightness.[97] This took patience.

Cheng likened this experience to "a fish swallowing a hook" as the silver needle "punctured" a nerve, which initiated a pulling sensation.[98] Each primary meridian, Cheng reasoned, corresponded to a cranial nerve pair, which were different from the nine and thirty-one pairs in Chen Xiutang's image of the Nerve Man (**Figure 3.7**). Regardless, nerves had to be self-evident for Cheng. He could use the nerves and the brain to explain clinical observations that he did not understand. The brain was the seat of all sensation. It was the essence of all meridians, which explained why paraplegics could still survive with a missing leg or limb, or a missing set of *jingluo* paths.[99]

"[This] fundamentally undermines previous theories," Cheng announced. "They can no longer be sustained."[100]

But nerves would again offer a partial picture. According to Cheng, needling an individual meridian site would stimulate an individual cranial nerve. Yet, nerves were everywhere, which meant that needling sites were everywhere. "Nerves in the body are in abundance," he observed. "Thus, on every inch in the body, sites are in abundance."[101] In other words, nerves were not localized. They wrapped around capillaries, pressed on lymph nodes, and extended across every surface and depth. This observation would diminish the uniqueness of both needling sites and nerves.

Yet *jingluo* were not everywhere; unlike nerves, they were not always accessible. One had to wait to "capture" the site. Further, needles were not like fishhooks or galvanizing probes; they did not require metal. Wood, bamboo, or metal needles and burning moxa stick could also activate sites that seemed to process more than sensation alone. In the end, nerve structures were far too generalized and far too ubiquitous to sustain the idiosyncratic, dynamic, and ephemeral effects of acupuncture and moxibustion.

Like Henry Head, the more Cheng Dan'an worked with patient bodies, the more disenchanted he became with using nerves to explain therapeutic responses. Sensory organs engaged with more than sensation. They facilitated something else. Yet, Cheng soon learned that nerves fell short of their explanatory promise. They did not reveal how needling and heating mobilized healing in the body. As a result, these limits renewed Cheng's appreciation for *jingluo*. He eventually returned to the ancient and early modern texts that described a cavernous landscape filled with meridians.[102] This was the same language that he initially criticized as vague and incomprehensible—language that he thought technical objects like cameras could make clear.

Following the uneventful collapse of the National government in 1949, the Communist government honored Cheng as an officially recognized "medical innovator." Cheng continued studying and translating Japanese texts, focusing on books on *jingluo* such as Yoshio Nagahama's 1950 *Study of Meridians*. In the preface to his translation, Cheng abandoned all references to nerves. Rather than being relics of the past, *jingluo* were central to medical practice. Cheng urged, "Not only do [meridians] extend from the internal viscera, but they give order to the internal viscera; they mediate the rise and fall of meridian-related qi in the body; they influence fluctuations in health and disease."[103] For Cheng, meridians were ontologically unique. They were idiosyncratic objects expressed on idiosyncratic bodies.

Rather than one standard body making sense of all bodies, Cheng kept detailed individual case records of his patients. He sketched individual faces, recorded individual conditions, and captured individually relevant sites on wax paper (**Figure 3.12**). These portraits did

Figure 3.12. Faces, twentieth century. Cheng Dan'an (1899–1957), ink on wax paper. Selected illustrations from nearly 400 clinical moxibustion cases from Cheng's practice as a rural doctor. He used moxibustion to treat various conditions: nasal congestion, pertussis, asthma, hypertension. Illustrations from the works of Cheng Dan'an. Compiled by the author

not invoke a photography studio but a clinical encounter. Their expressive contortions represented nasal congestion, pertussis, asthma, hemoptysis, pneumonia, hypertension, pleurisy, cholelithiasis, colitis, cholera, enteritis, bloody stools, diabetes, insomnia, beriberi, malaria, epilepsy, nosebleeds, polyps, tonsillitis, diphtheria, blood clotting, leukorrhea pediatric meningitis, and more.

On display were not idealized meridian models, but women, men, children, and elders exhibiting specific symptoms. Cheng's descriptions guided readers on taking individual measurements on the body, explaining how to use a string wrapped around the ankle or neck to measure and find the appropriate site to moxa on the spine.[104] By the 1950s, when he published his collection of case studies, Cheng had been elected to the Chinese Academy of Sciences and made vice-chairman of the Chinese Medical Association. By then, he had long abandoned the modernist promises of photography.[105]

Shades of Sensation

Sensations remained hard to pin down. Nervous expressions did not always correspond to their physical composition. Objects that initiated tactile sensations—pins, wools, acids, ice cubes, and boiling water—did not always generate sensations that matched their shape. Sharp objects did not always feel sharp; dull objects did not always feel dull; hot objects did not always feel hot. Under experimental conditions, hot sometimes felt cold; localized sensations sometimes felt dispersed. Henry Head had severed his nerves, brushed the head of his penis, and pressed glass tubes into men's anuses. But his maps failed. Like Cheng Dan'an, Henry Head was frustrated with the inconsistent expression and empirically slippery nature of nervous sensation. Others who tried to replicate Head's experiments identified the many weaknesses of his conceptual framework. Neurosurgeons and psychologists tried to look for the hierarchy of "primitive" and "developed" forms of sensation but found none.

For instance, Cornell psychology student Elsie Murray (1878–1965) claimed that there was no hierarchy. Categories of sensation blended together. Light touch, which Head described as a more "developed" sensation, more resembled a tickle, which Charles Darwin had once reduced to something "purposeless [in] character."[106] Light touch was not a refined sensation but a useless one.

Murray also pointed out that Head and Rivers were limited by their experimental tools. The cotton wool brushes, pins, and tubes of water barely exhausted the large range of textures and intensities that someone could process. Many things could produce specific kinds of "touch" or "tickling," Murray mused. Thin bristles, glass hairs, and small spears made of carbon, wood, and cork could be fashioned into different lengths, diameters, and degrees of sharpness, which initiated a variety of vivid textures on the skin.[107]

"Each of these—tickle, contact or brightness, pressure and pain—was correlated with a fairly well-defined range of intensities in the stimulus scale," Murray explained.[108]

What worried Elsie Murray was the inconsistency between how Head perceived his sensations and how he described them. Murray insisted that sensations were "not distinct qualities, but shades . . . within the same continuum."[109] She observed that Head failed to be technically precise with his language, using words like "contact" and "pressure" interchangeably. Sensations required rhetorical clarity in order to have conceptual clarity. They required consistent, precise, and elaborate definitions. Murray explained that "contact" was a superficial sensation that abruptly appeared and faded. Meanwhile "pressure" lasted longer, had a distinct shape and texture, faded slowly, and left an impression. She even defined "contact pressure" as an "intense but quickly-fading variety" closer to the skin's surface alongside deeper, "more severe and lasting" feelings.[110] In contrast, "dull" was too complicated to describe.[111]

For Murray, sensations occupied a "tridimensional" space.[112] They were dynamic, bodied, and embodied. With this in mind, the continuum of sensation shades seemed to make sense. Murray had distinguished "contact," "pressure," "contact pressure," and "dull" as a way to critique Head's experiments. When other neurophysiologists investigated further, they encountered similar inconsistencies. British surgeon Wilfred Trotter (1872–1939) and his student Hugh M. Davies expanded on Head's and Rivers' work.[113] Instead of severing two nerves, the men severed several nerves in each other over the course of seven months.[114] They started with a nerve at the knee, then the neck, then below the elbow, then in the elbow crease, then between the elbow and elbow crease, and finally in the thigh.[115]

This was an overwhelmingly tedious process.[116] Despite being trained neurosurgeons, Trotter and Davies experienced intense postoperative pain that lasted days.[117] After each new operation, both men struggled to focus on their experiment. They were hurting. Less "precise" or "painful" sensations took much longer to locate, with cold stimuli requiring 45 to 50 hours to identify. At one point, they realized that they spent hours using the wrong way of collecting data. "[W]e were led to make and record in detail a series of 50,000 separate touches only to discover that the solution we had sought was not to be obtained by that method," Trotter repined.[118]

Trotter and Davies filled their diagrams with clusters of dots, arrows, and circles (**Figure 3.13**). Each point translated a different sensation, including touch, temperature, or pressure.[119] But the individual sites were imprecise. Some responses were more diffuse than others. Like Murray, Trotter and Davies failed to distinguish between more and less developed forms of sensation. They hoped that this presented alternative, not contradictory, results to Head's conclusions.[120] At the very least, their results expanded on Murray's "shades within the same continuum."

In 1915, Edwin G. Boring (1886–1968) at Cornell University also cut into himself.[121] Only, he was less polite about it. Boring objected to every aspect of Head's theory. To Boring, Head's description of sensations, mapping of sensations, and proposed theory of sensations was profoundly misleading.[122] Boring severed his intercutaneous nerve that innervated an area below his wrist. Then, he stamped a grid onto his skin to measure how his pain thresholds responded to weights

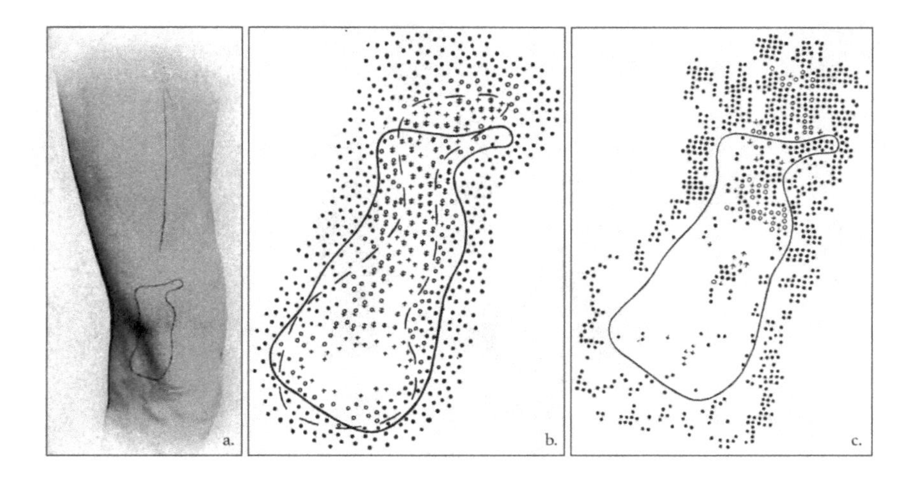

Figure 3.13. Inconvenient maps, twentieth century. Wilfred Trotter (1872–1939), print. These diagrams track areas of sensitivity and numbness on W. S. Trotter's right thigh two years after severing his nerve. Trotter and Davies used an electric current to measure sensitivity to touch, temperature, pressure, and hair stimulation, then transposed their findings from 31 diagrams and two photographic plates. From Trotter and Davies, "Experimental Studies of the Innervation of the Skin," 1909: plate 1, figs. 22 and 23

ranging from 1 to 6 grams (**Figure 3.14**). He didn't notice any changes to deep sensations of pain and pressure, but immediately noticed qualitative differences. Where Head had used words like "tingling" to describe "hair sensibility," Boring said that his sensations were more like a "flutter" and related to pressure rather than touch. The affected area felt "drier, more sticky, i.e. not slippery," he wrote.[123] Rather than a single nerve processing both fine and vague sensations via higher and lower sensibilities, Boring claimed that overlapping nerve roots caused overlapping skin sensations. Anatomical structure undergirded functional ambiguity.

When Edwin Boring published his critique of Head, the First World War had already altered Head's approach to empirically measuring sensation. While German neurologists like Maximilian von Frey (1852–1932) significantly influenced Head's conception of touch, pain,

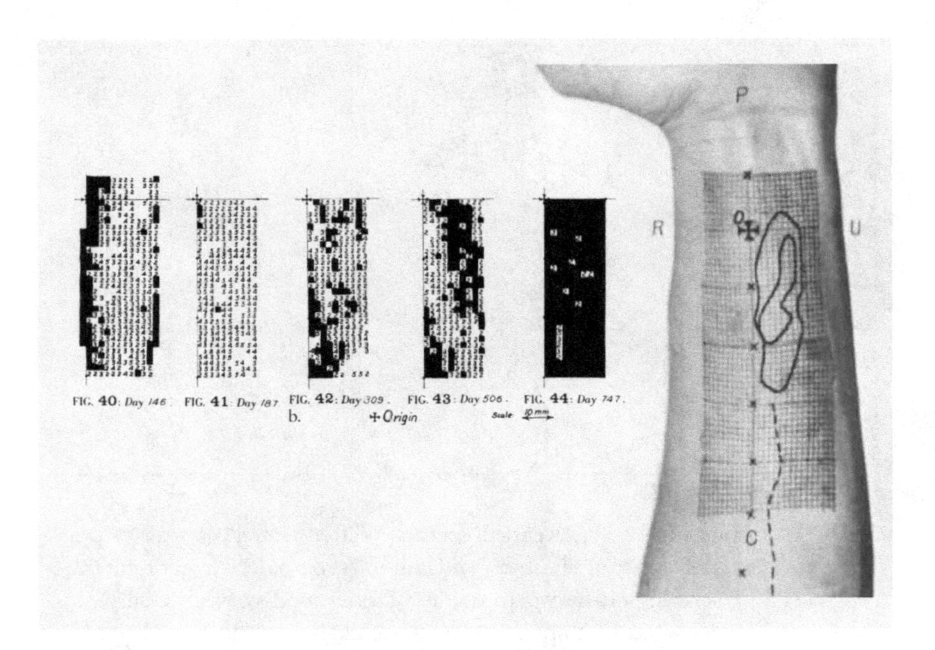

Figure 3.14. Overlapping grids, twentieth century. Edwin Boring (1886–1968), photograph. The image shows Boring's arm following nerve division. The affected area is outlined and overlaid with a stamped grid, where a square measures 2 millimeters. Boring stimulated points within the grid and measured his responses to pain in grams. Indicators from 1 to 5 grams are marked, whereas responses to 6 grams or more are shaded black. From Boring, "Cutaneous Sensation after Nerve Division," 1915: 43, plate 1, figs. 39–44

temperature, and pressure, he began to question the contributions of his German colleagues. Head resigned from his hospital post, abandoned his private practice, and worked to treat wounded soldiers.

Henry Head regularly confronted inconsistencies within and among patients. He needed a flexible imagination to navigate their constantly changing disease states. In his private life, Head expressed himself in poetry and prose, befriending authors and artists. Writer Robert Nichols, a close friend and former patient, described Head as an "artist physician."[124] Head shared with Nichols a belief that the self was highly malleable in form and in kind. He loved literature and music, understanding the constructed differences between symptom and cause, treatment and diagnosis. After joining the Council for Mental Hygiene in 1922, Head urged, "The hard and fast line so commonly drawn between organic and functional conditions [is] grossly fallacious."[125]

In 1925, Head was diagnosed with Parkinson's Disease and forced to retire. He continued to observe changes in his own body for insights on how to help patients. After recovering from influenza, he told his wife Ruth, "I have learnt a great deal by being ill that I hope to practice for the good of my patients when I get well."[126] Head's later work moved from mapping peripheral sensation to speech disorders, increasingly aware of people's situated, "neurological contrivance."[127] People were thoroughly subjective.

Yet few colleagues appreciated Head's empirical and speculative work. After Head's death in 1940, neurologist Francis Walshe (1885–1973) dismissed Head's earlier research on sensation. He echoed Boring's comments on anatomical structure and claimed that Head's approach had "manifest dangers," for theories of function "can never be regarded as soundly based until they rest upon an assured foundation of ascertained structure."[128] Other neurophysiologists later described the results of Head's work as "diffuse," "repetitive," and "outmoded."[129] To them, Head failed to make any meaningful contributions.

As psychologists, neurologists, and surgeons experimented on themselves, they puzzled over the blank spaces between dots, lines, and grids. Much seemed left to be desired in the absence of acupuncture-moxibustion practices or meridian maps. Researchers relied on self-reports, speculated on underlying structures, and grappled with the many textures that inconsistently registered in words and in memory. Rather than reinforcing a protopathic and epicritic hierarchy, experimenters like Trotter, Davies, and Boring described "shades" of feeling that more closely aligned with Elsie Murray's continuum. Too many variables affected responses to physical stimuli. Henry Head had tried to sort through these ambiguous sensations that were at once embodied and elusive, which epitomized the longstanding empirical tensions within neurophysiology. While Cheng Dan'an abandoned the neurophysiology to resurrect and reinforce the authority of classical texts, Henry Head abandoned peripheral neurophysiology to retreat into the black box of a distressed brain.

Conclusion

Nerves offered an uneasy foundation for articulating sensation and for explaining meridians. In the nineteenth century, physicians increasingly assumed that the stiff, solid, and stable quality of nerves could localize and explain the flexible, volatile, and ever-changing quality of sensations.[130] Capturing embodied sensations through conceptual, technological, and discursive means fell short of physiologists' broader epistemic ambitions. This chapter has shown these challenges manifesting in two cases, with Henry Head in London and Cheng Dan'an in Shanghai. Both cases confronted the practical hazards of assuming the anatomical stability of nerves. Head's case revealed how the promise of nerves failed to ground emerging theories of sensation. Cheng's case revealed how the promise of nerves failed to transform meridians into anatomical objects.

It was not that anatomy and physiology were at odds. Physicians continued to speculate on what nerves looked *like*, which generated more material metaphors (tree branches, water pipes, blood vessels) to supersede, presuppose, and justify what nerves *did*. In the eighteenth and nineteen centuries, these metaphors increasingly characterized brittle nerves and shriveled brains to house both higher and lower forms of cognition. The more physicians described mental disturbances as nerve damage, the more they uneasily collapsed the mind into the nerves. Under these circumstances, Henry Head hoped to use sensations to speculate on the nature of nervous anatomy, not the mind. Yet, the results only described tedious, distracting, unreliable experiences.

East Asian anatomy had long regarded the *xin* (心) as responsible for forming cognition. In the eighteenth and nineteenth centuries, translators and editors put forth new visions of anatomy amid political, economic, and intellectual transformations between Japan and China. Some amplified the role of nerves, others minimized it. Within two centuries, translators de-centered the *xin* in the body and assigned a new term for the brain in the form of *nao* (腦), which previously described the shell of the skull. Doing so reinforced meridians as capillaries that connected the *xin* as the heart and the brain. Neologisms like *naoqi* (brain Qi) then animated pairs of cranial nerves and pairs of meridian paths, approximating them so that they could be similar, if not the same thing.

Japan-trained physicians like Cheng Dan'an initially appropriated nerves and the brain to present meridians as solid anatomical structures. Despite the diverse blending of familiar and foreign ideas in a vibrant medical marketplace, the politics of twentieth-century statecraft rigidly juxtaposed a body made of nerves and a body made of meridians. Cheng followed his predecessors in trying to adapt this binary, using photography to make his point. Ultimately, he abandoned both nerves and photography, which only briefly satisfied modernist anxieties of accuracy through a mechanical gaze.

Still, nerves were not completely useless. At the very least, they projected onto other material metaphors that could superficially explain the meridians. In his later lectures, Cheng Dan'an described burning moxa as "activating" paths in the body like the "flick of a switch." As he taught students in rural classrooms with limited resources and dirt floors, Cheng pointed to the exposed beams and cables, gesturing to the entrance, room, and doorway lights. He explained, "Needling and heating sites are like levers on the switchboard [where] the nerve is the electric cable that links the internal organization of all the lights. If there were problems in one part of the system, if the lights were dimmed, then we can use needling and heating to turn on the switch that governs that problem area. So, for any problem that arises inside the body, needling and heating the exterior of the body will allow you to easily treat the disease. The more accurate the moxa sites, the more effective the treatment."[131]

Cheng aimed to convince others, not himself. Even though he had abandoned the explanatory power of nerves, he still understood the rhetorical force of material metaphors. Meridian paths were not nerves, but they could be activated like wires connected to an electric switchboard. This allowed him to describe switches that "governed" discrete areas of the body. The simplicity of the metaphor made the therapeutic effects seem self-evident. They could "easily treat" the ailment. Metaphors made meridians more visible and more compelling. Their simplicity allowed Cheng to reinforce the "accuracy" and "efficacy" of moxibustion.

Cheng and Head had both recognized the expanding distance between the external stimulation—the needle, pin, burning moxa, hot water tube, cotton brush—and the reported sensation—heavy, tight, sour, dispersed, distant, connected, shaded. These accounts exposed a paradox that underlay theories of sensation: even though physicians invoked nerves to explain the diagnostic techniques and therapeutic effects of acupuncture and moxibustion, nerves were hard to explain. Perhaps sensations clustered like branches. Perhaps nerves were like blood vessels. Perhaps meridians were as obvious as nerves—which were not obvious at all.

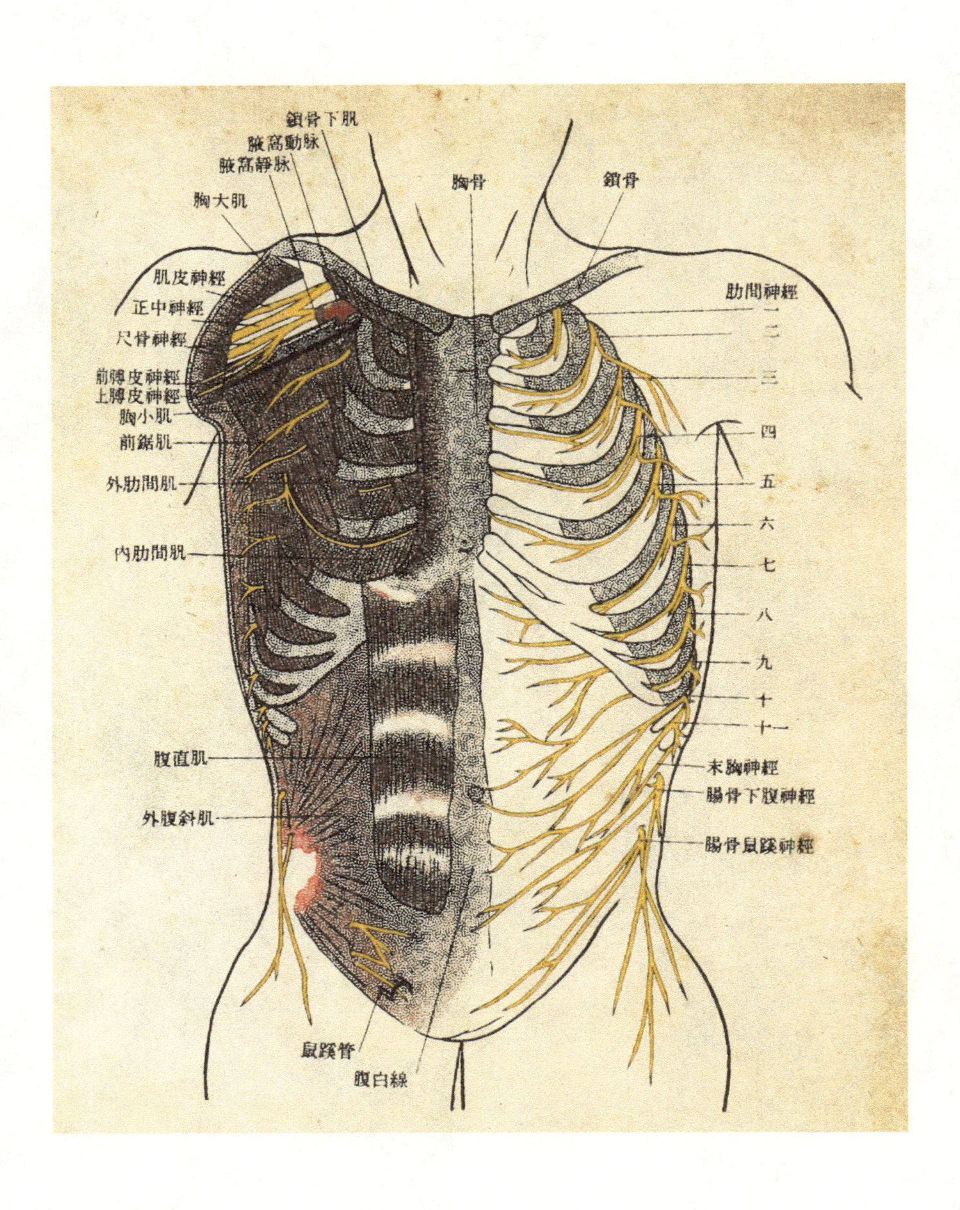

鎖骨下肌
腋窩動脉
腋窩靜脉
胸大肌
胸骨
鎖骨
肌皮神經
正中神經
尺骨神經
前膊皮神經
上膊皮神經
胸小肌
前鋸肌
外肋間肌
內肋間肌
腹直肌
外腹斜肌
鼠蹊管
腹白線
肋間神經
一
二
三
四
五
六
七
八
九
十
十一
末胸神經
腸骨下腹神經
腸骨鼠蹊神經

CHAPTER FOUR

Generic Maps and the Failure of Standardization

The Color of Nerves

They were yellow and sharp (**Figure 4.1**). To the left of the image, the nerves receded from view where they burrowed under layers of muscle and hid beneath a collar, a sternum, and a stack of cartilage. These bright yellow tentacles penetrated the grey and pink mass of crosshatched and stippled flesh. Captions labeled the dense nerve clusters in the shoulder, along with Sinographic translations of Latin terms for veins in the neck and abdomen. Flesh and bone acted as a sheath enveloping the yellow tubes to anchor them in place.

To the right of the image, the nerves looked liberated. They curved around the ribs. The yellow arms now reached over, under, and alongside bone. They connected to a row of numbers that descended in neat succession: 一, 二, 三, 四, 五, 六, 七, 八, 九, 十, 十一. These numbers did not count the ribs, but instead counted the 11 pairs of

Figure 4.1. (opposite) Yellow nerves, twentieth century. Wang Xuetai (1925–2008), chromolithograph. Wang worked with a limited palette of yellow, red, blue, and black. This view of the torso contrasts muscles and skeleton (*left*) with yellow-outlined nerves innervating the chest and stomach (*right*). From Zhu Lian, *Xin zhenjiu xue*, illustrated by Wang, 1950, fig. 6

individual nerve branches. These 11 pairs, also known as "intercostal nerves," grew from the spine and clung to the torso. Historically, illustrators varied in their nerve pairings, sometimes drawing 12, 16, or 30 pairs. For now, in this image, the fat yellow branches were easy to spot, easy to count, and easy to imagine.

Printed in 1951, this chromolithograph was credited to a young medical school graduate named Wang Xuetai (1925–2008). Wang had been recently radicalized by Zhu Lian (1909–1978), the vice minister of health for the Communist Party. United by a common cause under uncommon circumstances, Zhu recruited Wang to join her small team of medical experts. Zhu had been writing a book and needed editorial support to publish what would be known as *New Approaches to Acupuncture-Moxibustion* (*Xin zhenjiu xue*). There was nothing explicitly new about *New Approaches*; acu-moxa texts had been in circulation since the Republican period (1912–1949). Rather, the newness of *New Approaches* was meant to signal a fissure between opposing political factions, Communist efforts, and its Nationalist enemies. Zhu's book was a Communist text by a Communist Party author. In it, she addressed her comrades, promising to reveal simple and effective therapeutic techniques with the help of Wang's detailed images.

Introduction

This chapter closely reads Wang Xuetai's illustrations to understand the ways in which the similar, conflicting, and diverging conceptions of meridians and nerves manifested in the early Communist Party—a party that discursively upheld a persistent dialectical materialist discourse of the body. It explores Wang's illustrations over two decades to understand how his visual epistemology explicitly underpinned his ontological commitment to meridian paths. This chapter further contends that despite uniting Zhu and Wang, Communist materialist discourses also undergirded further materialist contradictions in anatomical standardization.

Although Zhu and Wang both studied classical medical texts, they eventually diverged in how they applied materialist discourse to justify the nature and need of meridians.[1] Zhu firmly maintained that nerves could explain and replace meridians. She considered meridian lines as unremarkable visual referents. If pressed further, Zhu simply stated that meridians were material products of nervous sensation. Wang, meanwhile, would uphold that meridians possessed their own unique ontology. The more Wang created meridian *tu* (圖), the more invested he became in them. Meridians would need to be understood on their own terms in their own form. The author and illustrator were at odds.

Zhu's and Wang's diverging visions of meridian paths were further grounded by a transnational Communist materialist ideology. In the mid-twentieth century, Chinese scientists had turned to their Soviet counterparts in Russia to supply alternative physiological models. Zhu Lian began reading translated experiments by Ivan Pavlov (1849–1936) and his lab, focusing on Pavlov's links between observable behavior and unseen physiological processes. Although Pavlov's research predated the Russian Communist Party, politically minded scientists appropriated his work into a materialist Communist framework.

Zhu Lian borrowed Pavlov's term of "higher nervous activity," or *gaoji shenjing huodong*, to explain the therapeutic effects of acupuncture and moxibustion as physical products of the mind. In theory, the higher activity of nerves manifested through the inhibition and excitation of nerves, which Pavlov and his collaborators could not empirically demonstrate. Here, again, was the longstanding contradiction that characterized neurophysiological research. The physical products of nerves manifested in discrete and measurable behaviors, but the physical products of the mind remained unseen and unmeasurable. Nevertheless, Zhu was unbothered as long as the invisible potential of nerves was *more* material than the invisible extension of meridians.

Zhu Lian's and Wang Xuetai's diverging perspective of meridians was subtle at first. Some of this subtlety emerged in Wang's selection of color pigments to illustrate anatomical structures. For instance, in the first edition of *New Approaches*, nerves and meridians were the same color. Yellow meridians floated above bone and distributed across an empty stomach (**Figure 4.2**). By taking the same color, perhaps

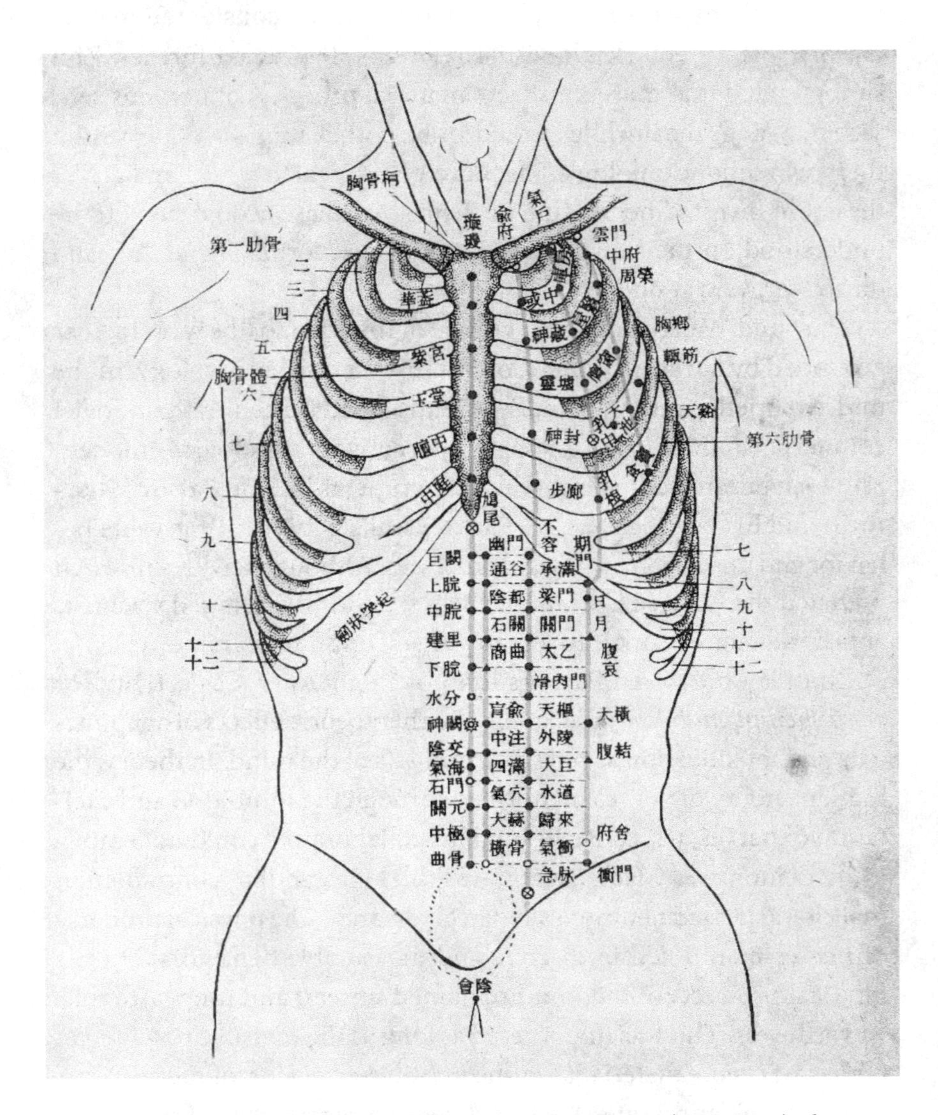

Figure 4.2. Yellow meridians, twentieth century. Wang Xuetai (1925–2008), chromolithograph. Five meridian fragments on the left side of the body extend from the collarbone toward the pelvis. The yellow meridians match the color of nerves in **Figure 4.1.** From Zhu Lian, *Xin zhenjiu xue*, illustrated by Wang, 1950, fig. 7

yellow nerves and meridians carried the same materiality. Wang had access to red, yellow, blue, and black pigments, which could have been combined in over two dozen ways. Yet he assigned yellow to both nerves and meridians. Doing so suggested that meridians and nerves shared qualitative characteristics.

By the second edition of *New Approaches published* in 1954, meridians and nerves became different colors. Meridians turned red; nerves turned black. Wang's new images rendered meridians and nerves as fundamentally different kinds of things. To be clear, nerves and meridians were different to Zhu and Wang. They simply disagreed on the nature of that difference. Zhu held that meridians were superfluous—that they need not exist. Wang suggested that nerves and meridians were not mutually exclusive—that there existed a meaningful ontological difference between the two.

This presented an ontological problem: were meridians anatomical structures or physiological expressions? If they were physiological expressions, then it was unclear what they expressed. Stimulating one meridian site did not always follow one meridian path. Needling the hand could treat the head or chest; burning moxa on the shoulder could treat the lungs or liver. With these observations, Zhu realized that Pavlov's "higher nervous activity" was insufficient to explain the therapeutic effects of sensory transduction. Neuroanatomy was a useful starting point but not the final answer.

At the same time, it was also possible that meridians were neither anatomical nor physiological. As Wang continued his career as writer, researcher, illustrator, and policy advisor, he increasingly appreciated how graphic genres sustained the ontological ambiguity of meridians. His bodies became more *tu*-like and less artistically rendered, losing dimension and shading. He inscribed meridians on transparent muscular bodies rather than on dense tissue. Meridians lost their color. Body diagrams looked less and less like bodies. Wang understood that his illustrations did not show a comprehensive view. They did not show the many connections between distressed areas and therapeutic sites, or variations in depth and destination. Instead, they suggested that meridians took their own form—as both an organizing schema and as an expressive network of flexible paths.

Un-making Meridians

Zhu Lian aimed to be pragmatic.[2] The ongoing civil war between the Communist and Nationalist parties had left her medical team in extreme isolation. By the late 1940s, political tensions prevented access to already limited resources. It was around this time that Zhu called upon a committee of comrades to help her write what would become *New Approaches to Acupuncture-Moxibustion*. She first began by revising her lecture notes to reconfigure medical training based on a close study of anatomy and physiology. In her view, accurately locating needling and heating sites depended on identifying precise physical locations on the body, specifically on her terms. In the introduction, Zhu explained: "A few years ago, we were in the old, liberated areas of the rural North. There, we used acu-moxa to successfully treat many diseases, which conserved our limited supply of drugs while preventing [diseases] . . . we also encountered practitioners who did not sterilize needles and had little understanding of anatomy and physiology. Despite their experience, this led to infection and other adverse consequences, so we felt compelled to enhance their understanding of the body."[3] Zhu was afraid of ignorance. The local medics and their years of "experience" still led to infections and unintended harms. Zhu felt that her own training and experience could offer a corrective. She had started her medical training at 16, apprenticed in Beijing and Shanghai, and studied with senior physicians. Zhu held a privileged view. Her medical arsenal demonstrated a skilled attention to the body. It demonstrated medical expertise undergirded by an "understanding of anatomy and physiology" that her rural counterparts lacked. All she needed was a textbook to fix her vision of the body.

Enter Wang Xuetai (1925–2008). He served as Zhu Lian's resident illustrator and created a series of elaborate images that promised to correct misconceptions about the body. In the first edition of *New*

Approaches, Wang Xuetai produced 14 full-page color prints showing the body in fragmented sections, such as the head, chest, arms, and legs.[4] Despite limited publishing resources, the first run printed 12,000 copies of *New Approaches* in 1951. The press had been strategic with its use of raw materials. The body of the text appeared on wood pulp paper that discolored over time, whereas Wang's lithographs appeared on sturdy white paper made to withstand processing up to four rounds of colored ink.[5] Printing images required negotiating resources.

Wang created his images with care. Lines were not just lines, and dots were not just dots. In **Figure 4.3**, a closer look at the torso reveals that individual therapeutic sites were graphically unequal. Some appeared as solid dots, while others did not. For instance, ❂ at the center of the navel marked the *shenque* (神闕) site. Meanwhile, five hollow sites marked with ⊙, designated *shuifen* (水分), *shimen* (石門), *henggu* (横骨), *qijiu* (氣僦), and *fushe* (府舍). Two crossed sites marked with ⊗ labeled *jiuwei* (鳩尾) and *jimai* (急脈). Triangle sites, marked with ▲, included *riyue* (日月), *fujie* (腹結), and *shangqu* (商曲).

These different sites did different things, although Wang did not explain how they differed until 1954.[6] The second edition of *New Approaches*, issued in hardback, had new images with more detail and text. Wang created these images from scratch, redrawing the head, chest, arms, and legs. The lithographic meridian torso now featured shaded skin and bone with an almost watercolor quality. Wang reversed the image, moving labels from the right to the left, which nearly doubled paths and sites on both sides (**Figure 4.4**). This time, meridians were red.

Among the 97 sites, some were more sensitive or more sacred. Wang included a legend explaining that ● sites were open to needling and moxa, ○ sites were forbidden for needling, ▲ were forbidden for moxa, and ⊗ were forbidden for both needling and moxa. For instance, *shenque* (previously marked with ❂) appeared in the center of the abdomen and now prohibited needling. *Jimai*, at the base of the third red path, was marked with ⊗ and forbidden for needling and moxa.[7] These suggested qualitative and ontological differences among

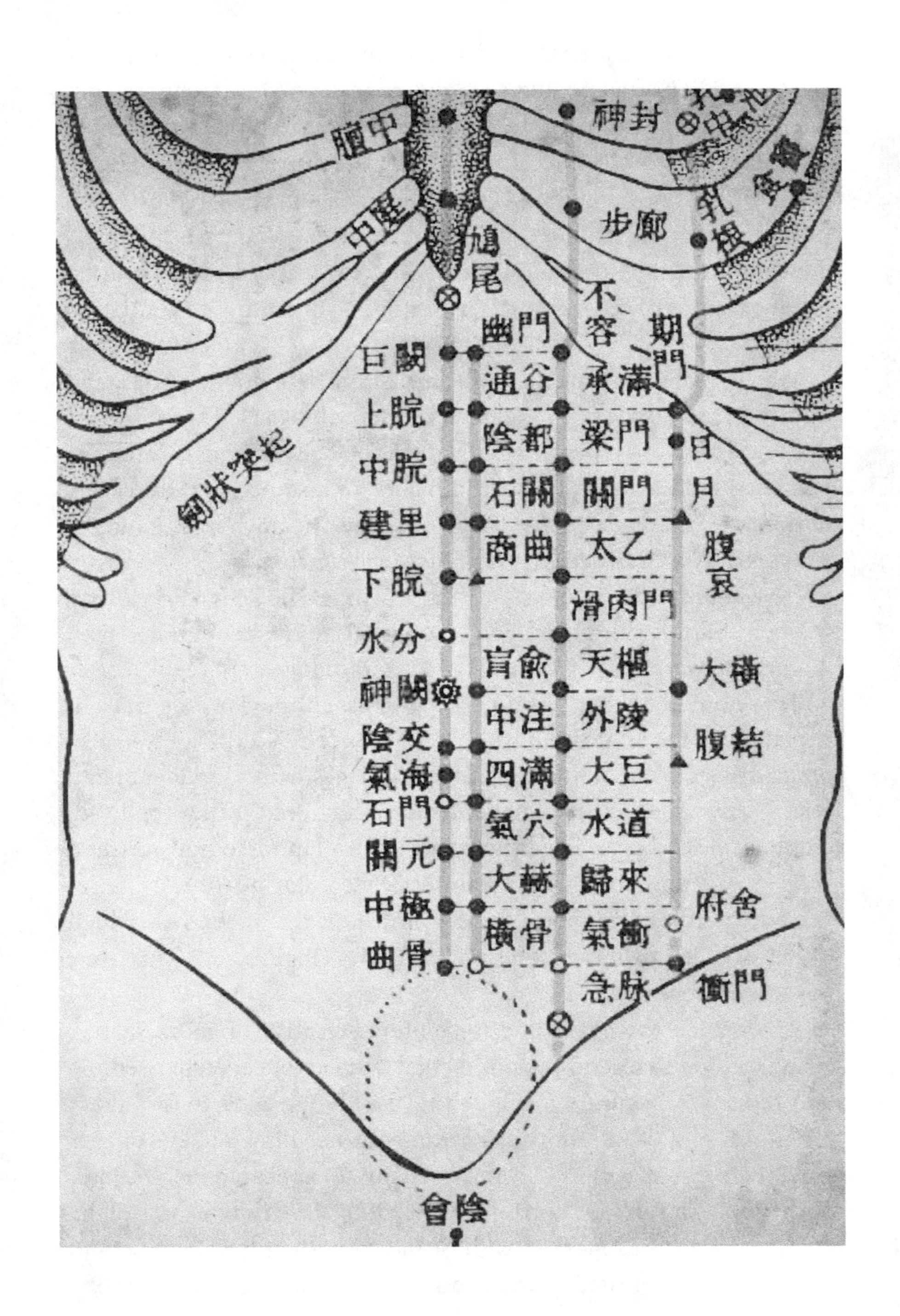

膻中
中庭
神封
步廊
乳根
食竇
鳩尾
不容
幽門
期門
巨闕
通谷
承滿
上脘
陰都
梁門
日月
中脘
石關
關門
建里
商曲
太乙
腹哀
下脘
滑肉門
水分
肓兪
天樞
大橫
神闕
中注
外陵
腹結
陰交
四滿
大巨
氣海
石門
氣穴
水道
關元
大赫
歸來
中極
府舍
曲骨
橫骨
氣衝
急脉
衝門
會陰
劍狀突起

sites—inappropriately stimulating one could severely injure the recipient.

The accompanying text explained that understanding acu-moxa required knowing the appropriate needling depths. Sites could rise and sink, resting at different depths—some closer to the skin surface, others farther away. These variations were measured in units of *fen*, which was a tenth of a *cun*.[8] For instance, sites at the center of the chest could be reached at three *fen* deep. Further down the same path, meridian sites could be accessed at six *fen*, five *fen*, and eight *fen* deep. The vertical action of a needle suggested that it moved to search for the exact location of an active site—that the site lay deeper in the body.

Wang's images did not register depth. The dots rested at the surface of the skin. Meridians settled on a single plane. To the right of the image, sites floated in space. What would have been a dense stack of organs appeared empty. In its absence, horizontal tracks dragged across the belly and anchored sites that were not visually defined by tissue or viscera. Like a weft securing a warp, these lines lifted individual sites from their depths, connecting them as a cohesive unit. In other words, Wang's image failed to show the full dimension of its lines and dots. It existed as a hybrid of graphic realism on the one hand and a schematic *tu* on the other.

There were other graphic limitations. Meridian sites not only dipped in space but also reached multiple destinations. Neighboring sites on one path could treat different disorders and areas in the body. For instance, two adjacent sites on the shoulder named *yunmen* (雲門) and *shidou* (食竇) accessed different disease states of different intensities. *Yunmen*, found beneath the collar bone, could treat coughing,

Figure 4.3. (opposite) Unequal sites (detail), twentieth century. Wang Xuetai (1925–2008), chromolithograph. In Wang's anatomical illustrations, sites are graphically differentiated. The symbol ❁ at the center of the navel marks *shenque* (神闕). The symbol ⭘ marks sites including *shuifen* (水分), *shimen* (石門), *henggu* (橫骨), *qijiu* (氣僦), and *fushe* (府舍). Two sites marked with ⊗ are *jiuwei* (鳩尾) and *jimai* (急脈). Sites marked with ▲ include *riyue* (日月), *fujie* (腹結), and *shangqu* (商曲). Wang did not, however, explain how these sets of sites differed. From Zhu Lian, *Xin zhenjiu xue*, illustrated by Wang X., 1950, fig. 6

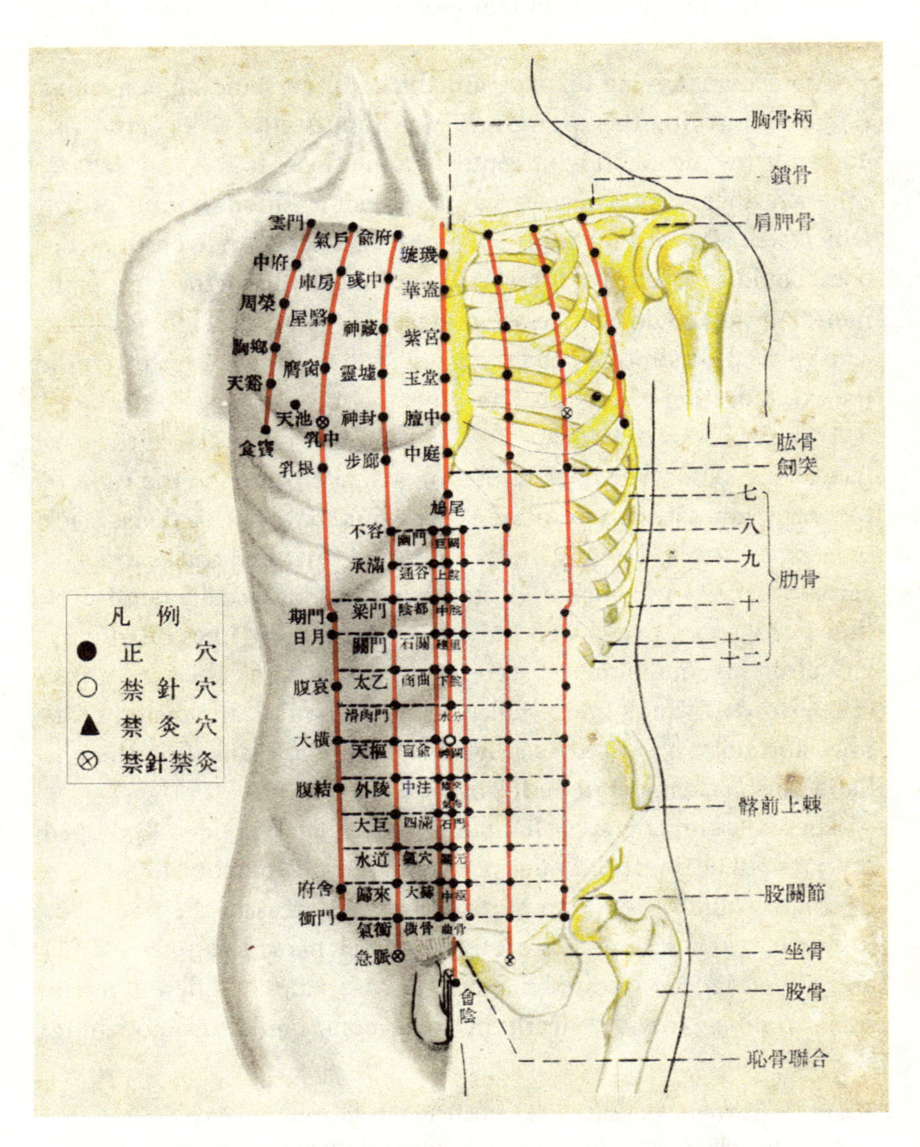

Figure 4.4. Red meridians, twentieth century. Wang Xuetai (1925–2008), chromolithograph. Wang added a legend in the 1954 edition of *New Approaches* to disambiguate differently marked sites. The symbol ● sites can be needled and treated with moxa, ○ are forbidden for needling, ▲ are forbidden for moxa, and ⊗ are forbidden for both needling and moxa. From Zhu Lian, *Xin zhenjiu xue*, 2nd ed., 1954, fig. 6

asthma, tonsillitis, and shoulder numbness, among various heart and lung issues. *Shidou*, further below on the same path, could treat pulmonary congestion, pneumonia, pleurisy, liver pain, and more.[9] One site addressed infections in the lungs. The other addressed disturbances in the liver. One site reached the throat. The other reached the chest. The red line connecting these sites only measured superficial distance, while more invisible paths extended from each site to other therapeutic areas. Beyond the red lines lay another complex set of unseen therapeutic paths.

These discrepancies between concealed depths and imagined destinations perplexed Wang Xuetai. He had illustrated a heuristic; he had not represented therapeutic possibility. The therapeutic effects of acu-moxa were too dynamic to illustrate on paper. They had never been illustrated on paper. Meanwhile, Zhu Lian took this as a reason to abandon any discussion of meridians. Her aim, again, was to educate readers in acupuncture and moxibustion through anatomy. Whereas previous books on acu-moxa introduced therapeutic sites based on how they were organized along individual meridian paths, Zhu took a different approach. Each chapter in *New Approaches* focused on a different region of the body: the side of the head, top of the head, area around the eyes, the mouth and nose, the face and cheeks, the back of the shoulders, the front of the shoulders, the chest, abdomen, the upper limbs, and the lower limbs. She had fractured the full lengths of multiple meridian paths and gathered them based on her own organizing principle of body parts. In her chapter on the chest, for instance, Zhu discussed different sites based on whether they were found on the "middle," "inner," or "outer" lines. Now, the lines simply organized individual points, nothing more. For Zhu, lines were no longer meridians.

The Ideological and the Embodied

Zhu Lian's decisions were ideologically grounded. She took seriously her responsibility to incorporate anatomy into her revolutionary cause. Medical education was part of her political work. Zhu had been born to a family of revolutionaries—her parents and grandparents had been involved in anti-foreigner raids during the Boxer Rebellion at the turn of the twentieth century. During her medical training in Shanghai, Zhu sponsored and married a political fugitive.[10] In 1935, she joined the Communist Party, being its first female member at 26 years old. She soon set up a clinic in Shijiazhuang and took on duties as the chief editor of two regional newspapers.[11] As she advanced in her position in the Party, Zhu assumed a masculine persona. She cut her hair short and wore a Mao suit.[12] She became her comrades by looking like them. Zhu was formidable. She later joined the Eighth Route Army and served as deputy health minister and clinical director of the 18th Army before becoming vice president of China Medical University.[13]

In the 1940s, Zhu and other core Communist Party members found themselves in the mountains of Yan'an. The chaos of civil war had strained already limited resources.[14] Party leaders like Mao Zedong (1893–1976) discussed establishing an ideological framework to advance their political cause, examining ways to reform and re-evaluate Chinese medicine. They criticized Nationalist campaigns as enemies of Chinese medical knowledge even when the Nationalists also tried to modernize medical knowledge. Republican-era physicians similarly wrestled with the absolute and ambiguous binary of "tradition" and "modernity" in creating new state-sanctioned medical pedagogy. But this history mattered little at the momentous United Front in Cultural Work in 1944. There, Yan'an officials devised a slogan to render Chinese medicine "scientific" and Western medicine "popular." The newly christened "scientific" Chinese medicine would them be incorporated into Communist identity.[15]

Zhu carried out this task to render Chinese medicine scientific. To begin her mission, she joined a group of surgeons, pathologists, and anatomists to learn acu-moxa from a local physician named Ren Zuotian. Initially engrossed in administrative duties, Zhu paid little attention to her training. However, on a frigid night, crossing a frozen river triggered a sharp sciatica pain that radiated from her hip down her leg.[16] Zhu was left in intense agony. Injections, medications, and heat brought no relief. When Zhu finally turned to acupuncture, the pain vanished after only one application.[17] This immediate recovery inspired Zhu to study acu-moxa with care. She shadowed senior physicians, studied classical texts, and compiled acu-moxa handouts to distribute as teaching material.[18] Her embodied experience of acu-moxa convinced her of its efficacy. Yet, efficacy was not quite the kind of "scientific" evidence that Zhu needed to fulfill her political duties. Historical records were full of case studies; hers was only one among many.

Zhu Lian gradually became known for her talent as a practitioner. She served as the personal physician to other influential Party cadres, including high-profile patients like Zhang Panshi (1905–2000), one of the founding members of *People's Daily (Renmin Ribao),* the official newspaper of the Party. When they first met, Zhang was in terrible shape. He had excessive fatigue, night sweats, chronic fever, and insomnia. "I was suffering from a nervous breakdown," Zhang recalled.[19] He had no appetite and took an assortment of drugs for sleep and digestion. Barely able to walk, Zhang finally asked Zhu for help. Still early in her acu-moxa training, Zhu wanted to test her new skills. Zhang reluctantly agreed. In a few days, his sleep improved, though he still sweated excessively. Gradually, he had the energy to work again. Over the next years, Zhang saw no other doctors, took no other medication, and only occasionally received treatments from Zhu or her assistant.[20]

Not everyone sympathized with Zhang's recovery. Although he had confidently attested to his own improvement, some of his colleagues dismissed his story. To them, acupuncture remained "unscientific." To them, Zhang's testimonial did not contribute any empirical evidence. His experience proved nothing. Zhang was sensitive to this critique and explained that he was not writing as a medical expert but as a patient for whom any effective treatment demonstrated "scientific value." Despite his own political standing, his words failed to impress.

Like Zhang Panshi, the eminent military general Zhu De (1886–1976) echoed a similar sentiment. He claimed "[A]cupuncture and moxibustion can effectively treat illnesses—this represents science."[21] For Zhu De, any form of observable phenomenon, anecdotal or not, established the fact that acu-moxa worked. Dong Biwu (1886–1975), the head of the Supreme Court, agreed. Dong had met Zhu Lian in his late 60s, when he suffered chronic shoulder pain. Dong was so impressed with Zhu that he would later bring her to Beijing and shield her from the Cultural Revolution to treat his severe facial trigeminal neuralgia. Dong even dedicated a poem to Zhu Lian that appeared in the third and final edition of *New Approaches to Acupuncture-Moxibustion* in 1980. Building on the testaments of Zhang Panshi and Zhu De, Dong Biwu identified Zhu's training in biomedicine as the source of her scientific authority. This, along with his own experience, confirmed the epistemic virtues of Zhu's practice.

Party cadres would disagree on what counted as scientific evidence. Neither skeptics nor supporters understood how acu-moxa worked, but they diverged on whether they had personally experienced its effects.[22] Embodied knowledge was useful knowledge. Either you had it, or you did not.

An Abundance of Metaphors

Zhu Lian had little to offer in explaining how acu-moxa treated her patients. She did not think meridians were relevant. Instead, she turned to new neurological theories. Based on her study of classical texts and clinical experience, Zhu added to her repertoire the recently translated works of Ivan Pavlov (1849–1936), whose long reputation in Russia had earned him recognition as a model Soviet scientist.[23] China's political alliance with the Soviet Union was at its height in the 1950s, which facilitated a flood of interest in Soviet literature, music, social theory, and scientific publications. Within a decade, nearly all of Pavlov's work had been translated into Chinese. In these papers, Pavlov and his colleagues discussed using needles to stimulate the central nervous system.[24] Observations from animal studies suggested that needling at nerves could cause diseases through eliciting inhibitory or excitatory signals.

Zhu Lian picked up on Ivan Pavlov's theories of sensation. Yet, his theories remained partial. The idea of exciting and inhibiting nerves to induce disease states seemed simple enough—perhaps too simple. It did not account for the numerous physiological inconsistencies that tormented Pavlov and his collaborators. Like physiologists wrestling with meridians, Pavlov was equally perplexed with what he described as "higher nervous activity," the same idea that Zhu and her colleagues translated as *gaoji shenjing huodong*. When Pavlov placed "higher nervous activity" under experimental conditions, he found the results increasingly difficult to interpret. They demanded that the experimenter draw inferences based on her own experiences.

The inhibition and excitation of nerves extended from a commitment to the idea that behaviors were physical products of the mind. Pavlov had collected thousands of drops of saliva from hundreds of dogs to access their inner "psychical experience."[25] Their gastric juices were products of their interior world. Pavlov's laboratory took to studying the mind through these observable behaviors. In other words, gastric juices expressed behaviors, and these behaviors stood in for the mind.[26] Pavlov took inspiration from authors who published in French, English, and Russian, developing an appreciation for the entangled relationship between the mind, the brain, and the gut.[27] Reflexes that radiated from the brain determined the fundamental composition of an organism. Reflexes emerged in an organism's preferences, tastes, and passions.[28] These passions, tastes, and preferences also manifested in the gut that grounded all movements in the body. In studying behaviors as expressions of the mind, Pavlov searched for a direct relationship between personality and physiology. He tried to isolate this direct relationship with his famous army of laboratory-bound dogs, who were each sorted by personality.[29]

But in Pavlov's research, no single path could map out a single response. A measurable behavior seemed to connect to numerous factors, each one as invisible as the next. For instance, saliva production was not the result of a single reflex, but a reaction emerging through a "chain" of responses. Each chain was like a thread that connected to a shifting "center" that existed only in relation to external forces.

As the Russian physiologist Ivan Sechenov (1829–1905) put it, the center of a reflex chain was like the center of gravity.[30] It moved when other factors weighed in. This explained why it persistently evaded experimenters.

Metaphors again shaped the limits of epistemology. Sechenov had borrowed language from physics, suggesting that physiological responses were like magnetic and electric fields. And when the metaphor of fields no longer applied, Sechenov turned to the metaphor of the machine. He presumed that the reflex chain was a series of mechanical, causal events. "[A] psychical phenomenon becomes part of a chain of machine-like processes," Sechenov wrote when explaining a fear response.[31] One event triggered the next, like links on a chain. These links were mechanical yet flexible, metal yet magnetic, electrical yet gravitational. Similarly, Pavlov's student Nikolai Tikhomirov also preferred to describe reflexes as something akin to gravitational forces. The gravitational tugging from many centers allowed Tikhomirov to explain the observation of "actions at a distance" that occurred without a visible stimulus.[32] These simultaneous forces allowed the body to function as a coherent unit.

Pavlov was not impressed. He had trained in physics and mathematics but avoided using the metaphor of a gravitational field. Invisible actions at a distance led to a slippery slope of immaterial reflexes. Invisibility and materiality were at odds. Pavlov instead preferred to reference visible agents that carried material signals—external elements that pressed on a dog's nose, ears, and eyes.[33] To validate these corporeal interactions, Pavlov described the interaction between centers and peripheries in the body as something more like Russian factory-cottage industries, or *kustarnyi lad*.[34] Communication occurred within communities; activity registered through action. This industrial metaphor was another stretch of the imagination that grounded the materiality of reflexes.

Still, invisible movements remained at work, and Pavlov eventually gave in to using gravitational metaphors.[35] His team would describe reflexes as "waves" that collided into one another.[36] A stronger force subdued a weaker one; a weaker force diffused a larger one.[37] Explanatory models grew more elaborate. The nature of "higher nervous activity" moved from chains with discrete centers to industrial interactions to interfering waves. This way, exciting and inhibiting stimuli generated other forms of dis-inhibiting, irradiating, concentrating, and inhibiting responses. A single response fractured into many forms, which Pavlov named the "law of dispersion." It was a force of nature. It was a law of nature. He wrote, "A stimulation entering the cerebral cortex which is not directed to a certain active nervous point begins to spread and disperse itself over the surface of the brain. If now a strongly excited point arises later, it will attract to itself cortical stimulations not only from the point originally irritated, but from all other points over which the stimulation had gradually spread. This is the law of dispersion, of irradiation of the-stimulation in the cerebral cortex."[38]

In this description, waves of inhibition could spread through the body. As they spread, they changed. This presented practical experimental challenges. Dispersed responses were hard to follow. Locating the "center" of a reflex as it transformed meant isolating a phenomenon that Pavlov assumed to be *both* discrete and dispersed. Laboratory conditions further complicated seemingly discrete processes of the "excitation" and "inhibition" of nerves. If a dog's straps were too tight or if he had been accidentally burned, then salivation would stop. Pavlov described these accidents as "destructive irradiation," or a kind of irradiation that destroyed the path that signaled excitation. Paths could vanish. With a destructive stimulus, they could die.

This is where needles came in. A scientist named Maria Erofeeva in Pavlov's lab ran the study of "destructive irradiation" by physically poking or electrically shocking dogs to initiate a "destructive" signal. In her experiments, Erofeeva paired unpleasant sensations with pleasant ones. She would first introduce a pleasant smell that caused the dog to salivate. Then, she would either poke or stun the dog to measure how one sensation overwhelmed the other. Initially, salivation decreased. But over repeated experiments, the dog associated the pleasant odor with the unpleasant jab and the unpleasant jab with food. Eventually, the shock no longer diminished salivation Instead, the shock intensified salivation. Paired sensations generated stronger responses over time rather than weaker ones. The simple "destructive" signal turned out to be unstable. It revealed a false binary between pleasant and unpleasant sensations because together, they intensified salivation. "Destructive irradiation" did not make sense.

Despite numerous experimental challenges, Pavlov's team persisted in researching nervous excitation and inhibition. They used fragrances, buzzers, and electric shocks in the laboratory. Still, they struggled to discursively fix what they observed as dispersed signals. Metaphors like chains, threads, mechanical sequences, industrial relationships, gravitational forces, and sound waves only provided partial explanations.

In the 1920s, Pavlov's former student Dmitrii Fursikov officially established the Communist Academy's Institute of Higher Nervous Activity. Members of the Academy used approaches in biophysics, biochemistry, histology, cytology, comparative anatomy, and physiology to explain how signals transformed in the body.[39] Rather than a chain of reflexes that pulled at each other, Pavlov later described signals entering the body as a cerebral "mosaic."[40] He wrote that sensory processes concentrated in "fine" and "definite" locations in the brain. Again, they were both discrete and dispersed. Signals were simultaneously singular and connected.

Pavlov did not make reflex maps because he could not. No single path traced a stimulus to a response. Responses, again, relied on multiple invisible factors. If an electric shock induced a "wave" of sensation, that wave could spread. As it spread, it could change. Multiple inputs competed, collided, and combined with each other. According to Pavlov's "law of dispersion," a reflex could be either a single chain or many chains. They could be competing gravitational forces or an elaborate mosaic of things tugging on each other and interfering with new sensory inputs. Responses in the body responded to too many conditions. Rather than conditioning the animal, the effects of stimuli were conditional, inconsistent, and compounding.

Appropriating Pavlov

In the 1920s, translations of Ivan Pavlov's ongoing work made their way to China. Pavlov had become a mascot for Communist science despite his own political ambivalence. It was the Communist cause that generated a language about materialist ideology that sustained the many of the unanswered questions left in Pavlov's research. Politically minded scientists held that Pavlov's work embodied "dialectical materialism," a concept that allowed differences and contradictions to exist within the same system. Dialectical materialism was critical to Communist philosophy and largely credited to Karl Marx and Friedrich Engels, who articulated and aimed to resolve tensions between social classes. In other words, dialectical materialism was a social theory that promised to make sense of all kinds of contradictions, establish universal rules of engagement, and delineate individual scales of difference.

This also applied to science.[41] Mao Zedong famously announced in a 1937 speech that a dialectical worldview described a dynamic system. It was accommodating, flexible, and adaptive so that internal contradictions could exist. Every system involved a movement of opposites, Mao reflected, and each individual entity within the system was guided by its own contradictions.[42] True to a dialectical world, "higher nervous activity" in theory and practice was full of contradictions.[43] Many scientists recognized these contradictions and committed themselves to investigating conditional reflexes. They applied the idea of opposing inhibition and excitation forces in their own work. They assumed that these opposing forces would manifest observable and embodied phenomena, even if sensations were not always observable or felt. Here was the contradiction.

Pavlov's popularity in China continued to grow. In 1934, the new Chinese Physiological Society elected Pavlov an honorary member. In 1935, over a dozen Chinese scholars attended the 15th International Conference on Physiology in Leningrad, hosted by Pavlov before his death the following year.[44] Chinese students abroad like Pan Shu (1897–1988), Gao Juefu (1896–1993), and Chen Hanbiao (1906–1982) took on Soviet science and translated Pavlov's papers.[45]

The connection between political ideology and epistemic practices was both innovative and pragmatic. For instance, future Pavlovian scientist Siegen K. Chou (1903–1996) first learned of the clinical implications of "higher nervous activity" at one of Pavlov's lectures delivered at Yale University in 1929. Intrigued, Chou decided to apply this idea to his own body.[46] He returned to Beijing and suffered from insomnia, which he interpreted as a manifestation of disrupted inhibitory functions. Chou reasoned that the sleep reflex was an inhibitory reflex that had been destructively interrupted by an excitatory reflex. In other words, insomnia was a kind of destructive excitement, like the "destructive irradiation" that Maria Erofeeva hypothesized in her dogs. When combined sensations intensified an animal's physiological response, it generated an excessive reaction that became pathological. To treat his insomnia, Chou self-medicated with sleeping pills.[47] This would re-calibrate his inhibition function and he recovered in two months.

Although Chou did not need to validate the use of sleeping pills, he still felt compelled to offer a systematic explanation for why they worked. He insisted that his experience justified Pavlov's description of "higher nervous activity," which remained a mere theory. Nevertheless, Chou was emboldened. In 1948, he established his own Pavlovian lab in China after working with William Horsley Gantt (1892–1980), one of the last US collaborators to study with Pavlov. In 1952, the People's Health Publishing House enlisted Chou to distribute Pavlovian materials en masse and reinforce the language of Communist dialectical materialism.[48] Around the same time, Peking University invited Soviet experts to design the curriculum for their department of biology and animal physiology.[49] In 1953, the Health Ministry established the Pavlovian Learning Society, supporting Pavlov-inspired departments at medical institutions in Tianjin, Shanghai, Chongqing, Hangzhou, Datong, Jinan, Chengdu, Qingdao, Wuhan, Yan'an, Yunnan, Henan, and Guangxi, among others. Pavlov was everywhere.

Even with its popularity, Pavlovian theory had its limits. Scientists had to deal with these limits. Soon after Pavlov's death in 1936, one scientist and translator named Guo Yicen (1894–1977) wrote in Pavlov's obituary, "In remembering Pavlov, we should of course recognize that he did not conceal his own shortcomings in his academic contributions. We do not idolize him as a leader who offers undisputed, absolute truths."[50] Guo recognized that Pavlov knew of the "shortcomings" in his own theories. Guo accepted that Soviet science and Communist materialist ideology did not offer "absolute truths."

Still, Guo himself invested heavily in Pavlovian science. He had studied physiology in Germany. He traveled to Moscow and consulted the physiologist Konstantin N. Kornilov (1879–1957) on how to use dialectical materialism to understand the mind. Kornilov pointed Guo to Pavlov's work on conditional reflexes. Guo held onto the promise of conditional reflexes even when he and other scientist-writers understood that Pavlov himself had moved on from his early ideas. They knew that there was nuance, complexity, and uncertainty in Pavlov's speculations. Guo, who had searched for physiological models of dialectical materialism in Moscow, recognized that Pavlov did not offer any external truths.

Disconnecting Meridians

Like her peers, Zhu Lian also embraced Pavlov. She almost had no other option; Pavlov was a Communist ally. His theories justified materialist discourses. The nervous system grounded the material process of sensory paths. It grounded the material nature of feelings and the emotions. Zhu also knew that Soviet science offered only a partial perspective. Like Guo Yichen, Zhu recognized that Pavlov's theories were not all-encompassing; his work did not offer universal laws. It could not. She echoed Guo's skepticism and instead emphasized that Pavlov's theories had "obvious" implications to clinical work and research on acupuncture and moxibustion.[51] "At the same time," Zhu added, "acupuncture and moxibustion can also offer important empirical evidence to enhance Pavlov's theory."[52]

Given the limits of "higher nervous activity," Zhu claimed that research on the effects of acupuncture and moxibustion stood to further "enhance" Soviet science. This was a bold claim. Zhu Lian recognized the many sensory inputs of poking, burning, and stunning dogs in Pavlov's lab. Like Maria Erofeeva's research in "destructive irradiation," the effects of acu-moxa could also be combined and intensified. Zhu Lian described that acu-moxa generated the "same clinical effects."[53]

However, acu-moxa went further. Acu-moxa marked the limits of "destructive irradiation" because it reached specific parts of the body to deliver specific therapeutic responses. In other words, the clinical effects of acu-moxa were not just like those in Pavlov's labs, but superior to them. It offered a kind of precision that Maria Erofeeva did not manage with her combination of fragrances, shocks, and jabs. In Pavlov's lab, biological explanations failed to clearly describe the interior spaces that mediated health and disease. Meanwhile, acu-moxa mapped the metaphorical mosaic; it reached a specific target for a specific effect.

By focusing her attention on the functional characteristics of acu-moxa sites, Zhu displaced the functional significance of meridians.[54] She was more concerned with explaining what Pavlov's "higher activity of nerves" could not, which was how acu-moxa reached discrete targets in the chest or in the gut. Meridians did not matter. They existed as part of diagnostic techniques and were otherwise irrelevant. Zhu's fixation on correcting Pavlov undergirded her pedagogical mission and research agenda. Specific acu-moxa sites allowed Zhu to work with certainty rather than with abstractions, as Chou had done when explaining why sleeping pills treated his insomnia. Individual sites induced individual effects.

When Zhu Lian established her own lab for acu-moxa research in 1952, she also oversaw the production of new anatomical *tu* (圖). In keeping with her interest with the clinical effects of acu-moxa, Zhu eliminated meridian lines. When a group of practitioners at the Beijing Traditional College of Chinese Medicine designed their own meridian *tu* for Zhu's approval, she dramatically revised the images. Zhu corrected what she perceived as anatomical errors and added thick outlines of nerves and muscles. She eliminated the 14 meridians, which the image was supposed to illustrate.[55]

For Zhu, meridian *tu* did not need meridians. In 1956, she published with her illustrator Wang Xuetai an abridged version of *New Approaches* called *The Handbook on Acupuncture and Moxibustion* (*Zhenjiu xue shouce*). The new *Handbook* was filled with 38 black and white illustrations without referencing *jingluo*. The body again appeared in fragmented parts with dots marking skin, muscle, and bone. Zhu reiterated that acu-moxa acted through the excitation and inhibition of nerves and neurons in the cerebral cortex. This was the accurate use of Pavlovian theory and dialectical materialism.

As a state-sanctioned publication, the *Handbook* circulated widely. The first round of printing produced 70,000 copies and resulted in a 1956 Korean translation.[56] This was followed by a Russian translation in 1959, which editors used as a template for translating the *Handbook* into even more languages.[57]

Yet, Zhu's retooling of "higher nervous activity" did not fully translate into Russian. Few authors emphasized the effects of needling at nerves with the same enthusiasm. They instead focused on Chinese *materia medica*. Books that introduced Chinese medicine to the Soviet Union reproduced Wang's lithographs but not always Zhu's text. Translations were partial. For instance, *Chinese Folk Medicine* (Китайская Народная Медицина), published in 1959, offered an overview of the recently liberated Chinese "folk" and Chinese "folk medicine" based on lectures from Jin Xin-Zhong (Qian Xinzhong, 1911–2009), the then Minister of Health. Qian had addressed a gathering of the All-Union Society for the Dissemination of Political and Scientific Knowledge in Moscow. His audience was already familiar with acupuncture-moxibustion.

Yet, without the same political context, learning acu-moxa in Moscow lacked the same sense of urgency as it had in Beijing.[58] Associating contemporary medical practices with classical texts was unproblematic. The translators of *Folk Medicine* described needling and heating therapies as a kind of "reflexotherapy" (рефлексотерапия), almost as an afterthought before turning to the more interesting "arsenal" of over 2,000 medicinal plants and animal parts.[59]

When it came to meridians, *Chinese Folk Medicine* presented new versions of Wang Xuetai's images (**Figure 4.5**). One image of the torso showed nine vertical lines inscribed over bone. It featured similar graphic contouring from Wang's work, such as the notch in the neck and the details in the abdominal muscles. The dots and individual sites remained roughly the same. Two sites on the nipple and one site at the base of the third tract were still forbidden from needling and heating. The ⊗ from Wang's lithograph was rotated to ⊕. Without a legend, these distinctions evaded readers.

The most obvious difference between this image and Wang's original was the 71 numbers positioned next to each dot. Each number marked a needling site that referenced a long list of names. The numbers zigzagged in sequence, ascending and descending on the image. They connected meridian sites across different meridian paths. From left to right, the numbers started at 1 and moved up from the base of the first time, then descended to number 20 at the base of the second line before moving back up to 40 at the top of the third line. Meanwhile, some numbers on the vertical lines were out of place, such as a floating number 7 on the shoulder and a wandering 35 above the left nipple. The modified image presented a new logic, a new order, a new sequence of sites. Lines were no longer meridian paths but lines for reference.

Chinese Folk Medicine transformed, simplified, and rearranged Wang's illustrations. Sites and paths flattened further; many disappeared. When the names of the sites were transliterated, they also sounded different. The Cyrillic transliteration in the image caption adopted two different writing conventions. Chinese texts appeared in both Pinyin and Wade-Giles. For instance, the site *tianchi* (天池) on the torso followed the Pinyin convention (Тянь-чи). In contrast, *riyue* (日月) followed the Wade-Giles convention in the form of *zhiyue* (Жи-юе). This subtle mixing of linguistic standards reflected ongoing attempts to develop standards for reforming Chinese Romanization.[60] Transliteration was not so straightforward.

Romanizing Chinese itself was a political task given the numerous Northern and Southern dialects. In 1949, the Beijing dialect became

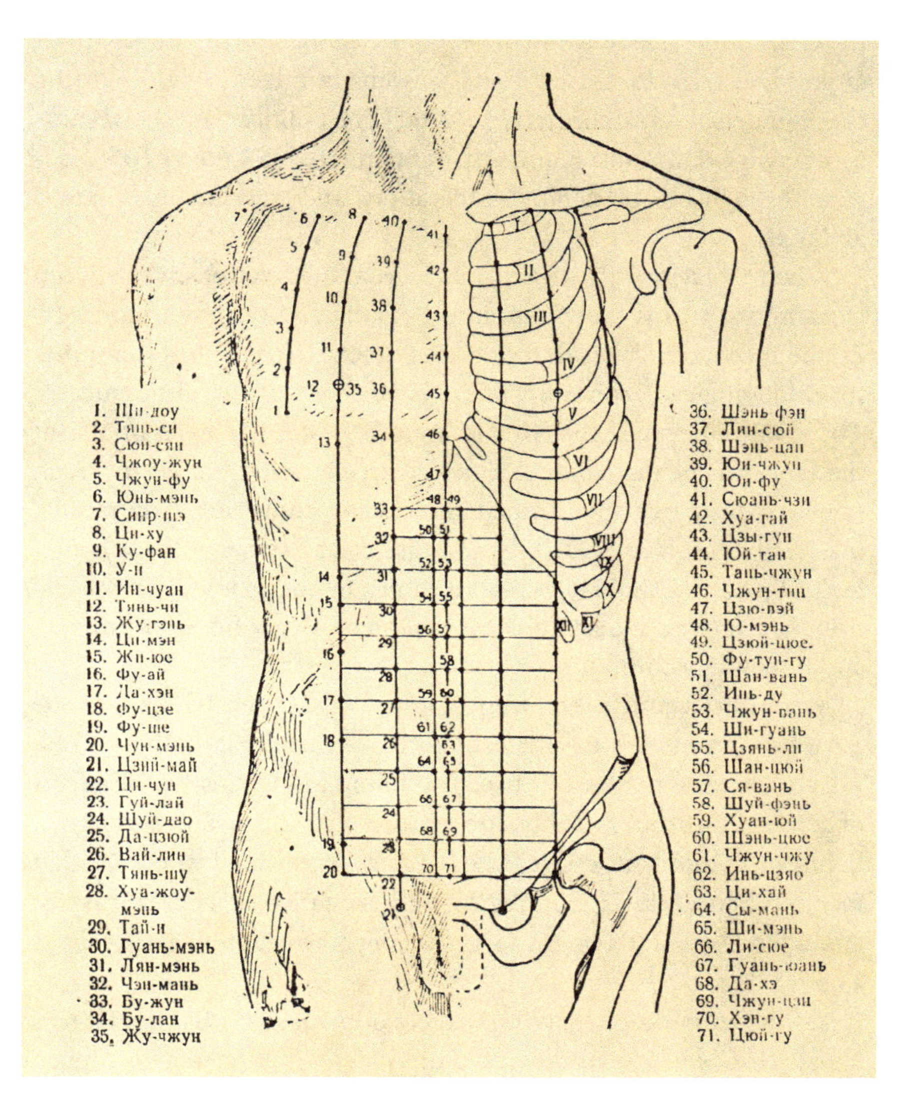

Figure 4.5. Translated fragments, twentieth century. Unknown artist, illustration. This image, from a volume that translates as *Chinese Folk Medicine*, re-presents Wang's lithograph of meridian fragments on the chest. Sites are numbered along the path with their names listed on either side of the torso: 71 arabic numerals indicate the position of meridian sites; 12 roman numerals mark the ribs. From Qian, *Kitayskaya Narodnaya Meditsina*, 1959, 34

the national standard in China, replacing all others after the rise of the Communist Party. Meanwhile, ethnic Chinese living in the Soviet Union had tried to use Latinxua Sinwenz, a mixed style of writing that blended several Northern dialects.[61] The political and social shifts across two Communist regimes then manifested in *Chinese Folk Medicine* through inconsistent transliterations and translations of image and text.

When *Chinese Folk Medicine* appeared, the relationship between Communist China and Communist Russia had been deteriorating. By the 1960s, observers described the Sino–Soviet split as stemming from ideological, economic, political, and personality differences.[62] The alliance between the two Communist states had been brief. But the scope of such an alliance significantly shaped the ways in which scientists sustained and justified knowledge production. Communist materialist discourse aided the popularity of Pavlovian theory. Dialectical materialism, which placed things in contrast and contradiction, had created a black box for claims on many forms of unseen, and potentially conflicting phenomena.

More importantly, the political alliance between China and the Soviet Union created an intellectual environment that motivated Zhu Lian to fixate on nerves. She used this foundation to justify the effects of acupuncture and moxibustion without meridians. She also hoped to improve, supplant, and replace the same Pavlovian framework that inspired her work. The inhibition and excitation of nerves was not enough to explain the many targeted effects of needling and heating. Zhu wanted to push further.

Yet, when parts of *New Approaches* appeared in Russian, editors focused on *materia medica* as a uniquely Chinese folk medicine, rather than on acu-moxa as a uniquely Chinese science. Translators had rearranged and simplified Wang's images. These transformations held ontological consequences. Zhu Lian's fixation on scientific innovation and Wang's dedication to detailed images were historically and contextually contingent. Despite the trans-national Communist discourse, using acu-moxa to improve on Pavlov only mattered in China.

Wang Xuetai was also invested in dialectical materialism. He too was dedicated to the Communist cause. After receiving his medical degree, Wang turned down a position at a large urban hospital to join Zhu Lian's faculty in the outskirts of northern China. He chose a life of revolution. Like comrades who studied Mao's lectures on contradiction, Wang's first single-author paper argued that Chinese medicine, or "national" medicine, embodied the law of contradiction. Chinese medicine grounded a dialectical worldview.

Titled, when translated, "Embodying the Law of Contradiction in a Dialectical Treatment of Chinese Medicine," Wang's paper quoted common Communist adages where life was decided through "mutual struggle" or *xianghu douzhen*. This struggle ensured progress. He wrote: "Without exception, all things subjected to medical science are governed by the law of contradiction. As Engels indicated long ago, 'Life is a contradiction which is present in things and processes themselves, and which constantly originates and resolves itself; and as soon as the contradiction ceases, life, too, comes to an end, and death steps in.' Our task as health care workers is to ensure the health of the living and requires us to engage with disease through the law of contradiction and resolve all other contradictions that threaten life."[63] Like his mentor Zhu Lian, Wang applied Communist social theory to medical practice. He took literally Engel's famous refrain that life was sustained through contradiction and that the absence of contradiction resulted in death.

But unlike Zhu Lian, Wang gravitated toward historical medical texts. Learning classical techniques and how they evolved over time was critical to the study of contradiction. In Wang's view, this made classical Chinese medicine "rational" because it embodied universal contradictions. Its longevity and efficacy secured its objectivity. It offered, in Wang's words, "a true reflection of the laws that guided physical movements in the objective world."[64] Chinese medical theory and practice was ideologically cogent, epistemically sound, and therapeutically effective.

When Wang first began his work on *New Approaches*, he realized that his role as an illustrator was to help translate early modern texts for a modern reader. Zhu assigned Wang with studying sources from the premodern and early modern period.[65] These included the third-century classic on acupuncture moxibustion (*Zhenjiu jiayi jing*) and Sun Simiao's *Essential Formulas Worth a Thousand in Gold for Emergencies* (*Qianjin yifang*) from the sixth century. Wang also read annotated versions of *Simple Questions* (*Suwen*) and *The Medical Secrets of an Official* (*Waitai miyao*) from the Tang dynasty, in addition to *Bronze Images of Acupuncture and Moxibustion Sites* (*Tongren shuxue zhenjiu tu jing*) and *Xi Fangzi's Illuminated Moxibustion Classic* (*Xifang zi mingtang jiu jing*) from the Song dynasty.

Rather than promoting a new Communist Chinese science, Wang invoked classical continuity. He copied meridian men that graphically conformed to modernist visions but did not supplant pre-modern ones. Here was an already coherent system that rendered nerves redundant. Classical texts had already described meridians as an articulate system with numerous branches and sub-branches. Meridians already embodied a dialectical world view. They were flexible objects; they were dynamic; they operated based on the opposing relationship of Yin and Yang body parts. With this commitment in mind, Wang did not attach himself to the inhibition and excitation of nerves as an explanatory mechanism for acupuncture and moxibustion. Wang did not need nerves to explain meridians.

Wang continued his work as a researcher, assistant, physician, and illustrator. While working with Zhu Lian, he undertook on his own experimental studies. In 1956, Wang and his collaborators observed the effects of moxibustion in over a dozen patients. He reported that applying moxa heat to one part of the body generated repeatable patterns of sensation in other parts of the body. For instance, when participants received moxibustion above their tailbone, they felt warming sensations traveling to their abdomen, lower limbs, and feet. These sensations lasted for about half an hour. "The effects of moxibustion are sometimes very fast," Wang reported, "It can immediately help insomniacs find sleep."[66] These curious effects occupied a temporal space, traveling at varying speeds, sometimes offering immediate relief that defied physiological standards. They worked faster than Chou's sleeping pills. "Even protein denaturation does not happen as quickly," Wang added.[67]

Wang did not ascribe these sensations to the "higher activity of nerves." The more he studied classical sources, the less he translated the meridians as nerves.

But despite his growing fixation on meridians, Wang knew better than to openly contradict his mentor. Zhu Lian had established powerful ties within the Communist Party through her early work as a military doctor.[68] Wang Xuetai was in no position to contradict her. Instead, they continued to collaborate and published an abridged edition of *New Approaches*. For years, they worked toward establishing modernist visions of the body. Both were invested in the recent success of the Chinese Communist Party. Political ideology dictated the terms of knowledge production, even if the knowledge itself was not entirely new.

Wang used his images to establish a sense of continuity. His early illustrations of meridian men for the first edition of *New Approaches* looked distinct, awkward, and off-balance—most likely copied from the classic *The Great Compendium of Acupuncture and Moxibustion* (*Zhenjiu dacheng*). For instance, there was a striking similarity between Wang's foot-Yang-*ming*-stomach meridian man and the early modern stomach meridian man (**Figure 4.6**).[69] Both men sported similar head buns, a bent right arm, a hidden left arm, and a robe that partially concealed the front of the waist. Yet upon closer examination, Wang's image was no direct copy. He had taken his time to imitate the image with his own hand and added his own flourishes. There were noticeable differences, such as the position of the inscriptions for each site, the disinterested facial expression, and the addition of a left shoe.

Four years later, Wang offered his own set of illustrations. In the second edition of *New Approaches*, he discarded his original prints and re-created new ones that resembled the men from *Treatise on the Fourteen Meridians* (**Figure 4.7**). This time, the stomach meridian man

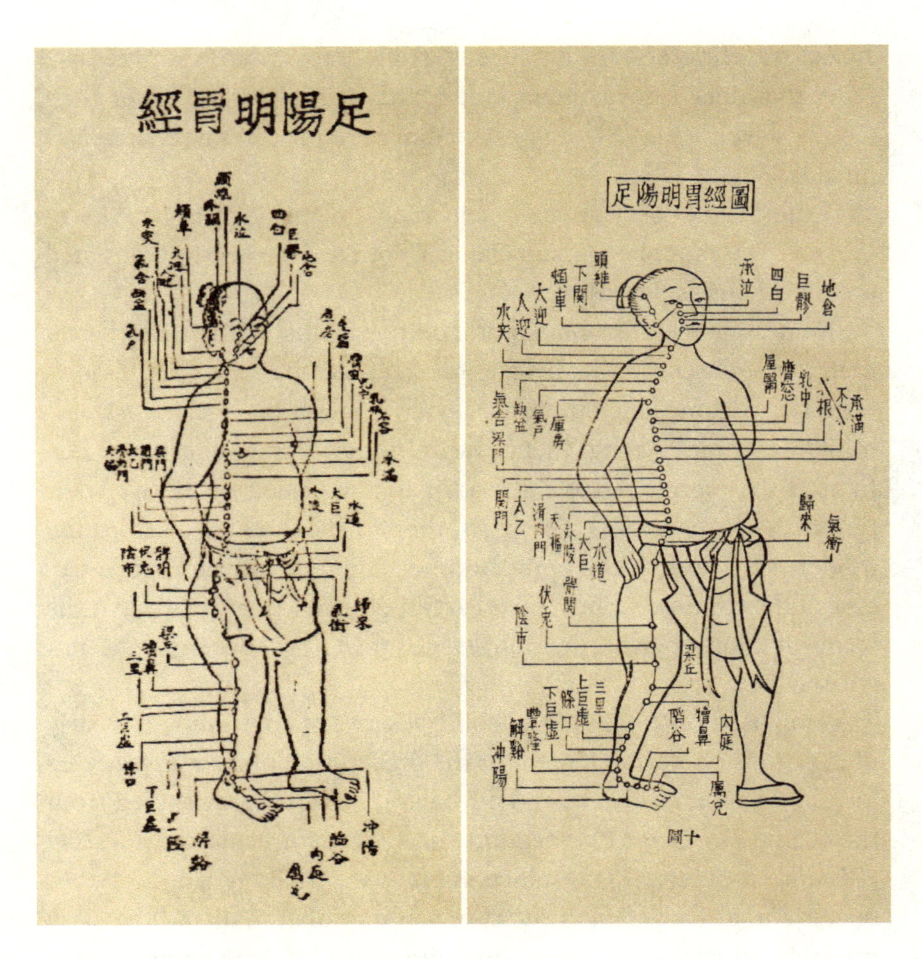

Figure 4.6. Stomach meridian men, twentieth century. Woodblock print and lithograph. Wang Xuetai's first copy of meridian men most likely came from *Zhenjiu dacheng* (*left*). His meridian men, though, were distinct, being slightly awkward and off-balance (*right*). Both meridian men sported similar head buns, a bent right arm, a concealed left arm, and a robe that partially concealed the front of the waist. Wang had imitated, not copied, the image. From Zheng and Liu, *Zhenjiu dacheng*, 1936, 15; Zhu Lian, *Xin zhenjiu xue*, illustrated by Wang, 1950, fig. 10

had facial hair. His right hand lifted his robe exposing bare feet; his left hand pinched an invisible beard. Wang used a complementary combination of pigments and adorned stomach man with blue, green, and teal on his hat and robes, accented by a grass-like pattern and

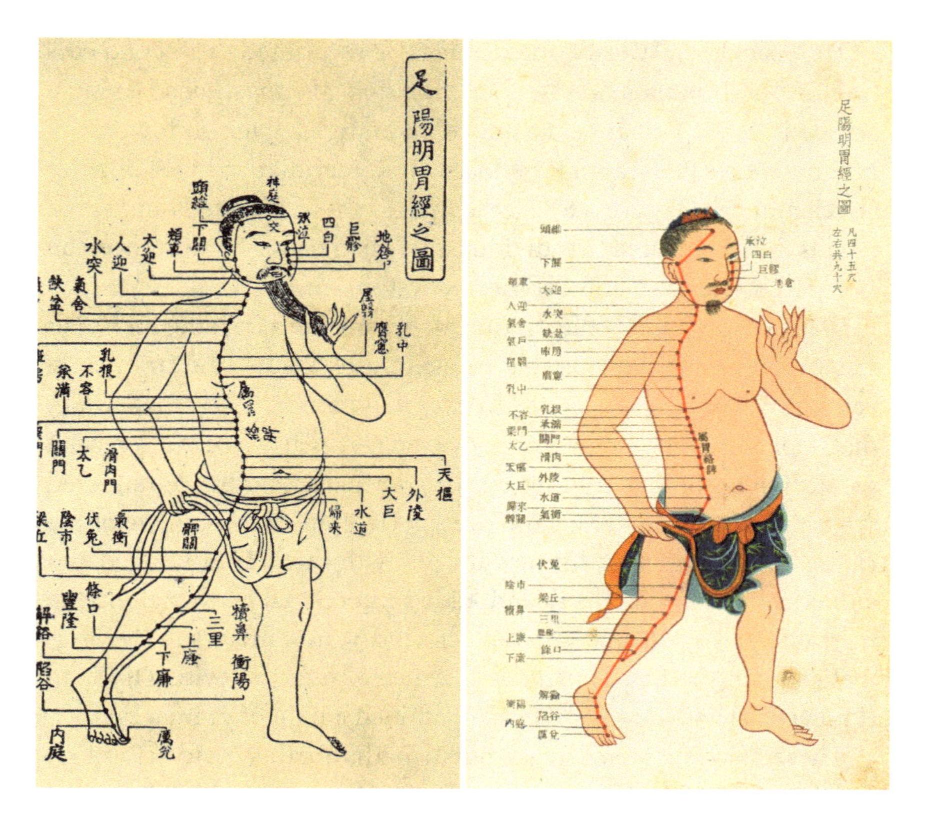

Figure 4.7. Stomach meridian men, twentieth century. Wang Xuetai (1925–2008), chromolithograph. Wang's illustrations in the second edition, of 1954, resembled the men from *Shisi jing fahui*, in the treatise on 14 meridians. The stomach meridian man now had facial hair, a left hand holding an invisible beard, a right hand lifting his robe, and bare feet. From Zhu L., *Xin zhenjiu xue*, 2nd ed., 1954, fig. 64

bright orange trim. Other meridian men in Wang's set wore even more elaborate headpieces and refined textiles in brown, gold, yellow, light blue, dark blue, red, and light orange. Their robes featured elaborate patterns with clouds, flowers, phoenixes, and pendants.

Despite the effort Wang exerted to make these images, Zhu Lian relegated them in her book. They appeared toward the end of *New Approaches*, over 300 pages away from the first set of color plates that displayed meridian fragments in the head, torso, arms, and legs. For Zhu Lian, Wang's efforts to resurrect all 14 meridians were unnecessary.

Ten years later, Wang worked alone. He re-printed the second edition of *The Handbook on Acupuncture and Moxibustion* as a single author in 1966. This time, he added an entire chapter on the 14 meridians. He created muscular, short-haired meridian men resembling Greek athletes (**Figure 4.8**). These images showed dotted outlines of organs that connected to individual meridian paths. For instance, the stomach meridian man now showed a stomach.

Wang's technique of illustrating an organ along with the meridian was not new. It is possible that he again took inspiration from early versions of *The Great Compendium of Acupuncture and Moxibustion* that depicted an organ floating above the meridian man. Wang placed an outline of the stomach inside the transparent body. He rendered the organ visible, material, and present. Despite the small size of the *Handbook*, Wang made his new images with care. He numbered and named each individual site and added directional arrows along the meridian paths. Ultimately, Wang created 33 new images to the original 38 from the first *Handbook* that he co-authored with Zhu Lian (**Figure 4.9**). His new *Handbook* contained a total of 71 images.

These new chapters on meridian paths further paid tribute to classical texts. Wang assigned each meridian path with an excerpt from an early modern source.[70] For instance, he opened his chapter on meridians with instructions from the sixteenth century: "*Jing* is the trunk and *luo* are the branches, but they emerge from the same source. If you are a doctor and do not [know] the *jingluo*, then you will be moving in darkness."[71] For Wang, meridians were not superfluous or arbitrary. Classical texts said so. Meridians were at the core of medical knowledge that needed new kinds of epistemic and ontological packaging.

Figure 4.8. (opposite) Stomach meridian men, twentieth century. Lithographs. Wang Xuetai reprinted the second edition of the *Acupuncture-Moxibustion Handbook* with him as a single author in 1966. This time, his new stomach meridian man featured a stomach (*right*). Again mimicking early versions of *Zhenjiu dacheng* (*left*) that depicted the meridian organ above the meridian, Wang outlined the stomach in the transparent body, floating in the gut, visible, material, present. From Wang, *Zhenjiu xue shouce*, 1966, fig. 23; Zheng and Liu, *Zhenjiu dacheng*, 1936, 15

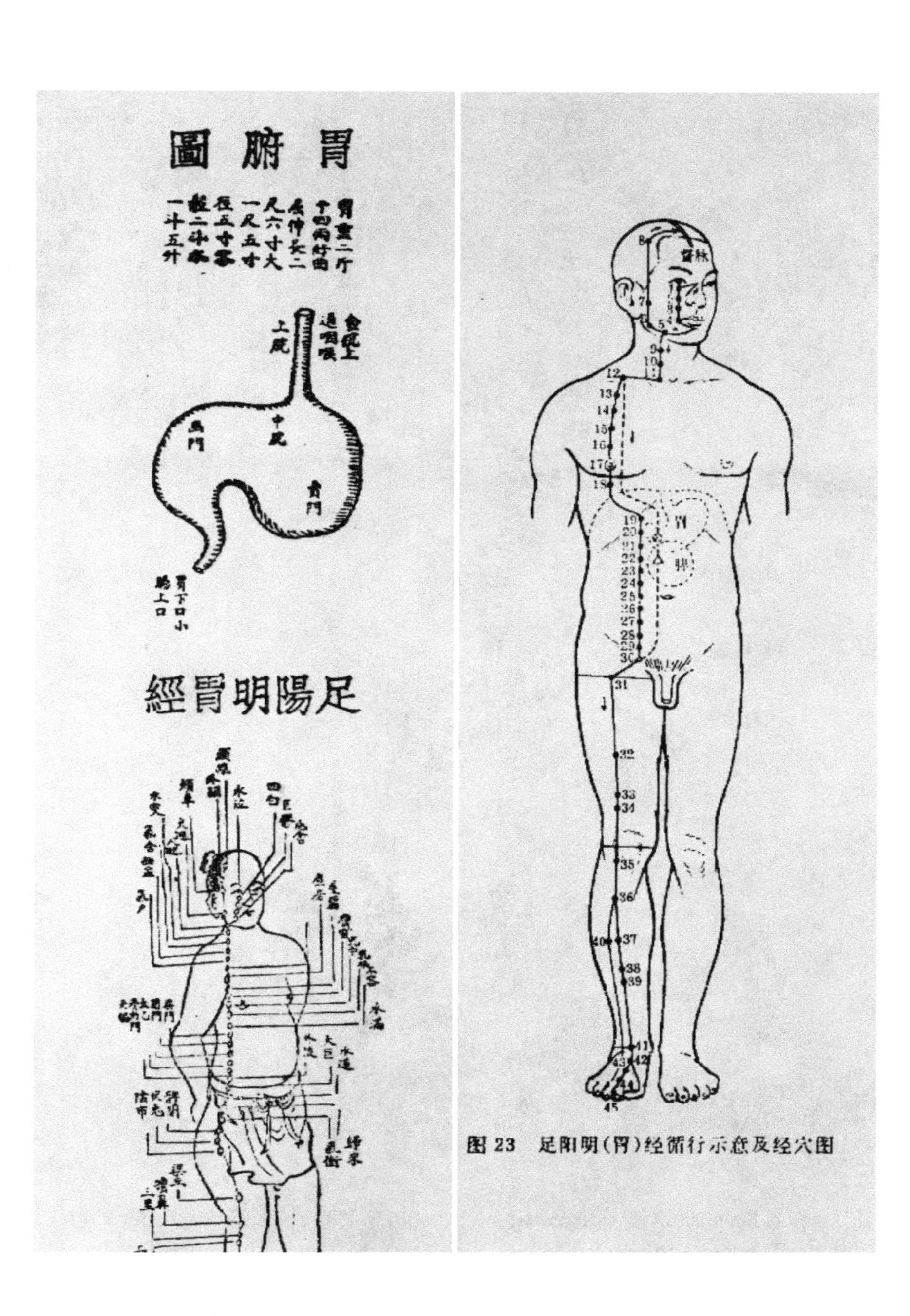

图 23 足阳明(胃)经循行示意及经穴图

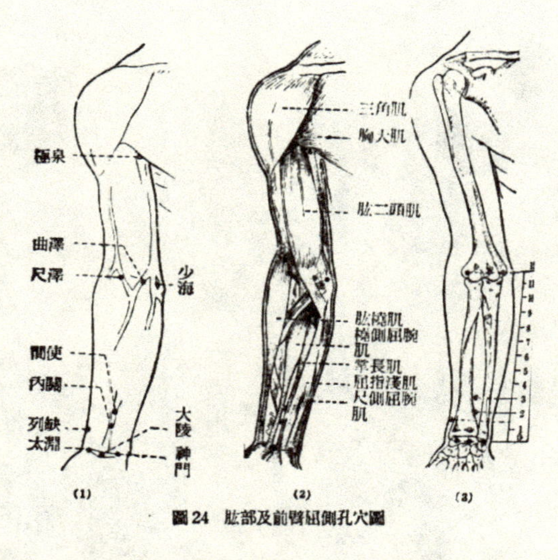

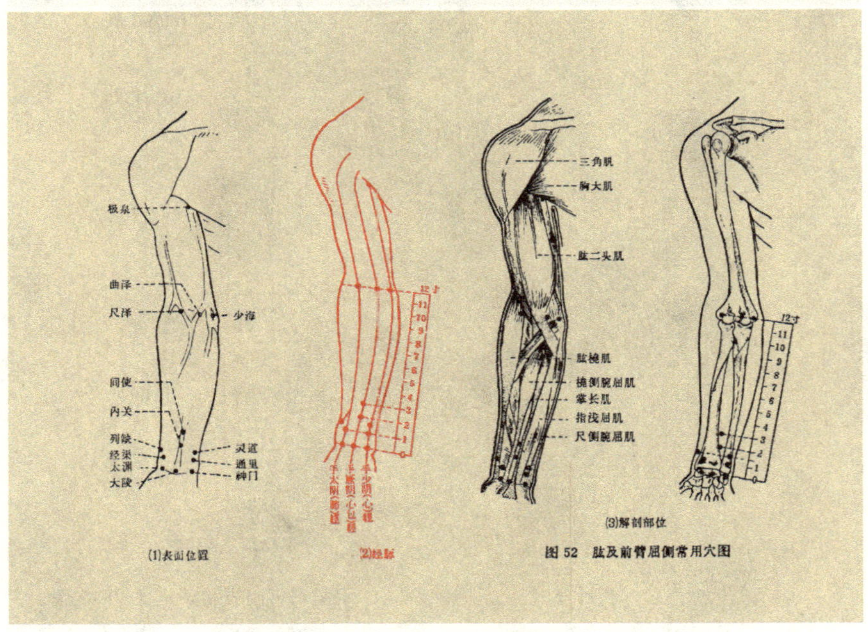

Figure 4.9. Re-placing meridians, twentieth century. Wang Xuetai (1925–2008), lithographs. The 1956 edition (*top*) showed acupuncture sites without meridians, while the 1966 edition (*bottom*) reintroduced meridian lines, highlighted in red. From Wang, *Zhenjiu xue shouce*, 1956, fig. 24; Wang, *Zhenjiu xue shouce*, 1966, fig. 52

Instead of focusing on any single system, Wang urged the reader to grasp how physiological systems coordinated with one another. Nerves were a small part of a complex body. "For example," he wrote, "the segments of tissue between the spinal cord allow for the connection between somatic nerves and internal organs."[72] Wang highlighted the importance of intercostal tissues over nerves that connected with larger structures in the body. He added, "as early as two years ago, we also discovered directional relationships among body parts. For instance, the *hoku* site on the hands affects the mouth and nose, the *neiguan* on the forearm affects the ears, and the *zusanli* on the leg affects the gastrointestinal system, the *yinxue* on the toes affects the uterus. . . . [T]he effects of needling follow these relationships."[73] In this excerpt, Wang introduced *jingluo* through a process of discovery. Classical texts already established connections between the *hoku* site and the face, *neiguan* and the ears, *zusanli* and the stomach, and *yinxue* and the uterus. But Wang's patients felt these connections for the first time, and their reports epistemically validated these sites. Their reports of embodied knowledge were useful insofar as they could be assigned biomedical vernacular had established grounds for truth claims.

Wang further paired embodied knowledge with historical inquiry. He again cited medieval texts to describe the morphology of *jingluo* in the biomedical body. He distinguished the varieties of structures that constituted meridians, including large *luomai*, slightly smaller *mai*, and even smaller *sunmai*. Larger paths like *jing* were known and named, whereas smaller types of *sunmai* were known and unnamed.[74] The premodern medical texts that Wang explored compared *jingluo* with irrigation canals with trunks and branches, varying in size and shape. Some structures were identified, others anonymous. Wang mapped these structures trunk-by-trunk, labeling points and assigning direction to each major meridian. On paper, major meridians branched into smaller, unnamed, and elusive forms.

Rather than representing the many trunks, branches, and twigs in the body, a different kind of puzzle preoccupied Wang. The paths were like rivers and their courses were not fixed.[75] Moreover, depending on the state of the body, some paths were more apparent, and sometimes less apparent. This was the same issue that exasperated early-twentieth-century medical reformers who tried to standardize on paper the location of meridian paths. Channels moved, sometimes draining and sometimes flooding across layers of tissue. Wang faced the same challenge as his predecessors: How should he capture objects in constant motion?

Wang turned to the genre of a *tu*. Like the thirteenth-century scholar Wang Haogu (1200–1264?) and the sixteenth-century physician Manase Dōsan (c. 1507–1594), Wang Xuetai created a new image that only showed the 12 primary meridians.[76] He first tested out a small version of his meridian *tu* in a 1960 article on meridians. The image was barely legible, and roughly took the shape of a circle (**Figure 4.10**). The text offered no instructions on how to make sense of this image. The reader could barely discern three tracts with light and dark patterns. Each track was punctuated by arrows that separated four quadrants. This apparently showed the 12 regular meridians.[77] What appealed to Wang Xuetai was not visualizing the exact location of meridian paths and their associated parts, but their broader structural orientation.

Wang continued to build on this drawing. In 1966, he published a significantly modified version of the looping meridian tracts in the second edition of the *Handbook*. Meridians now appeared as one continuous loop (**Figure 4.11**). The circle looked like a compass where the horizontal and vertical axes divided the tracks into two hemispheres and four quadrants. At the top were the hands (*shou*), at the bottom were the feet (*zu*), to the right was the head (*tou*), and to the left was the torso composed of the chest (*xiong*) and abdomen (*fu*). The hands and feet formed the vertical axis, and the head and chest formed the horizontal axis. This oriented the body by rolling the head down to the side.[78]

For Wang, a shifting orientation offered clarity. These quadrants had meaning. On the left hemisphere, Wang drew Yin meridians with a solid line. On the right hemisphere, he drew Yang meridians with double lines. Along the horizontal axis, the upper half of the circle

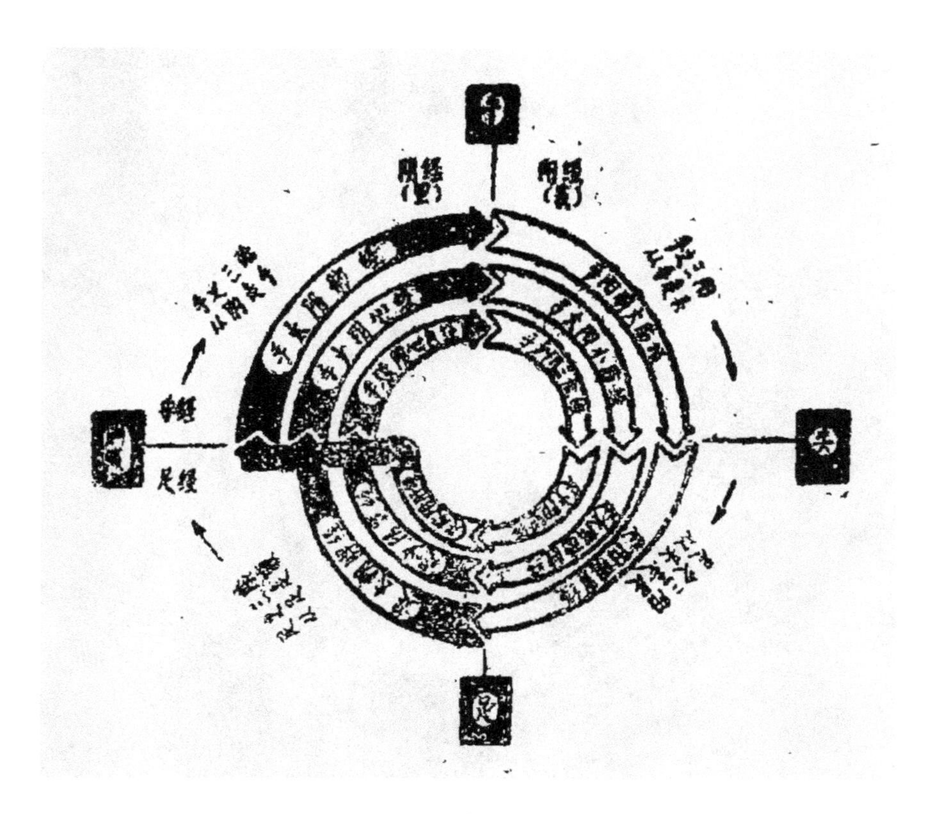

Figure 4.10. Meridians in rotation, twentieth century. Wang Xuetai (1925–2008), lithograph. The more Wang studied *jingluo*, the more he came to imagine them as connecting in a single system. This circular map orients the body to meridians, moving from the hands (*top*) to the head (*right*) to the feet (*bottom*) and through the torso (*left*). The dark arrows on the left show Yin meridians, and the transparent arrows on the right show Yang meridians. From Wang, "*Zhenjiu jiangzuo (xu)*," 1960, 56

indicated hand meridians that moved from left to right, connecting organs in the torso to the head. The lower half of the circle featured the feet meridians that moved from right to left, connecting the head to the internal viscera.

This image guided readers. It showed how meridians derived their names—that they were based on the hands, feet, Yin, and Yang orientations. The top of the image showed hand meridians (*shou jing*), and the bottom half of the image showed feet meridians (*zu jing*). For instance, the top-right quadrant showed Yang-hand meridians; the

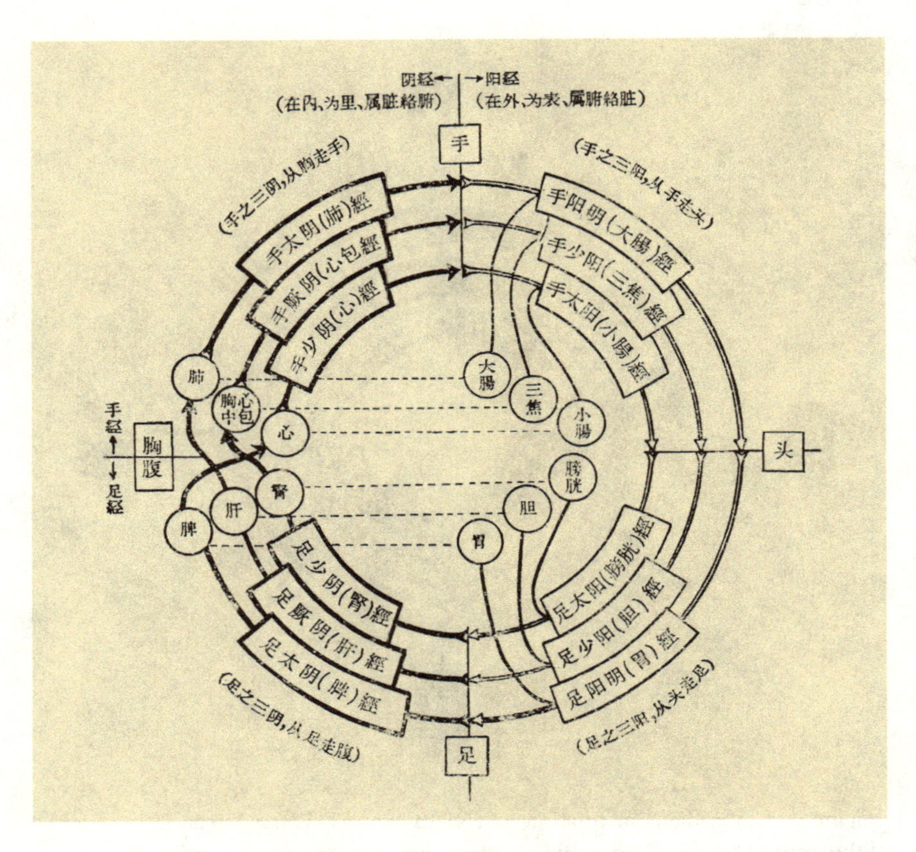

Figure 4.11. Meridian wheel, twentieth century. Wang Xuetai (1925–2008), lithograph. Wang modified his circular diagram and added the 12 inner organs with Yin organs connecting to Yin meridians (*left*) and Yang organs linking to Yang paths (*right*). From Wang, *Zhenjiu xue shouce*, 1966, fig. 20

lower-right quadrant showed Yang-feet meridians. In contrast, the top-left quadrant showed Yin-hands meridians; the lower-left quadrant showed Yin-feet meridians.

The image also showed how individual meridian path names arose from individual organs. Across the Yin–Yang divide stretched lines connecting 12 circles. These were the 12 internal viscera—lung, pericardium, heart, kidney, liver, and spleen on the left (Yin), and large intestine, triple burner, small intestine, gallbladder, bladder, and stomach on the right (Yang). For instance, Wang Xuetai's stomach meridian man showed a path called *zuyangming* that corresponded to the stomach. The same meridian in **Figure 4.11** was found in the outer tract in the lower left quadrant of the diagram.

The more Wang created medical images, the more his illustrations became *tu*-like. His final image of the meridian paths did not appear on a human form. Instead, he arranged the body according to the orientation of the meridians, placing meridians at the center of a meridian map. This system of the 12 major meridians that arose from 12 major organs presented a complete and coherent system. Tracing any part of the image would follow the arrows in a continuous loop.

This image dramatically deviated from the yellow fragments that Wang produced for Zhu Lian in 1948. As Wang Xuetai closely copied images from classical medical books, meridian lines no longer connected a series of dots on the page. They represented a full system. For Wang, learning about the meridians had increasingly become essential to understanding the Chinese, and Chinese Communist, medical body. He was not the first physician to turn to classical texts, as numerous scholars over centuries revisited and reinterpreted classical texts and images. But when Wang urged the reader to consider the wisdom of Ming dynasty writers, he further insisted that this was because meridian paths aligned with Communist ideals. They operated through opposing pairs to make legible patterns in the body. This mode of pattern recognition embodied a contradiction that sustained life. Meridian paths represented the ideal dialectical system.

Conclusion

This chapter has explored the conceptual relationship early Communist physiologists established between meridians and nerves. In this period, discourses of dialectical materialism allowed physicians and researchers to assemble biological objects and embodied experiences into a kind of Communist science. As a social theory, conflict and contradiction applied to meridians and nerves as systems that were independent, mutually exclusive, and mutually inclusive.

These conditions resulted in a variety of interpretive and improvisatory practices. Zhu Lian and Wang Xuetai both took seriously the political and ideological applications of dialectical materialism and emerged with their own diverging conclusions. For instance, Zhu Lian adopted Pavlovian theory so that that acupuncture and moxibustion would test its limits. Keenly aware of these limits, Pavlov and his collaborators had encountered numerous obstacles, counting thousands of drops of saliva in dogs only to find that joining physical and conditional stimuli produced confusing results. A single observable response could have extended from multiple unseen factors.

Pavlov aimed to apply discrete, measurable, controllable inputs like buzzers, shocks, and perfume sprays, yet remained unable to explain the effects of these stimuli without relying on metaphors from physics, hydraulics, and economics. He described contradictory sensations as though they were chains, threads, sequences, relationships, gravitational forces, and sound waves. Perhaps reflexes spread like waves; perhaps they tugged like chains; perhaps they were mosaic fragments. For Zhu Lian, these conceptual limits were exactly the kinds of effects that acupuncture and moxibustion not only explained, but targeted in the body with distinct accuracy.

Meanwhile, Wang Xuetai already remained unconvinced that nerves could explain everything. He would eventually dedicate his career to studying meridian *tu* (圖), producing and re-producing copies of early modern meridian men. Years after being Zhu's assistant, Wang eventually founded his own research lab.[79] The more Wang studied meridian *tu*, the more he recognized the utility of meridians. He would place meridians at the center of his illustrations, creating versions meridian *tu* that did not resemble a body. The more Wang created images, the more *tu*-like they became. Body parts conformed to the meridians, not the other way around.

Wang copied early modern meridian men to fit modernist visions without supplanting pre-modern ones. He relied on a new epistemic packaging to join quotations from classical texts with new experimental reports and graphic images. But beyond the image, Wang struggled to legitimate historical texts on their own terms. Toward the end of his decades-long career, Wang surveyed over 9,000 publications on acupuncture and moxibustion, lamenting that barely a handful drew on classical sources.[80] He remained frustrated that contemporaries dismissed early modern meridian men.

Communist-materialist discourses facilitated Zhu's and Wang's diverging perspectives while also failing to establish what counted as science. This left Zhu and Wang to defend their ideas on their own. Despite their reputation as trained physicians, their comrades doubted the efficacy of acu-moxa by rejecting embodied experience as a way of knowing. They refused to recognize the importance of first-hand experience, even when Communist discourses celebrated rural epistemologies that were idiosyncratic, improvisatory, and innovative.

When translators introduced Zhu's text and Wang's images to a Soviet audience in Russian, their sense of urgency had been diminished; their images were simplified, and concerns diffused. Despite the unifying aspirations of transnational Communist discourses, the language of dialectical materialism to legitimate acupuncture and moxibustion only mattered in China. Dialectical materialism had created a politically specific plurality that embraced conflicting empirical standards. Zhu and Wang could expand beyond Pavlov. But beyond China, the ideological stakes of their debates no longer mattered.

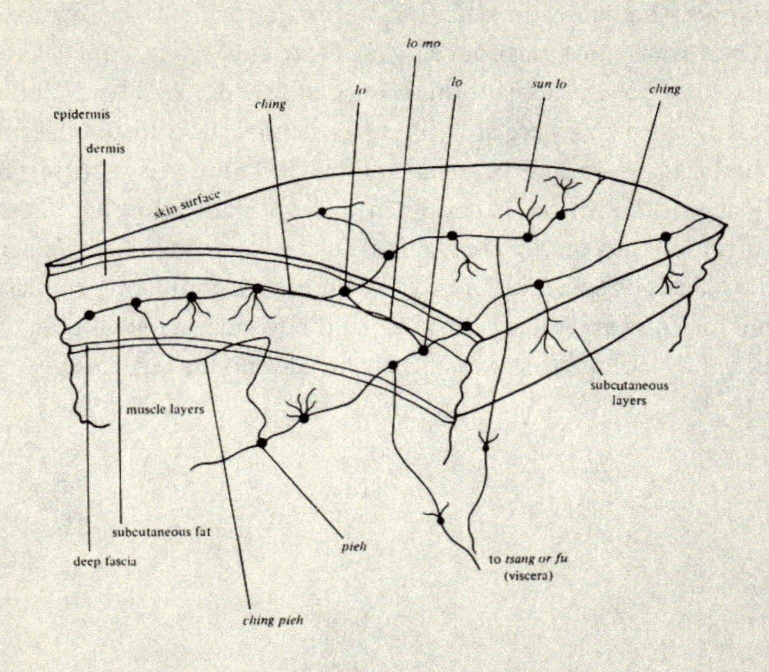
lo mo
lo
lo
sun lo
ching
ching
epidermis
dermis
skin surface
subcutaneous
layers
muscle layers
subcutaneous fat
deep fascia
pieh
to tsang or fu
(viscera)
ching pieh

CHAPTER FIVE

Modern Mediations in Difference and Diplomacy

The Electric Grid

The dots hung together in a loose configuration (**Figure 5.1**). Lines gathered at each node, tethered in a single unit. Its wiry arms reached out like tentacles probing their surroundings. Other lines plunged downward, disappearing into an invisible foundation. It was difficult to know what the lines were. These delicate filaments remained ambiguous because the accompanying text did not specify the thickness of the lines or the size of each node. The image made no reference to scale, although it seemed to show a very thin thing.

Supporting this network of lines was a cross-section of skin. The image indicated five layers: epidermis, dermis, subcutaneous fat, deep fascia, and muscle. More lines neatly marked the boundaries of each

Figure 5.1. (opposite) Rationalized depths, twentieth century. Lu Gwei-djen (1904–1991) and Joseph Needham (1900–1995), illustration. The skin appears as a solid block with transparent layers that support paths that look like nerves but are not. The original caption reads, "Diagram to elucidate *ching* (acu-tracts), *lo* (acu-junctions), tract-connecting branches (*lo mo*), dendritic 'capillary' micro-branches (*sun lo*), acu-junctions for long tract-connecting cross-channel branches (*pieh*), and these long connections themselves (*ching pieh*)." From Lu and Needham, *Celestial Lancets*, 17, fig. 3

layer, positioning the image in a three-point perspective. Unlike most images that illustrated skin morphology in the 1970s, this image did not show texture. It did not show a dense cluster of hair follicles, sweat glands, capillaries, or nerve endings.[1] It did not resemble a geological wedge stacked with rigid skin cells. The image merely mimicked illustrations of skin morphology and left blank what would have been tissue, fat, and muscle. In doing so, it evoked the whimsical, cartoonish qualities of a *tu* (圖). Flesh was left to the viewer's imagination. All that mattered were the lines and dots.

The set of wiry branches was credited in part to Lu Gwei-djen (1904–1991), a biochemist who would become one of the most influential historians of medicine in the twentieth century. Known for her magnum opus, *Celestial Lancets* (1980), Lu offered one of the first comprehensive histories of acupuncture and moxibustion that was written and published in English. Lu's aim was simple: introduce meridians as anatomical objects and their therapeutic applications as epistemically informative. She relied on the Wade-Giles Romanization system to present meridian fragments as *ching*, *ching pieh*, *lo mo*, and *sun lo.* Yet Lu avoided explicitly translating these terms as "meridians." Instead, she described them as branches, tracts, and connections. In her caption, Lu explained that the image showed "acu-tracts," "long connections," "tract-connecting branches," and "micro-branches."[2] She introduced the dots more elaborately as "acu-junctions for long tract-connecting cross-channel branches."

Lu's use of branches, tracts, and connections served as metaphors for meridians. Meridians became complex structures that could split, cross, and merge. Each possible connection registered a new name. Each new line registered a new relationship. Every new line signified a distinct relationship, necessitating its own label. Lu elaborated in her image caption that a dot (•) could be a *lo* or a *pieh*.[3] Paths between dots (•—•) could be *ching*, *ching pieh*, *lo mo*, and *sun lo*. One dot registered two names. One line had four identities. Even though Lu's image, like many before it, strove to elaborate the numerous identities of therapeutic pathways, it seemed to strain against its own contradictions.

There existed an asymmetry between the signified and its sign. Although the abundance of names suggested ontological diversity, the untextured lines and dots flattened this multiplicity. Unable to resolve this tension, Lu's work failed to initiate an enthusiastic adoption of anatomical branches, tracts, and connections. It did not encourage a steady acceptance of *lo*, *pieh*, *ching*, *ching pieh*, *lo mo*, and *sun lo* into a medical vocabulary. It did not establish meridians as visually distinct anatomical structures.

Introduction

This chapter situates Lu Gwei-djen's intellectual contributions amidst global conversations assessing the relationship between meridians and nerves in the mid-twentieth century. During this period, acupuncture had been promoted explicitly as a Chinese invention, although varieties of acu-moxa had long histories across Korea, Japan, Vietnam, Mongolia, and elsewhere. Yet acupuncture was branded as a product of China—a creation of the state that functioned as a diplomatic object.

Specifically, acu-moxa manifested as *zhenjiu mazui*, loosely translated as "acupuncture anesthesia" or "acupuncture analgesia."[4] In the 1970s, acupuncture analgesia circulated through reports describing anesthetists using needles to induce localized numbness during surgery. This was a highly performative procedure. Hospitals in China welcomed foreign visitors, dignitaries, and journalists to observe patients under the effects of acupuncture analgesia. The patients remained awake, alert, and at ease while surgeons undertook a cardiac bypass, drained a lung abscess, or performed a caesarian section.

Many such reports found an engaged readership, inspiring physicians to experiment on themselves. They, too, plugged needles into electric boxes to feel its effects on their hands and feet. Researchers in Berlin, London, Singapore, Michigan, and Colombo speculated on the mechanisms behind acupuncture analgesia, intrigued and perplexed by its effects on chronic and acute pain. They proposed new theories to explain these effects, with Lu Gwei-djen finding herself at the center of these scholarly exchanges. She corresponded with the Canadian neurophysiologist Ronald Melzack (1929–2019), who cited physical responses to needling as justification for his famous "gate control" theory of pain, and the English physician Felix Mann (1931–2014), who invoked neurophysiology to supplant classical forms of medical cosmology. Among these many correspondences, meridians and neurophysiology seemed to share an intimate ontological proximity.

However, this ostensible closeness belied an uneasy relationship. The ontological encounter between meridians and nerves had been a diplomatic one, with reports heralding acu-moxa as a distinctly "Chinese" practice while simultaneously arousing skepticism from readers wary of endorsing a "Chinese" medicine. At the same time, researchers with a deep interest in acupuncture analgesia remained unsettled by their own findings—neurophysiology could not account for all the effects of acupuncture and moxibustion, just as meridians did not consistently engage with neurophysiology. Meridians were not nerves. Lu Gwei-djen often said as much. Even in illustrating meridians as branch-like structures reminiscent of neural pathways, she maintained that they were distinct entities.

To understand the nuances of diplomatically navigating ontological difference, this chapter presents a close analysis of Lu's writings alongside published accounts of acupuncture analgesia. In doing so, it highlights the degrees of divergence that practitioners upheld between meridians and nerves.

Profound tensions underlay Lu Gwei-djen's translation efforts. For her, what we call meridians were not "meridians." Lu refused to apply this term to the branches, tracts, and connections she illustrated, fearing it would undermine her attempt to establish them as anatomical entities. In her view, anatomical structures were defined as visible and material substances. She positioned invisibility and materiality as being mutually exclusive. Structures that evaded the naked eye were not anatomical. Lu and her collaborator Joseph Needham (1900–1995) explained that because "meridian" denoted an invisible thing, it presented a non-material object.[5] In a footnote, they wrote: "It has become customary for Western writers to translate *ching* [*jing*] as 'meridians,' but the analogy with astronomical hour-circles or terrestrial longitude is so far-fetched that we do not adopt the term. 'Tract' is as nearly non-committal as possible for the supposed channel of the *chhi* [*qi*], and perhaps the established use of the word for the bundles of neuron fibres in the spinal cord may be a recommendation for our use."[6] Lu and Needham emphasized the inherent physicality of paths and lines. Meridians didn't exist; longitudes were a stretch of the imagination. But "bundles of neuron fibres" were real; they were physical, stable, and structural.[7] "Tracts" could also be a kind of anatomical structure. For them, "tracts" presented materiality in a way that "meridians" did not. The material implications of these two metaphors had tremendous consequences.

Lu Gwei-djen insisted tracts were neither meridians nor nerves. Tracts were tracts. *Ching* were *ching*. This tautology was not born out of ignorance, but out of a close study of hundreds of classical and contemporary texts. Lu often worked in isolation, annotating Cheng Dan'an's early books, collecting English translations of Zhu Lian's articles, and analyzing comparisons of acu-moxa images to Henry Head's sensation men.[8] Even as the novelty of acupuncture anesthesia intensified the imperative to rationalize the effects of needling, Lu only linked needling to nerves when it suited her aims. When visiting hospitals in China, she used acupuncture to ask surgeons whether tracts could be material objects in their own right. She wanted them to concede that *ching*, *ching pieh*, *lo mo*, and *sun lo* were physical things.

In other words, Lu recognized the political stakes of her work. She wanted to define the boundaries of a distinctly "Chinese" body. Yet the more she oriented toward difference, the harder it became to approximate similarity. In a way, Lu had found herself trapped between degrees of difference. Defining "Chinese" medicine as distinctly "Chinese" rendered it fundamentally different from a neurophysiological body. Orientalist tropes of difference inhibited commensurability. Lu had hoped to both justify difference and diminish it. She would engage with neuroscience at a distance, refusing to allow acu-moxa to be explained away by nerves.

Bodies in Migration

Born in Nanjing in 1904, Lu Gwei-djen came of age amid vibrant intellectual communities that drew on new models of modernity. Chinese feminism translated radical ideologies that theorized and disrupted Confucian hierarchies embedded in legal, ritual, and social institutions.[9] In this fragmented political landscape, campaigns to standardize medical practice promoted integrating native and foreign knowledge systems.[10] But rather than establishing a unified course of medical understanding, these projects intensified the growing plurality among empiricists and practitioners. As we saw with Cheng Dan'an's photographs, associating biomedicine with science and science with modernity resulted in a dramatic break from earlier attempts to converge two approaches to the body.[11] This divergence was at once conceptual, social, and financial. Compared with Chinese-run institutions, foreign-run schools were often better funded and educated more elites.[12] Physicians abroad felt embarrassed by this discrepancy and blamed local practices for hindering Chinese economic and political advancement.[13]

Like her peers, Lu Gwei-djen developed a complex relationship with native and non-native knowledge. At first, she fought against foreign intellectual, social, and political norms. She protested as a teenager during the May Fourth Movement in 1919 and apparently dismissed English learning as a task for "traitors and fools."[14] But rather than resist English, she soon mastered it. In 1922, Lu enrolled in Ginling College, a Christian school taught by American women affiliated with Smith College.[15] There, she majored in English and chemistry. Her classmates were politically active and missed weeks of instruction to participate in public demonstrations.[16] After graduating from Ginling, Lu continued training in Beijing before becoming a visiting scholar in Shanghai, where she studied pathology and pharmacology. A year later, she taught biology and biochemistry at St. John's College before working as a biochemical researcher at the newly established Henry Lester Institute.[17]

For Lu, education enacted politics. With the support of her British mentors in Shanghai, Lu continued her studies at Cambridge University in 1937, which required dodging Japanese bombs and enduring a difficult three-month journey by sea. At Cambridge she formally studied metabolism with Dorothy Needham (1896–1987), an expert in muscle biochemistry.[18] Lu then met Dorothy's husband, Joseph Needham, who was known for his research in embryology.[19] Even before arriving in Cambridge, Lu had admired the Needham's progressive politics, such as their work as delegates of the Association of Scientific Workers and their financial contributions to the families of soldiers who died fighting Francisco Franco's Nationalist forces in Spain.[20]

Lu Gwei-djen knew that she was an anomaly. She was a Chinese woman at an institution that welcomed few women in science and fewer scientists from East Asia. She also grasped her influence on her Cambridge mentors. She shared their sense of political urgency and became lifelong friends with Dorothy and later collaborator and lover of Joseph Needham. Dorothy and Joseph supported Lu's research. They encouraged her to undertake a history of nutrition in China and analyze the patient data that she gathered over four years in Shanghai.

While at Cambridge, Lu became a historian. In her graduate research, she specialized in metabolic disorders and vitamin B1 deficiency, which she linked to China's long history of treating beri-beri, or foot Qi.[21] By way of introducing her thesis, she described the expansive knowledge of Chinese *materia medica* that remained unknown to her readers: "Such knowledge has certainly existed at least since the 8th century in China . . . and the candidate proposes, therefore, to devote the major part of this portion to a chapter on the history of biochemistry, nutrition and medicine which has hitherto escaped attention. This has been largely due to lack of a proper index to the Chinese Classics, and also to the fact that the extreme divergence of the Chinese language from other languages."[22] For Lu, the perceived ignorance of nutrition in China was based on two factors: first, no one had tried to catalog all of the historical texts in China; second, Chinese was just too different from other languages for people to meaningfully engage with it.

Neither was entirely accurate, yet as Lu prepared her 1939 thesis, she compiled her own catalogue of medical treatises and authors. She introduced beri-beri as "Chio Ch'i" (*jiaoqi*) and then described the contributions of Shen Nung (*shen nong*), Chou Li (*zhou li*), Chou Kung (*zhou gong*), Chen Wang (*cheng wang*), Chang Chi (*zhang ji*), Han Yü (*han yu*), Hu Se-Huei (*hui sihui*), and Ching Tsong (*jing zong*). In her drafts, Lu tested out different kinds of transliterations. She was new to Romanizing Chinese names with Wade-Giles, crossing out some spellings and re-writing new ones. In the margins, she calculated publication dates and the authors' lifespans.[23]

Early on, Lu aimed to translate one material approach into another. Biochemistry relied on the interaction of organic molecules. Organic molecules articulated the effects of plant matter. Knowledge of plant matter constituted classical Chinese *materia medica*. The logic of this relationship allowed Lu to position *materia medica* as the intellectual impetus of her scientific inquiry.

Lu completed her doctoral research in just two years. She planned on returning to China until World War II left her stranded in California.[24] During this period, Joseph Needham received a private telegram from China's United Kingdom ambassador inviting him to help rebuild its scientific programs.[25] Lu had taught Needham enough Nanjing-style Mandarin for him to take the job. In the mid-1940s, Needham established the Sino-British Science Co-operation Office at the Chongqing embassy, befriending the future minister of culture, Guo Moruo (1892–1978), before his rise to fame.[26] Lu became bound to this alliance. When Lu's mother and younger brother died, Guo Moruo and Communist leader Zhou Enlai (1898–1976) convinced her to remain in Cambridge and assist the Needhams.[27] Homesick yet dutiful, Lu agreed to stay.[28]

After finally returning to China, Lu soon left home again without plans to return. She worked at UNESCO in Paris before retiring in 1956 to Cambridge University. There, she trained herself as a Sinologist and historian of science. Through self-study, she mastered classical Chinese, collected sources on the history of medicine in China, and advised Joseph Needham on managing his finances.[29] Joseph Needham further reported Lu's contributions to the Wellcome Trust, detailing her close study of technical terminology in classical Chinese and her study of the history of epilepsy.[30] This transition defined the rest of her career.

Lu kept busy. Upon returning to Cambridge, she indexed over 17,000 Chinese scientists, engineers, mathematicians, and physicians. Needham described Lu as "uniquely well equipped" to support his work as she joined him on extended trips to China.[31] Now visiting as a research ambassador, Lu facilitated meetings with officials and practitioners. She established institutional contacts and interviewed researchers on their ideas about neurophysiology and gathered material for her historical research.[32] Alongside her ongoing autodidactic training as a classicist, Lu had become formidable.

Physiological Orientations: Yin and Yang

Lu Gwei-djen seldom made images. Instead, she focused on compiling dictionaries. She surveyed hundreds of classical and contemporary Chinese texts to establish standards for translation. At times, she maintained Wade-Giles's Romanizations; other times, she offered short English descriptions or, when persuaded by Joseph Needham, settled on Latin terms. The precision of words mattered, even if she had to adjust her expectations for accuracy. For instance, Lu preserved the transliteration of *ching*, *ching pieh*, *lo mo*, and *sun lo*, explaining that these structures represented "acu-tracts," "connections," or simply "tracts." She insisted that these paths were not "meridians," which she believed undermined their basic materiality. To Lu, the simple lines of longitude presented no material qualities in the world, at least none that Lu could see.

Translation enacted ontological claims. Establishing the materiality of paths meant establishing a foundation for all aspects of Chinese medical theory. It meant defining Qi (氣) and Blood (血), which were at the heart of debates preoccupying major figures like Tang Zonghai (1851–1908) and Yu Yan (1879–1954) in the late nineteenth and early twentieth centuries.[33] On the one hand, those who cared about the material attributes of Qi and Blood asserted that meridians were distinct from neurological and circulatory networks.[34] On the other hand, those who did not care about the material attributes of Qi and Blood claimed that meridians were merely misguided and crude representations of blood vessels.[35]

Lu reviewed these debates with a critical eye. Therapeutic paths like *ching*, *ching pieh*, *lo mo*, and *sun lo* were material, but they were not entirely material without some comparison to nerves. This was a tricky position. Lu did not want nerves to overtake the basic integrity of a Chinese medical body. She scoffed at contemporaries who surrendered to the descriptive power of nerves and noticed that Cheng Dan'an's 1931 book equated the 12 meridians with the 12 cranial nerves. Cheng even suggested that the 31 pairs of primary nerves could explain the entire system of therapeutic tracts. According to Cheng, his pathologist friend said that muscle cells generated enough electric friction to induce healing currents in the body. "We need to advance our knowledge about this," Cheng declared.[36]

In the margin, Lu wrote: ?

Clearly, cranial nerves did not explain the healing effects of therapeutic paths. Perhaps it was a problem of translation, an issue of the imagination, or the limits of inexperience. Nerves appeared to be an obvious explanation, but for Lu, such an explanation was incomplete and incorrect.

In 1964, Lu compiled a personal dictionary with hundreds of translations. For some entries, Lu wrote that characters were "so full of meaning that they should be kept as Romanized Chinese terms as much as possible."[37] She often told Joseph Needham that English translations of classical terms were misleading, while he accepted the limits of translation.[38]

Lu dedicated one section of her dictionary to the physiological structure of the 12 *jingluo* (經絡). She portrayed them as a network concerned with "chhi [*qi*] circulation." Qi appeared as a conceptual foundation that sustained therapeutic paths. They traveled "deep inside the body" and were "rather long."[39] Lu echoed classical texts describing paths extending from the hands and feet. They penetrated "the visera [*sic*]" before turning to "parts of the neck and head."[40] Lu elaborated further, adding that the paths connected classical anatomy to contemporary physiology. They represented a "close relationship" to "both physiology and . . . basic theories of traditional medicine."[41]

Like many of her predecessors and peers, Lu then characterized classical anatomy based on their Yin and Yang orientations. The Yin and Yang binary was a relational orientation in that Yin organs were Yin relative to Yang organs, and some organs could be more Yin than others. This was a stable and longstanding mode of classification. They were stable enough for Lu Gwei-djen to present a particular perspective of anatomy. Yin and Yang oriented body parts, the organs, and the meridians. Yin body parts gave rise to Yin meridians that connected to Yin organs. Yang body parts gave rise to Yang meridians that connected to Yang organs. Yin and Yang anchored varieties of *ching*, *ching pieh*, *lo mo*, and *sun lo*.

Lu Gwei-djen and Joseph Needham pressed further. They then described the physiological patterns of Yin and Yang as born out of cellular differentiation. Cellular differentiation extended from embryology, which Lu and Needham considered as a universal and fundamental biological process. It gave rise to different kinds of muscular, cardiovascular, fascial, and nervous tissue. Lu and Needham positioned what they understood as the most basic element of biological development to explain East Asian anatomy. It was in this embryonic body, the earliest manifestation of the head, hands, feet, and torso, that offered the earliest manifestation of Yin and Yang body parts.

Per Lu's instructions, Joseph Needham sketched Yin and Yang body parts on an embryological body (**Figure 5.2**). He started with a bulge differentiated by two arrows pointing in opposite directions. A narrow paddle-like leg sprouted from one side, accompanied by an arrow. In the surrounding text, English inscriptions appeared in black; Chinese descriptions appeared in blue. In black, Needham labeled the head, tail, belly, back, and limbs as "caudal," "cephalic," "ventral," "anterior," "dorsal," "posterior," "proximal," and "distal."[42] In blue, he marked Yin and Yang body parts. The head, back, and limbs were Yang (陽); the belly, tail, and interior were Yin (陰). A dashed line leading to the

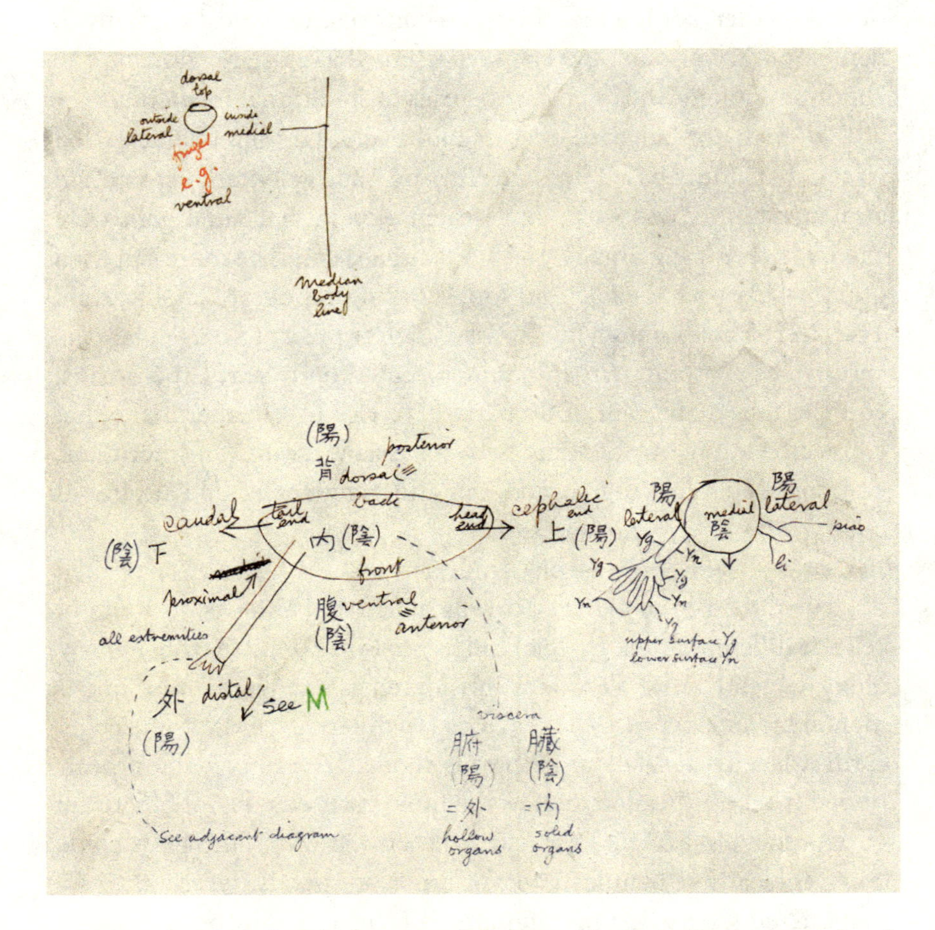

Figure 5.2. Ontological projections, twentieth century. Lu Gwei-djen (1904–1991) and Joseph Needham (1900–1995), illustration. In black, Lu labeled the caudal (tail end), cephalic (head end), dorsal and posterior on the back, ventral and anterior on the front, and proximal and distal on the extended limb. In blue, Needham marked each area as Yin (陰) and Yang (陽), or *nei* (內) and *wai* (外). This kind of "ontological projection" required distortions that traveled in one direction, where caudal and cephalic could give rise to Yin and Yang parts of the body, but Yin and Yang did not define the head and tail. From Lu and Needham, "Index Termini Technologici Medici Sinensio-Anglici Lu-Needhameinsii," 1964, 251. Reproduced courtesy of the Needham Research Institute SCC2/321/4

belly points to Yang "hollow organs" (also known as *wai*, "outer") and Yin "solid organs" (also known as *nei*, "inner"). This orientation grounded a kind of "Chinese" physiology.

Needham's inscriptions grew more elaborate. He added a cross-section with a front view of the arm in a smaller adjacent diagram. He drew a circle labeled "medial" at the center and "lateral" on either side in black. Different sides of body parts required different names. In blue, the same arm extended with articulated fingers. Here, Needham wrote "upper surface Yg" and "lower surface Yn," to show that the upper part of the limb was Yang, and the lower part was Yin. Moving digit-by-digit, Needham clustered arrows labeling the upper side Yang and lower Yin.

The relative positions of Yin and Yang as inner (*nei*) and outer (*wai*) seemed conceptually straightforward. Yet their graphic positions on the page were ornate. They necessitated an abundance of arrows to address every possible dimension. Yin and Yang did not proceed in a basic alternating pattern but took into account the relative interiority and exteriority of body parts. Accounting for these additional interior and exterior dimensions (*nei* and *wai*), front and back, left and right, registered numerous inconsistencies in Needham's image.

Translating body parts did not mean these new names and labels were interchangeable. For instance, the dorsal posterior could be assigned as Yang, but Yang could not be translated as the dorsal posterior. Similarly, the cephalic end was also Yang, but Yang was not the cephalic end. Yin and Yang were relative to each other. They only represented relative positions on the page. They connected the head, back, belly, and limbs. Together, these labels made sense as long as the body remained vaguely body-like.

Lu insisted on the physical quality of meridians. Needham's diagram allowed for that argument. A physiologically informed Yin and Yang orientation offered a template for everything else in the body. Lu and Needham rallied every embryological label—the head as the cephalic end, the tail as the caudal end—to apply Yin and Yang qualities. Caudal, cephalic, ventral, anterior, dorsal, posterior, proximal, and distal parts reinforced a directional body already on the page. Yin and Yang, *nei* and *wai*, extended the meaning of these different directions to characterize the biological Yin and Yang qualities of *ching*, *ching pieh*, *lo mo*, and *sun lo*.

In other words, Yin and Yang exceeded directional orientations. Lu's conceptual, textual, and visual translation presented an example of what I have described as a kind of "ontological projection."[43] Lu and Needham leveraged the embryonic body to articulate Yin and Yang orientations, which then articulated meridian paths. These levels of transformation and translation were projected from one level to the next. Like mathematical projections, translations only went in one direction. Proximal parts could be hands. Hands could be labeled Yin or Yang, *nei* or *wai*. However, Yin and Yang could not be translated as hands. They could not replace hands. *Nei* and *wai* did not supersede the proximal extremities. Instead, Yin and Yang, *nei* and *wai*, relied on embryonic vocabulary to undergird a physical and physiological body. Classical texts described Yin/Yang and *nei/wai* as complex and systematic patterns that gave rise to meridians. These patterns in an embryological context established *jingluo* as real, material things.

Unsettling Inner Alchemy

Lu Gwei-djen encountered practical and interpretive challenges to translating Yin/Yang and *nei/wai* parts of the body. There were no English terms for Yin and Yang, but there were English equivalents for *nei* and *wai*. *Nei* often represented "inner," and *wai* as "outer." Lu did not appreciate the simplicity of this dichotomy. *Nei* as "inner" and *wai* as "outer" alone was not enough. They were more than relational descriptions. *Nei* and *wai* were also material things. *Nei* represented corporeal matter, whereas *wai* was something beyond corporeal. Lu privately insisted that *nei* and *wai* held further implications in Chinese anatomy. When colleagues contacted Joseph Needham with questions regarding medical texts and terms, Lu liberally advised Needham to defend their superfluous translations. Her notes to Needham advocated a materialist position where classical medical theory dealt with a corporeal body. As a result, everything else in this body was similarly corporeal.

Lu's dedication to the corporeal implications of *nei* and *wai* also shaped her treatment of the first-century BCE compendium, *Huangdi neijing*. By the twentieth century, this compendium had come to represent the timeless qualities of Chinese medical theory. Lu aimed to rectify perceptions of this text, starting with the title. To her, *Huangdi neijing* translated as "The Yellow Emperor's Manual of Corporeal [Medicine]." This may seem like a cumbersome rendition of what is now known as the *Yellow Emperor's Inner Canon*. When historian of science Nathan Sivin (1931–2022) inquired about the translation of the title, he pointed out that Needham and Lu translated *nei* as "corporeal." Sivin wrote to Needham and suggested that rather than translating *nei jing* as "corporeal medicine," it would be better translated as "the inner canon." This would place *Huangdi neijing* alongside other medical texts that discussed the "outer canon." For Sivin, *nei* and *wai* oriented things inside and outside the body.

Lu wrote, "Only not true!"[44]

She insisted on abandoning references to inner and outer in the title. Rather than translating *nei* to indicate the body's interior, Lu maintained that *nei* connoted "corporeal medicine." She emphasized to Needham that translating *nei* as "corporeal medicine" accounted for the historical uses of *nei* and *wai*. Historical accounts corresponded *nei* and *wai* to Yin and Yang areas of the body that gave rise to meridian paths. For her, *nei* had to explicitly indicate a physical, corporeal space, which needed to appear in the translated title of the most famous Chinese medical text. Again, the materiality of meridians preoccupied Lu.

In classical corpora, many things happened inside the body. *Nei* appeared frequently in these descriptions and Lu studied them closely. She collected numerous sources detailing internal processes. In 1971, she drafted an essay titled "The Inner Elixir (*Nei Tan*); Chinese Physiological Alchemy," in which she introduced a method for cultivating longevity and immortality. Lu translated *nei* as "inner" to represent the physicality of interior spaces, stressing that the inner elixir, or *nei tan*, involved corporeal immortality because "no other was conceivable."[45] Immortality was not an abstract state of being. Lu warned, "No greater mistake could be made than to analogise *nei tan* with the 'spiritual alchemy' of the West."[46] The body "was physiological through and through," she added.[47]

Lu argued that *nei tan* was different from the occult practices. It was not just any kind of alchemy, and it was especially unrelated to European conceptions of alchemy.[48] Instead, *nei tan* represented a "physiological alchemy." She explicitly avoided the term "inner alchemy," which generated false equivalences, among other inaccuracies. Neither *nei* nor *tan* referred to alchemical practices. Even worse, Lu's contemporaries considered "alchemy" as a shorthand for non-modern, non-scientific, and non-rational practices.[49] Although these attitudes later changed, Lu did not anticipate them. In the 1970s, she worked to separate *nei tan* from the non-material and irrational aspects of "alchemy," as she had known it then. Regardless, the translation stuck; *nei tan* as "inner alchemy" has persisted in describing a broad range of medical and ritualized healing practices in East Asian texts, whereas histories of alchemy in Europe have grown increasingly reflexive, revisionist, and rational.

Nei tan was a special kind of practice. It was like yoga, but not yoga. Lu explained that yogic practices shared similar intentions of self-preservation, only *nei tan* was more "moderate."[50] *Nei tan* aligned with body cultivation. It was a corporeal practice that involved "retracing one's steps along the road of bodily decay" to prevent atrophy of vital *ching* (精) fluids like semen and saliva that manifested physiological expressions of Yin and Yang.[51]

Lu's detailed descriptions detached *nei tan* from other Asian systems. She rendered it less radical and maintained that *nei tan* was an epistemic practice—a way of knowing and engaging the body. Lu added that *nei tan* was "in a *real* sense akin to the *optimistic and experimental* outlook of modern science, especially biochemistry, endocrinology, and geriatrics."[52] Here was a technique that belonged to the natural and modern sciences.

This may have seemed like a stretch. Lu worked hard to equate classical *nei tan* with biochemistry. Yet, examining her drafts reveals that she was attempting to convince herself of this comparison as well.

Lu claimed in her final paper that *nei tan* was a corporeal cultivation practice that aimed to attain longevity and immortality. But when she first started writing the paper, she quoted ancient texts that explained how "*chhi* [*qi*] can preserve ~~the invisible~~ life."[53] She had crossed out "the invisible," which suggested that *chhi* (*qi*) occupied a separate ontology in Lu's own conception of the body. In Lu's paper, Qi could not be material if it were invisible. Simply put, "physiological" meant "physical," and "physical" meant "visible." A physiological body needed to be seen in order to be real.

Lu labored to highlight the theoretical authority of physiological alchemy. Her original draft described the *nei tan* technique as a philosophy that represented the "sanity" and "sobriety" of Daoist philosophers. Later, she crossed out these descriptions and changed them to emphasize the "empiricism" and "rationality" of Daoist philosophers.[54] Perhaps "sanity" and "sobriety" diminished the ideological and practical aspects of *nei tan*. She avoided hints of oriental mysticism, hoping to link *nei tan* to "Western" science as a modern, logical system. In doing so, Lu curated an innovative vision of *nei tan*. This body manifested characteristics of Yin and Yang, *nei* and *wai*, in a way that was not abstract. It corresponded to the dorsal, ventral, and proximal regions. Its physiological orientation manifested in physiological alchemy (however problematic the translation) and made paths like *ching*, *ching pieh*, *lo mo*, and *sun lo* "physiological through and through."

By insisting that all forms of alchemy occurred in the same physical body, Lu attempted to close the conceptual distance between *nei* as a dynamic interior space and *nei* as a corporeal medicine. Still, the conceptual distances continued to expand. Such a distance became particularly difficult to breach when nerves came into the picture. Nerves did not explain internal alchemies, whether rational, irrational, corporeal, or noncorporeal. Nerves had little to do with the saliva and semen that excreted from Yin and Yang body parts. Nerves did not account for the reversal of cellular atrophy. Even though nerves were appealing as an explanatory mechanism, *nei tan* had no ontological anchor in neurophysiology. Lu had worked to convince herself and her reader of the material, universal, practical, and enduring qualities of *nei tan*. But she offered no comprehensive conceptual framework for these practices. Lu presented *nei tan* on its own terms but interrupted her own efforts by articulating it as biochemistry. Her desire to insist on a uniquely Chinese corporeal body conflicted with her desire to render *nei tan* as a uniquely modern scientific object. Her translation had limits, with too much at stake and too much left unexplained.

Material Metaphors

Lu Gwei-djen's translations aimed to liberate Yin/Yang, *nei/wai* from a reputation of being fanciful theoretical objects. She relied on the specialized metaphor of embryological orientation rather than on the popular metaphor of nerves to render Yin/Yang and *nei/wai* as territories on the body that one could point out, trace, and comprehend. Nevertheless, nerves entered discourses of needling with profound and relentless visibility.

In the 1970s, the growing popularity of *zhenjiu mazui*, or "acupuncture analgesia," appeared as a modified version of needling that accompanied surgical procedures.[55] It induced localized insensitivity to pain and allowed patients to remain conscious throughout an operation. Numerous reports from hospitals in Beijing, Guangzhou, Shanghai, Colombo, Macau, and Hong Kong featured photographs of men and women resting on operating tables. They smiled into the camera, awake and alert, as a team of surgical staff carefully cut into their necks, bellies, hearts, and heads (**Figure 5.3**).

Needling looked different in the operating room. Some needles were large, some were small, some were manipulated by hand, and most were connected to a battery-powered box. Electric currents pulsated

Figure 5.3. (opposite) Awake and alert, twentieth century. Unknown photographer, press photos. Reported cases of acupuncture analgesia, or *zhenjiu mazui*, showed patients conscious while undergoing surgeries. Here, a woman watches surgeons remove her ovarian cysts (*above*) and serve her tea (*below*). Such images supported emerging discourses on how needling operated on the nerves, generating new theories of pain sensation. From Lee, "Acupuncture Anaesthesia," 1971

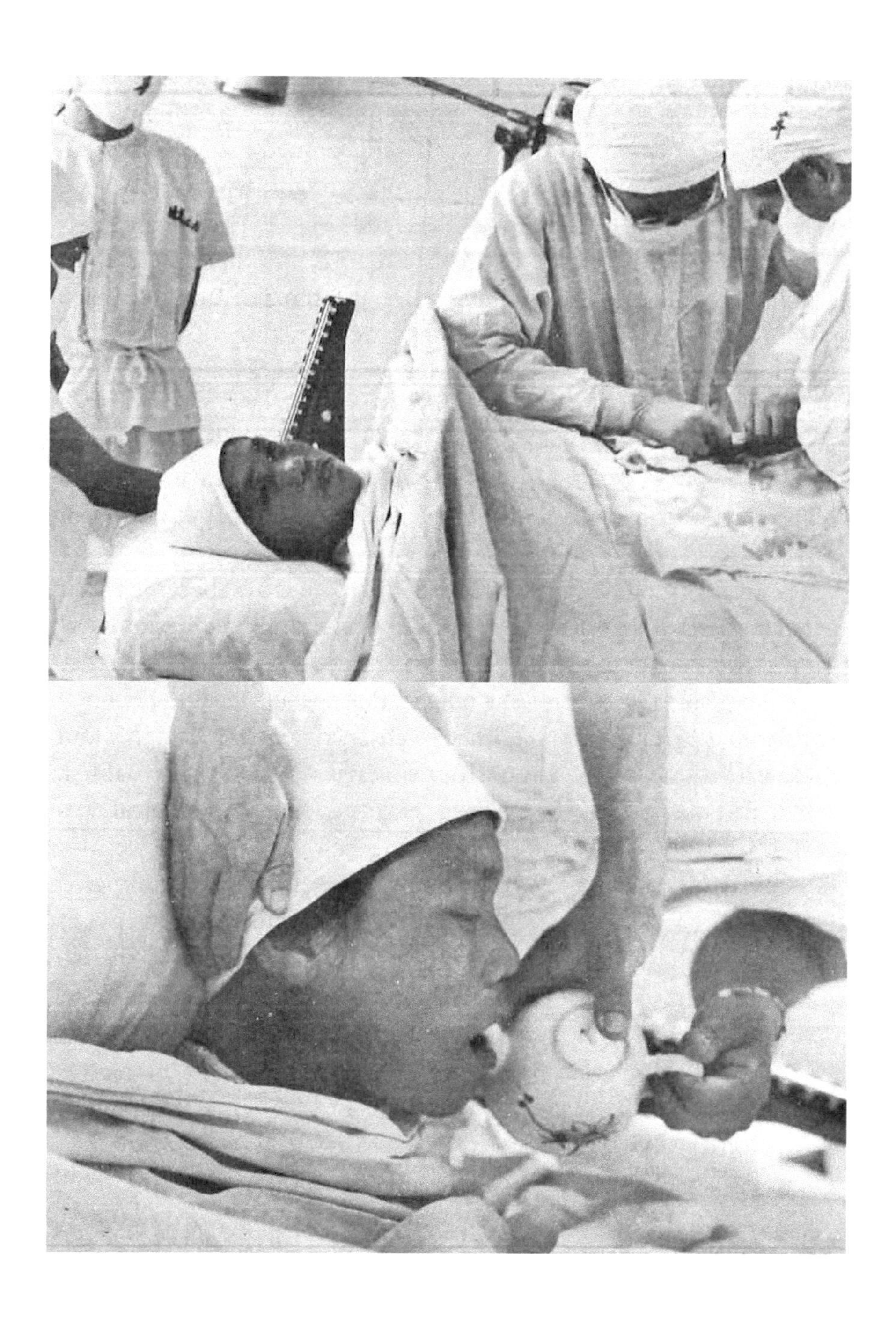

from the box to cause the needles to rotate over 100 cycles per minute.[56] Thicker needles rotated with so much speed that they often left marks on the skin, reminiscent of classical styles of needling. When connected to a wire, these new attachments pinned the patient to the table. They reinforced the idea that simple, straightforward electronic means could make therapeutic paths legible.

Lu Gwei-djen collected hundreds of reports on acupuncture analgesia. She knew that using acupuncture and moxibustion to induce numbness was not new. Classical literature frequently described different needling methods for addressing acute and chronic pain.[57] For instance, Cheng Dan'an used moxa to treat toothaches; Zhu Lian used needling to treat her sciatica. However, the new material environment of the operating room and the intensified political atmosphere of the Cold War generated new interest in the clinical applications of acu-moxa.

For instance, physicians like Han Suyin (1916–2012) became captivated by the skin's potential to register needles and the physiological implications of acupuncture. She allowed researchers to probe at her skin, skeptical at first. Then, she noticed that specific areas of her body registered different responses. "The electrical potential was higher and more varied than that of the surrounding tissue," she wrote.[58] Baffled, she turned to her collaborators, who suggested that physiological processes and emotional states could influence the skin's electrical potential. Han continued: "Unbelieving, I held out my hand, compared it with the chart, and asked [about] the "acupuncture point" between the thumb and the index finger . . . The point electrode was applied to the exact point, and the needle jumped, showing a bio-electric charge. A few millimeters away from the precise spot my skin was electrically inert. I tried many other points on my body, referring to the model, and felt mystified."[59]

To Han, the electric probe responded to areas on the skin with different levels of "bio-electric charge," whereas others remained "electrically inert." Like metal detectors, the machine unearthed sensitive, active sites. This was not an arbitrary or accidental discovery. One had to know the location and names of the sites to find them. The epistemic redundancy of knowing the location of the sites, searching for the sites, finding the sites, and then translating the sites as electrically active further reinforced the sites as registering electric conductance.

However compelling, writers like Han Suyin knew that probing the skin gave an incomplete picture. Her experience with the probes had left her feeling "mystified." The researchers explained that the site on her hand somehow connected to the larger intestine. Although Han had focused on the electrical potential of her skin, she would need to carry this logic from the body's surface to the body's interior. She stretched her imagination to her internal viscera and offered a full-page illustration. Accompanying her article was a meridian man that she generically described as "Acupuncture Points and Meridians on the Human Body" (**Figure 5.4**). Of course, this image represented just one of many paths, despite the plural "meridians" in the title. Specifically, it featured a single path: the major-Yang-foot-bladder-meridian seen in chapter 1.

This was a popular image. Writers often featured bladder-meridian men on the cover of acupuncture books.[60] Han Suyin had shared a similar impulse; her bladder-meridian man closely resembled other bladder-meridian men in the early modern classic *Treatise on the Fourteen Meridians* (*Shisi jing fahui*). It seemed to imitate Wang Xuetai's bladder-meridian man from Zhu Lian's 1950 book *New Approaches to Acupuncture and Moxibustion* (**Figure 5.5**).

But Han Suyin's bladder-meridian man diverged from his predecessors. He was different, uncanny. He resembled other major-Yang-foot-bladder-meridian men with a similarly pleated headpiece and a matching robe secured by a matching sash.[61] However, his headpiece was softer, and robe was more articulated. His mustache was fuller and cascaded around a delicate lip. His nostrils flared, and his sideburns hid a handsome ear. He stared directly at the viewer, eyebrows brooding and inquisitive. His legs spread apart, shifting his weight to one side to support a rounder belly and waist. He stood firmly with expressive toes and toenails. The freehand inscriptions of the site

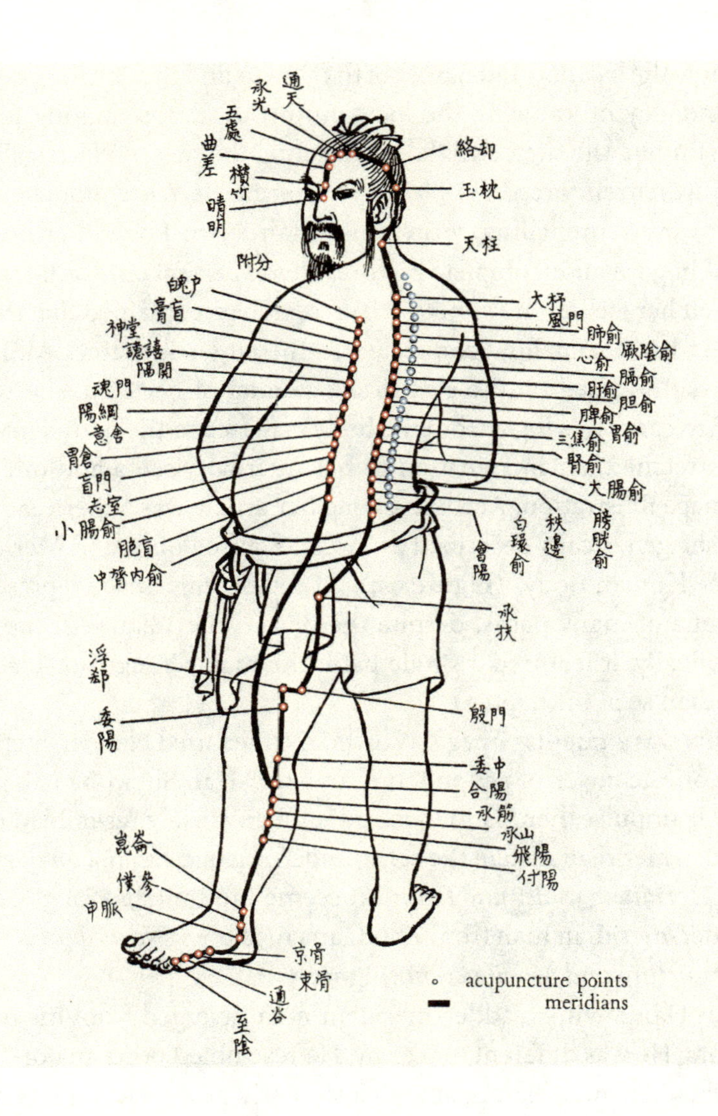

Figure 5.4. Bladder foot meridian man, twentieth century. Unknown artist, illustration. A modified freehand rendition by Han Suyin (1916–2012) of the *zu taiyin pangguang jing tu* (minor-Yin-bladder-foot meridian). Three rows on the back show two paths for needling (marked in red by the author) and one vertical row of 25 indentations (marked in blue by the author). From Han, "Scientific Evidence of Acupuncture," 1964, 9. Modified by the author

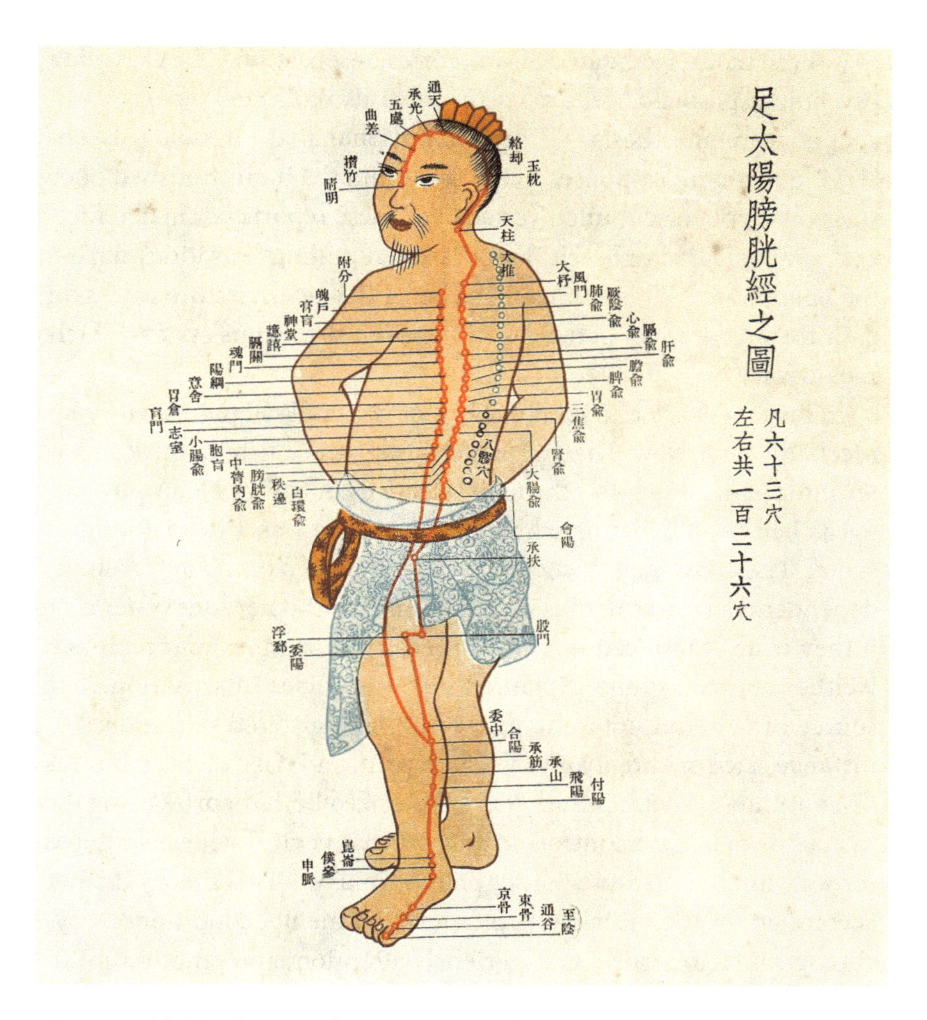

Figure 5.5. Bladder foot meridian man, twentieth century. Wang Xuetai (1925–2008), lithograph. Wang's meridian man from the second edition of *New Approaches* listed 63 meridian sites that, when doubled on both sides of the body, totaled 126 sites. From Zhu Lian, *Xin zhenjiu xue*, 2nd ed., 1954, fig. 8

names suggested that the illustrator took further liberties. Unlike vertically stacked characters reading top-down, Han's text read left-to-right. This was a modern meridian man.

Lu Gwei-djen noticed that as reports of acupuncture analgesia proliferated, so did discussions about nerves. Neurophysiologists felt that these reports validated new theories of sensation. The Canadian psychologist Ronald Melzack (1929–2019) also collected reports about acupuncture anesthesia. He read about Shanghai physiologists observing that pain responses in the spine and the brain vanished once surgeons applied acupuncture needles. These reports fascinated Melzack. Patients had only felt "an occasional pulling sensation" during the operation.[62] "This confirms the belief that acupuncture works on pain through known pathways in the central nervous system," Melzack wrote.[63]

In 1965, Melzack and his collaborator Patrick Wall published "Pain Mechanism: A New Theory," which proposed a gate control metaphor for pain perception through selective controls.[64] Neurophysiologists had been debating whether or not pain was a special kind of signal. They considered whether perceptions of pain operated on independent networks in the peripheral and central nervous systems or if they instead induced impulse patterns in non-specific receptors.[65] Neither approach fully explained inconsistencies in sensation. The source of pain did not quite map onto the registered sensation. For instance, a lesion did not always cause pain, and if it did, the pain was discontinuous. Melzack and Wall synthesized both theories using the metaphor of a "gate control" to suggest that central neurons filtered sensory input with and without pain or injury.[66] This theory further accounted for the individual's physical and mental conditions. "Psychological factors such as experience, attention, and emotion influence pain response and perception by acting on the gate control system," they mused.[67] Under the right conditions, patients could distract themselves from their pain.

Melzack's and Wall's theory aimed to be dynamic. The metaphor of "gates" was a powerful one. Gates could open wide. Gates could open partially. Gates could be controlled. They could be regulated for speed, intensity, and fluctuation. Indeed, when Ronald Melzack later reflected on the theory, he explained that the "gatelike mechanisms" at different centers in the body were shifting and changing. Building on this metaphor, he wrote: "The gates can be opened or closed to varying degrees by nerve impulses in the larger- and smaller-diameter fibers in each sensory nerve running from the body's surface to the spine and the brain. Activity in large fibers tends to close gates and lessen the pain, while small-fiber activity tends to open gates and increase pain. . . Fibers from the cortex of the brain—the center of the memory, attention, anxiety, and interpretive functions—also can either open or close gates."[68] For Melzack, gates theoretically responded to differently sized fibers in the body that facilitated memory, attention, and emotion. These were material and mental "impulses." Still, other neurophysiologists who considered the metaphor of sensory "gates" found it unappealing. Melzack and Wall's paper barely impacted the field, attracting only a handful of citations in the first ten years following publication.[69]

Initially, the gate control metaphor met little acclaim. But Ronald Melzack persisted, publishing *The Puzzle of Pain* in 1973. On the cover was a small meridian man. Melzack had developed an apparent fascination with acupuncture analgesia, seizing upon it to validate his theory. In the book, Melzack elaborated how gate control theory revealed the mechanisms for particular kinds of numbness. It accounted for the effects of a many kinds of analgesia from hyperstimulated analgesia to acupuncture analgesia. Having read Han Suyin's account of researchers probing at her skin, Melzack included a full-page spread of her major-Yang-foot-bladder-meridian man in his book (**Figure 5.6**). Bladder man again stood with his head turned, eyes searching, mustache wispy, belly bulging, and legs parted. Yet, this bladder man appeared uncanny in a different way. Its illustrator did not know how to read Chinese. The text surrounding the image made no sense. Some characters looked like characters, but with malformed radicals or extra sets of lines. Bladder man was accompanied by gibberish.

Like Han Suyin, Melzack captioned the image as a "typical acupuncture chart." But this bladder meridian was more exceptional and less representative of other paths. It was different. Among the 12 meridians, the bladder meridian was the only image that graphically featured indentations of the spine.[70] It was also the only among the 12 primary meridians to travel along the back and break into multiple paths.[71] It reached from the nose to the pinky toe and split below the

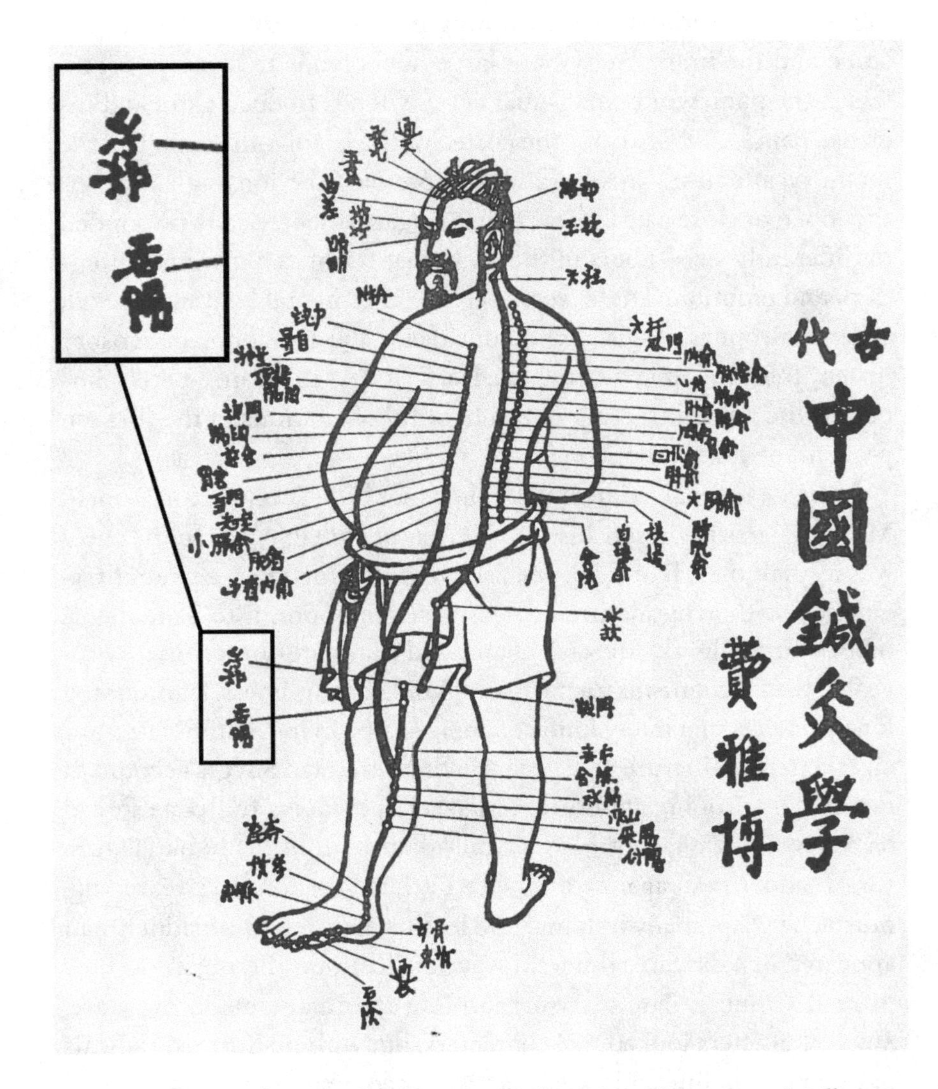

Figure 5.6. Bladder foot meridian man, twentieth century. Unknown artist, illustration. Ronald Melzack's attempt to copy Han Suyin's bladder meridian man is evident from the odd combination of vertical and horizontal text and graphic errors. The nonsense characters (*inset*) suggest that this was traced by someone illiterate in Chinese. In the image's caption, Melzack said it belonged to a set of "typical acupuncture charts, showing the sites for insertion of acupuncture needles." From Melzack, *The Puzzle of Pain*, 1973. Copyright © 1974. Reprinted by permission of Basic Books, an imprint of Hachette Book Group, Inc.

knee to include a second set of paths on the back.[72] This abundance of paths further meant that the bladder meridian featured almost twice as many sites as the other primary meridians. Whereas other paths averaged around 20 needling sites, the bladder meridian carried over 60.

The proximity of the bladder meridian to the spine implicitly suggested that all meridian paths engaged with spinal nerves or, at the very least, reflected patterns of spinal innervation. Melzack's version of a conceptually botched bladder meridian man suggested that the paths were symbolically Chinese, physiologically permanent, and practically legible.

Melzack continued publishing on how needling-induced numbness related to gate control theory. He collaborated with his McGill colleague Dr. Mary-Ellen Jeans on a short paper introducing a "physiological explanation" of acupuncture anesthesia.[73] As a gesture of cultural sensitivity, they included a section introducing Yin and Yang as "universal forces" and "spirits," to which Lu Gwei-djen would have objected.[74] With Yin and Yang out of the way, Melzack and Jeans compared sensations in the spine to something like a switchboard.[75] The "opening" and "closing" of gates addressed the temporal variability of acupuncture analgesia that happened before, during, and after surgery. The metaphor of a switchboard would account for the 20-minute "induction" time necessary for needling to induce analgesia, the hours of analgesia sustained during the procedure, the extended period of recovery after the operation, and further effects of reduced swelling and post-operative pain.

Even though the gate control theory presented an imperfect metaphor, "closing the gate"—any gate—seemed to explain all aspects of needling and all aspects of pain. As a theory, it allowed for multiple sources of sensation to converge on the body. In the context of acupuncture analgesia, it explained how the needling could "cancel out" the destructive sensation of a surgeon's scalpel or a dentist's drill. The gates worked because needles opened some channels and sealed others. After the surgeon removed the needles, the effects lingered on as "gates" gradually opened.

Gate control theory served as much to explain acupuncture analgesia as acupuncture analgesia helped to explain gate control theory. The more needling reports populated the Anglophone world, the more the gate control idea flourished. By the 1980s, its popularity had increased tenfold, with over 1,000 citations.[76] "Support from the gate-control theory has come from electrical stimulating devices," Melzack observed, pointing out the battery boxes that stimulated needles during surgery benefitted his work.[77] These nameless devices had registered their effects on the skin. Their effects further validated Melzack's otherwise obscure theory.

However, the exact placement of the electrode raised new epistemic questions. If one knew where to place the probe, did they need the probe to verify and validate meridian sites? Was knowing the location of the site enough?

Half a Million Operations

Lu Gwei-djen and Joseph Needham corresponded with Ronald Melzack, and the three soon became friends. Lu and Needham heavily relied on gate control theory to explain acupuncture, with Lu generously citing Melzack in *Celestial Lancets*. Just as Melzack leveraged acupuncture analgesia to explicate gate control, Lu recognized that gate control could reciprocally explicate acupuncture analgesia.

Addressing acupuncture analgesia was a political undertaking. It brought attention to China even though acupuncture analgesia was a global phenomenon.[78] For instance, when *New York Times* Vice President James Reston underwent an emergency appendectomy in Beijing in 1971, he famously reported how acupuncture and moxibustion relieved his post-operative pain. Reston described how his physician inserted three long needles in his right elbow and knees to ease his abdominal discomfort before lighting two moxa sticks on his abdomen.[79]

News of Reston's operation that appeared in the *Chicago Tribune* and the *Washington Post* garnered mixed reviews.[80] Some physicians supported Reston's experience in the *Journal of American Medical Association*, explaining how therapeutic needling substantially lowered cerebral induction voltage.[81] Other experimental physiologists attributed all therapeutic effects of needling to gate control theory. Meanwhile, American physicians demanded thorough studies of acupuncture analgesia on American bodies.[82] They essentialized biological and cultural differences emphasizing that experiences of pain were shaped by social, cultural, and psychological circumstances. In other words, "Chinese" needles could only work on "Chinese" bodies, even when acupuncture had long traveled beyond China and developed in Seoul, Sri Lanka, and São Paulo.

Although some American physicians racialized acupuncture analgesia to delegitimize its effects, members of the Chinese Communist Party popularized it as a legitimate national public health initiative.[83] Numerous reports explained that the first recorded acupuncture analgesia operation occurred during the Great Leap Forward in 1958. In the 1960s and 1970s, needling took further political urgency during the Cultural Revolution.[84] One writer claimed that the lack of supplies inspired physicians "the spirit of daring to think and daring to act."[85] Acupuncture needles were cheap, effective, and empowering. They facilitated a kind of rural epistemology and rural virtues that allowed the patient to remain awake during her operation and interact with the surgical team. It allowed medical students and amateur enthusiasts to experiment on their own bodies.[86] Acupuncture analgesia represented the non-elite knowledge that characterized its apparent Communist origins.

Lu Gwei-djen had to see the effects of acupuncture analgesia for herself. In 1972, she and Joseph Needham traveled to China to observe a series of surgeries. They were aware of the Chinese government's recent attempts to accelerate scientific development and economic growth, and despite observing China's successes and failures, Lu and Needham remained devoted to Communist Party rhetoric. In 1970, Needham famously announced, "Chinese Marxism is truly Marxism with a difference" in a Cambridge lecture celebrating Mao Zedong's legacy.[87] In private, Needham confessed that he recognized China's complicated reality to Lu and Nathan Sivin.[88] Despite the tragedies of the Great Leap Forward and the Cultural Revolution, losing Chinese contacts by condemning Mao was too risky.[89]

During their trip, Lu and Needham arrived at the Second General Hospital in Guangzhou.[90] Hosted by a team of surgeons, physicians, and scientists, they observed the procedure for a 57-year-old sailor who suffered from a kidney stone.[91] The physician applied four needles, two on his abdomen, one on the hand, and another on the foot. The team connected the stomach needles to a 9-volt 0.5-milliamp direct current that reached up to 500 milliamps. Meanwhile, a technician manually manipulated the needles on the hand and foot.

"Throughout the operation the patient was awake and calm, able to speak with those in the theatre, and to make any movements which the surgeon requested," Lu and Needham wrote.[92]

The second operation was a cesarean section on a 24-year-old woman, which required stimulating different points. The team applied one needle in her abdomen, left leg, right knee, and two in her ear. Later in the operation, they added two more needles in her ear. "She lay on her back throughout the procedure perfectly calm and impassive, not wincing or showing any signs of pain even during the extraction of the infant," Lu and Needham noticed. [93] The procedure had not affected the newborn, who cried and breathed normally.[94] This seemed to confirm many of the published reports. Acupuncture analgesia had worked. It had allowed the patient to remain awake during the operation and safely deliver her baby.

Lu and Needham recognized that both cases required very different types of surgery around the lower abdomen. The first case involved a middle-aged man with kidney stones, and the second involved a young woman giving birth. To safely cut into the kidney, physicians had to stimulate sites on the hand, foot, and abdomen. To cut into a pregnant belly, physicians stimulated sites on the leg, knee, ear, and stomach. These activated sites allowed the physician to cut into multiple layers of flesh and extract a kidney stone or an infant. This kind of operation was very different from the material body practices that Lu had studied in her work on the inner elixir, the *nei tan*. Rather than replenishing the body, needling served as a means of distraction, at least according to the gate control theory of Melzack and Wall.

From Guangzhou, Lu and Needham traveled north to Beijing. They learned from physicians at the Chinese Academy for Research in Traditional Chinese Medicine (the institute that Zhu Lian founded) that around 500,000 reports of operations assisted by acupuncture analgesia had been recorded in China.[95] Effects varied depending on the type of surgery, the patient's condition, and the selected needling sites. Abdominal operations were considerably more challenging to anesthetize with needling alone, often requiring surgeons in Guangzhou to supplement analgesic effects with a relaxing agent or turn to generalized chemical anesthesia as a last resort.[96]

Physicians reported that fewer needling sites were more effective than many. Surgeons in Beijing explained to Lu and Needham that reducing from 21 down to 3 sites produced better results. Too many needles generated "cancellation effects," they said.[97] Some physicians attributed this to Melzack-Wall's gate control theory, proposing that fewer central nervous system inputs amplified the intended anesthetic effects.[98]

Yet, Melzack's and Wall's gate control theory had obvious limits. For instance, it did not explain why acupuncture analgesia was more effective in the upper regions of the body compared with the lower regions. It did not explain why patients responded better to some sites than to others. To resolve these anomalies, one physician in Beijing suggested to Lu and Needham that the anesthetic effects operated through three independent networks: the autonomic and sympathetic aspects of the central nervous system, the lymphatic system, and the *jingluo* system. In response, Needham agreed that the central nervous system was not the only active network. But the weaker effects of needling on the abdomen indicated that it may not in fact operate through the autonomic and sympathetic systems.[99] "As for the [*jingluo*] system (whatever it is)," he added, "the effects go through it."[100]

This response did little to satisfy Lu Gwei-djen. She was more interested in how physicians who performed surgeries using acupuncture analgesia understood the meridians. On numerous occasions, she asked them:

"What could have been the origin of the *jingluo* system?"[101]

"How to interpret [*jingluo*] system and its origins?"[102]

"What about the origin of acupuncture?"[103]

The physicians replied:

"Trial and error probably."[104]

"Essentially clinical experience."[105]

"Probably mainly empirical, perhaps arose by following sensation after the introduction of needles."[106]

When one surgeon speculated that meridian maps developed from "a primitive knowledge of anatomy," Lu privately noted, "the ancients know more anatomy than usually credited."[107] Later, Needham redirected the conversation to nerves and referred pain. This prompted physicians to agree that referred pain likely contributed to needling therapeutics and meridian theory.[108]

Toward the end of their China trip, Joseph Needham received acupuncture from two different physicians to treat his swollen ankles and kidney issues.[109] At 71 years old, his health was in distress. The first doctor, Li Ta-Li, demonstrated needling on herself before treating Needham. She stimulated a point on her lower leg, describing numbness spreading from the needling site to her foot. Lingering feelings of heaviness followed by further numbness lasted longer in the arms and legs and felt more extreme in the torso, she explained. Li then inserted a needle into Needham's right leg at the *san li* point.[110] "Then a downward feeling of numbness (*ma*) ran along the long bones as far down as the upper surface of foot," he observed, adding, "twiddling renewed the sensation each time."[111]

The next day, Needham noticed his right leg hurt less than before the treatment. A few days later, he approached a second doctor, Liu Hêng-Fêng, to treat his left leg. Liu used a different needling technique, penetrating the point more deeply. Needham quickly felt numbness radiate down his leg, although his muscles contracted around the needle, making it hard to remove.[112] He did not elaborate much on this second encounter. In his diary, Needham seemed impressed with Li's willingness to help him anticipate the effects of needling, while his minimal reflection on Liu's treatment left Needham desiring more.

It was difficult to know how Lu and Needham integrated their research with their embodied experiences. When one physician asked their thoughts on *jingluo*, Needham replied, "[W]e feel undoubtedly it must represent some reality, perhaps more physiological than anatomical, but of course not yet clear exactly what!"[113] The therapeutic and anesthetic effects of needling the body worked, but the relationship among the sites remained ambiguous. When Li Ta-Li needled at Needham's leg, Lu Gwei-djen asked her if the path of sensation was, in fact, a meridian path.

"Yes," Li replied, "though not always so."[114]

Sensations could travel along different paths and produce a variety of responses among patients. Clinical encounters revealed idiosyncrasies in intensity and duration. Acupuncture worked, yet its techniques and outcomes varied. The physiological responses aligned with modern pain theories like Melzack's and Wall's gate control theory, while also evading the explanatory framework of a central nervous system. Few of the physicians Needham and Lu encountered reflected on the historicity of meridian paths. Modern doctors appeared ignorant of East Asian anatomical history and how meridians engaged with organs. They lacked appreciation for physiological practices like *nei tan*. To Lu Gwei-djen, much more needed to be done.

Failed Experiments

After returning to Cambridge, Lu continued following new reports on acupuncture analgesia.[115] Physiologists beyond China increasingly began to explore needling-induced numbness. Researchers in North Korea, Europe, Asia, and the United States tested a variety of needling techniques and outcomes. Some found that stimulating sites on the lower limb increased pain thresholds in rabbits. Others found that acupuncture altered sensations on the skin and in the spine.[116] A French American researcher reported electrochemical skin stimulation aided tissue repair.[117] Researchers in New York used the Melzack-Wall gate control theory to develop a technique called transcutaneous nerve stimulation that activated needles at 200 hertz to relieve pain.

Researchers also found that searching for sites could also deactivate them. Some reports stated that the probes identifying sites could neutralize those same sites.[118] Experimental procedures jeopardized clinical efficacy. For some, like Han Suyin, positive probe readings confirmed meridians through skin conductance. For others, negative readings despite skin conductance meant that meridians did not exist. Nevertheless, self-experimenters dragged electrodes across the surface of their skin to read its superficial voltage. Positive and negative readings required empiricists to make sense of their results. The idea that electrodes registered something meaningful—and more importantly, something *real*—promised more than it delivered.

The paradox that probes both validated and neutralized sites presented an epistemic puzzle. What were probes for? What did discourses of electricity overstate? What did it overshadow?

These questions manifested in how physicians both embraced acupuncture-moxibustion and rejected *jingluo*. For instance, London-based physician Felix Mann (1931–2012) had been an avid student of acu-moxa, working diligently to translate books from Chinese-language sources that introduced meridian paths. Mann had closely studied Wang Xuetai's meridian men and even featured a replica of Wang's bladder-meridian man on multiple editions of his book, *Acupuncture: The Chinese Art of Healing* (**Figure 5.7**). At the same time, Mann examined his skin with an electric probe that registered no positive readings. He eventually bought every probe on the market, none of which registered a single "active" meridian site on his skin. Mann was beyond disappointed. He later wrote: "[H]owever diligently I tried using this apparatus, I could detect neither acupuncture points, whether active or inactive, nor meridians. Subsequently I bought two more electrical acupuncture point detectors with the same dismal result. Over the years several dozen different models have been made by various manufacturers, some of which I have tested whilst looking at the manufacturers' displays at acupuncture conferences."[119] For Mann, the silence of the electric probes betrayed the wisdom of classical texts. If electrodes did not work on his skin, then meridians did not exist. The texts that he had meticulously translated into English had deceived him.[120] He blamed his uncritical, vulnerable, youthful enthusiasm.[121] He felt deceived.

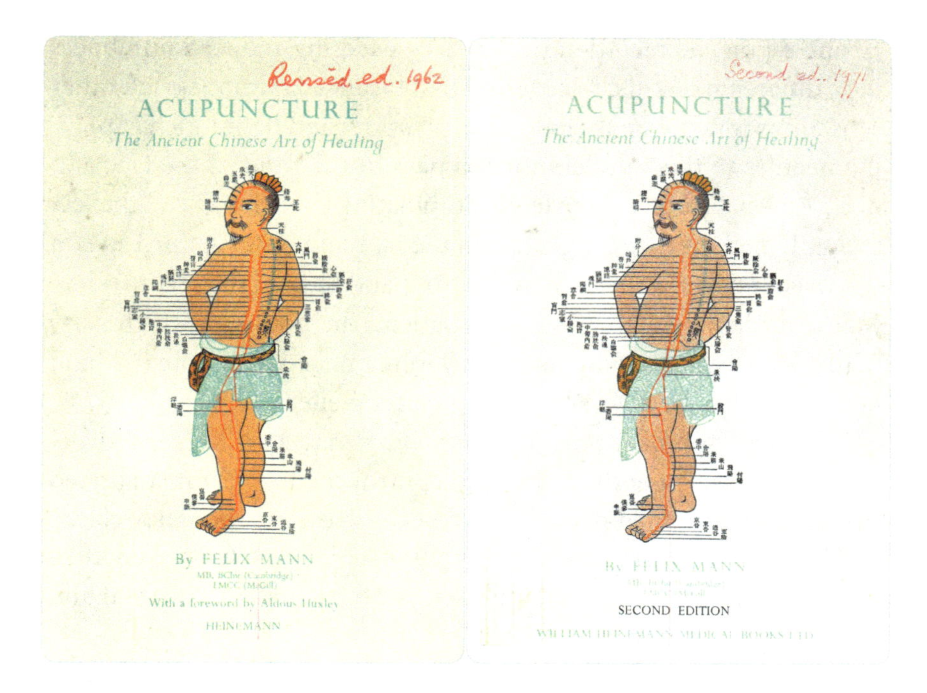

Figure 5.7. Bladder foot meridian men, twentieth century. Wang Xuetai (1925–2008), illustration. So visually compelling was Wang's 1954 bladder man in supporting discourses of needling at nerves that Felix Mann (1931–2014) featured it on the cover of multiple editions of his popular book introducing acupuncture. Mann regularly mailed his books to Joseph Needham and Lu Gwei-djen, who closely annotated his chapters to reference in *Celestial Lancets*. From Mann, *Acupuncture*, 1961 and 1972 eds. Reproduced Courtesy of the Needham Research Institute

Despite his disappointment, Mann still used acu-moxa to treat his patients. He started training at Cambridge University and Westminster Hospital in 1958 and eventually became friends with Lu Gwei-djen and Joseph Needham. Lu appreciated Mann's translations, citing them in *Celestial Lancets* as some of the best English-language resources on acu-moxa.[122]

Mann developed a complicated relationship with meridians. He wrote about them, rejected them, and taught them to students. In the early 1970s, he collaborated with scientists in neurobiology, anatomy, and dental surgery to contribute to the growing number of clinical observations demonstrating how acupuncture treated extreme pain.[123]

In one report, he recorded 100 cases of needling-induced numbness in 35 different patients.[124] He found that localized numbness left other sensations intact, like temperature, pressure, and light touch. Applying needles to the hand eliminated pain in the front of the face and neck, especially around the teeth. Stimulating a second site on the feet strengthened the numbing effect in the face and neck. A third site on the wrist induced analgesia on the front and back of the chest. When Mann needled these three sites, his patients did not know which parts of the body lost sensation unless they were pricked with a pin.[125] Mann had controlled for a possible strong placebo effect.

To explain this phenomenon, Mann turned to Melzack's and Wall's flawed yet conveniently vague gate control theory. Mann suggested that needles modified nervous control centers, even though Melzack and Wall were unsure how this was possible.[126] Mann tentatively concluded that, like asthma, the numbing effects combined mental and physical phenomena.[127]

Mann incorporated these theories into new editions of his books. He added modified sensation maps by German physicians who elaborated on Henry Head's early work on affected sensation. By then, neurophysiologists described Head's images as representing "dermatomes" that grounded descriptions of "referred" pain. Mann explicitly linked referred pain and acupuncture, insisting that they relied on the same physiological principles. Mann explained that this manifested through the immediate effects of needling. Stimulating an active site could reach the other side of the body in less than two seconds. "This speed of conduction excludes the blood and lymphatic systems," Mann wrote, "leav[ing] to my way of thinking, the nervous system as the only contender."[128] For him, this reaction conclusively "proved" that needling operated on nerves and nerves alone.

Meanwhile, other research teams did not fully buy into referred pain. They did not see nerves as the only explanatory mechanism for needling effects. For instance, in 1963, a North Korean team led by physiologist Kim Bong Han (1916–1966?) reported locating correlations between needling sites and capillary clusters.[129] Kim apparently used radioactive phosphorus to identify small oblong corpuscle cells near acupuncture sites in arms and legs, naming them "Bonghan corpuscles." Kim published color prints of the dark-red Bonghan corpuscles, graphically registering the needling site as an anatomical entity. He translated the findings into English, Russian, and Chinese, sending multiple copies to Lu and Needham. Yet, despite Kim's efforts, Bonghan corpuscles did not gain the same appeal as dermatomes and referred pain. Electric probes had already amplified the popularity of nervous stimulation and inhibited the possibility of other physiological mechanisms.

Whereas Kim's team had focused on defining the ontology of needling sites in the arms and legs, Mann turned to the spine. He noted that some needling sites, especially those on the back, linked to specific organs.[130] Given that East Asian anatomy only partially mapped onto biomedical anatomy, Mann used dermatomes to fill in the gaps. He modified a version of Jay Keegan's and Frederick Garrett's 1948 dermatome maps, which showed a neat row of reflex patterns. Mann then matched each reflex to individual needling sites.[131] These needling sites had come from the bladder meridian man. The bladder meridian had conveniently featured the back, which Han Suyin and Ronald Melzack had also appreciated. For Mann, the prominence of the back in the bladder meridian conveniently connected innervation patterns to meridian paths. It allowed him to anchor all meridian paths to the spine.

Mann illustrated the three parallel relationships between needling sites, organs, and dermatome levels (**Figure 5.8**). Needling sites appeared in sequence as B13, B14, B15, B18, B19, B20, B21, B23, B25, B27, B28 (where "B" stood for "bladder").[132] The column of corresponding organs featured eleven of the 12 viscera, listing in sequence the lungs, pericardium, heart, liver, gallbladder, spleen, stomach, kidney, large intestine, small intestine, and bladder. The third column presented dermatome levels as T2–T9, C8–T8, T6–T11, T6–T11, T7–T10, T5–T9, T9–L3, T9–L1, T6–T11, T11–S4. This juxtaposition made a direct argument. Visually comparing the stacked, neatly divided sites with organs and dermatomes forced a conceptual equivalence. It suggested that individual meridian sites were connected directly to spinal nerve

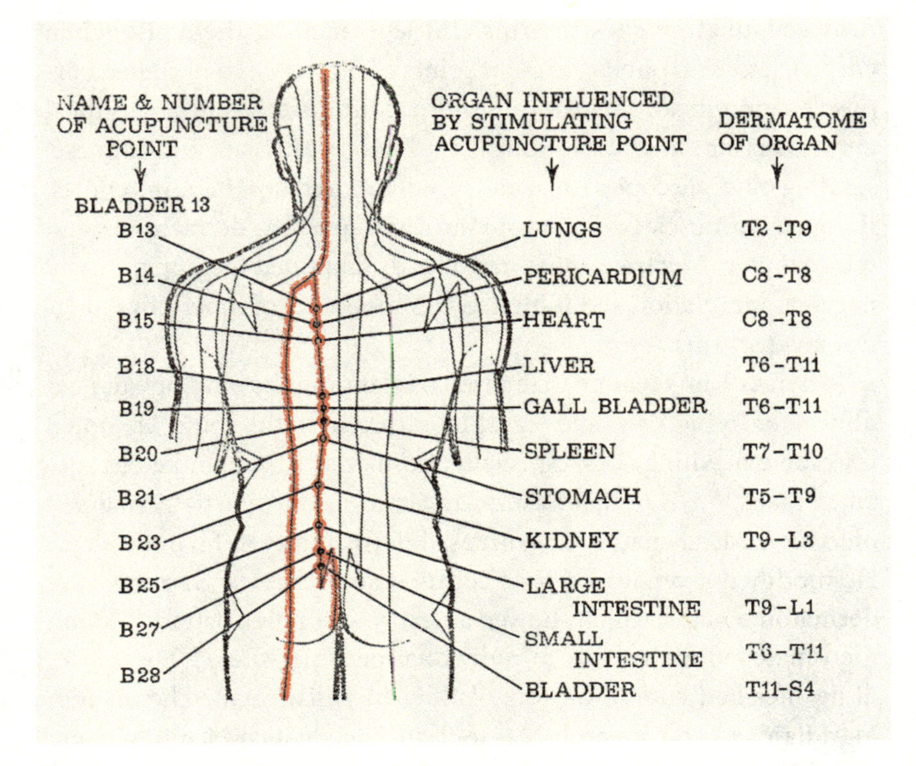

Figure 5.8. Bladders, backs, and dermatome organs, twentieth century. Unknown artist, illustration. Felix Mann paired the bladder meridian (marked in red by the author) with dermatomes, which he described as "organs" located at specific spinal levels, in addition to classical organs in the gut. From Mann, *Acupuncture*, 1971, fig. 1. Modified by the author

clusters. At the same time, the comparison presented a false equivalence. Mann had carefully selected an image to impose this connection between meridians and nerves. Specifically, he imposed a false equivalence between the bladder meridian and dermatome levels. Each were fragments of a larger system.

Mann's image was both persuasive and partial. The names of needling sites mirrored the names of spinal levels, even though each object had emerged from entirely different epistemic imperatives. Even though dermatomes appeared in the image as discrete regions on the spine, they overlapped across thoracic, cervical, and lumbar levels.

Dermatomes, like most areas of sensation, were unstable objects. They varied among individuals and occupied fluctuating regions of the upper and lower areas on the back.[133] "There is often considerable variation in the dermatome pattern according to the method of investigation," Mann acknowledged.[134] Dermatomes were ephemeral. They were not always obvious, and they often emerged in disease states. In some cases, they appeared on the body during viral infections of the nerve root. For patients with severed spinal nerves, they manifested as partial changes in the electrical resistance of their skin.

Despite the inconsistency of how dermatomes were expressed at the surface of the body, Mann still held onto nerves as both the anatomical basis and physiological expression of acupuncture. He studied maps illustrating idealized patterns of deep pain and observed, "It is interesting to note that the majority of acupuncture points on the abdomen, thorax, and back are near the mid-ventral and mid-dorsal lines."[135] This observation also motivated Mann to invoke embryological orientations, further noting that sensitive sites clustered around a dorsal-ventral axis. Mann relied on developmental biology to reject meridians along with its Yin and Yang orientations. For him, "acupuncture points" were only active sites on a neurophysiological atlas.

By 1977, Felix Mann fully embraced neurophysiological patterns to articulate how needling worked within and across the body. On the interior flap of his book *Scientific Aspects of Acupuncture*, he announced, "[This book] shows that the traditional 'acupuncture point' does not exist, that the 'meridians' are imaginary lines, and that the 'laws of acupuncture' are largely mythical. The reader will see that the traditional theoretical foundation of acupuncture is mostly incorrect."[136] Mann held that lines and points on a body map were fundamentally misleading. Rather than a single line that led to a single point, paths and points shifted across the body, sometimes corresponding to dermatomes, areas of deep pain, or reflex arcs.[137] These all explained how disorders operated through nervous and electric impulses. Although Mann still used classical meridian men as practical guides for teaching acupuncture and moxibustion, he denied them any material reality. Ontological resemblances were enough. Dermatomes resembled meridians enough to replace them entirely. Images of sensitive sites around the dorsal-ventral axis were conceptually reliable enough to replace Yin and Yang orientations.

Meanwhile, electrodes read superficial currents, sometimes correctly, sometimes incorrectly, sometimes failing entirely. Facing his own unsuccessful attempts to detect acupuncture sites electrically, Mann turned to incomplete, inconsistent clinical and experimental sensation data. Dermatomes, referred pain, and reflex arcs overlapped with meridian men, and rather than validating them, supplanted them.

In his private letters to Lu Gwei-djen and Joseph Needham, Felix Mann directed their attention to dermatomes and Henry Head's early work on referred pain.[138] This intrigued Needham. He researched further and discovered Head's original papers archived in the Royal Society of Medicine Library on cutaneous tenderness, herpes zoster, and disease-induced analgesia. Impressed, Needham described Head's work as "classical for its subtlety and originality."[139] Lu and Needham reproduced a modified version of Head's dermatome maps in their book *Celestial Lancets*, although they did not consider neurophysiology as a means for replacing meridians (**Figure 5.9**).

Mann often mailed updated editions of his books to Lu and Needham with handwritten notes highlighting new chapters on neural theory and needling techniques. When Lu Gwei-djen read through Mann's updated chapters on radiation and referred pain, she remained unimpressed. "I don't like his use of the word 'Radiation,'" she wrote in the margin.[140]

The politics that promoted needling at nerves sustained a volatile relationship. Lu and Needham urged their readers to appreciate the embodied reality of acupuncture analgesia even when variations of the procedure complicated their translation efforts. Meridians, which Lu and Needham did not translate as meridians, were not purely neurophysiological expressions but could support neurophysiological theories. Although acupuncture and moxibustion did not engage with nerves alone, they did engage with the unexplained elements of neurophysiology that neuroanatomy did not address.

Acupuncture analgesia had inspired serious epistemological inter-

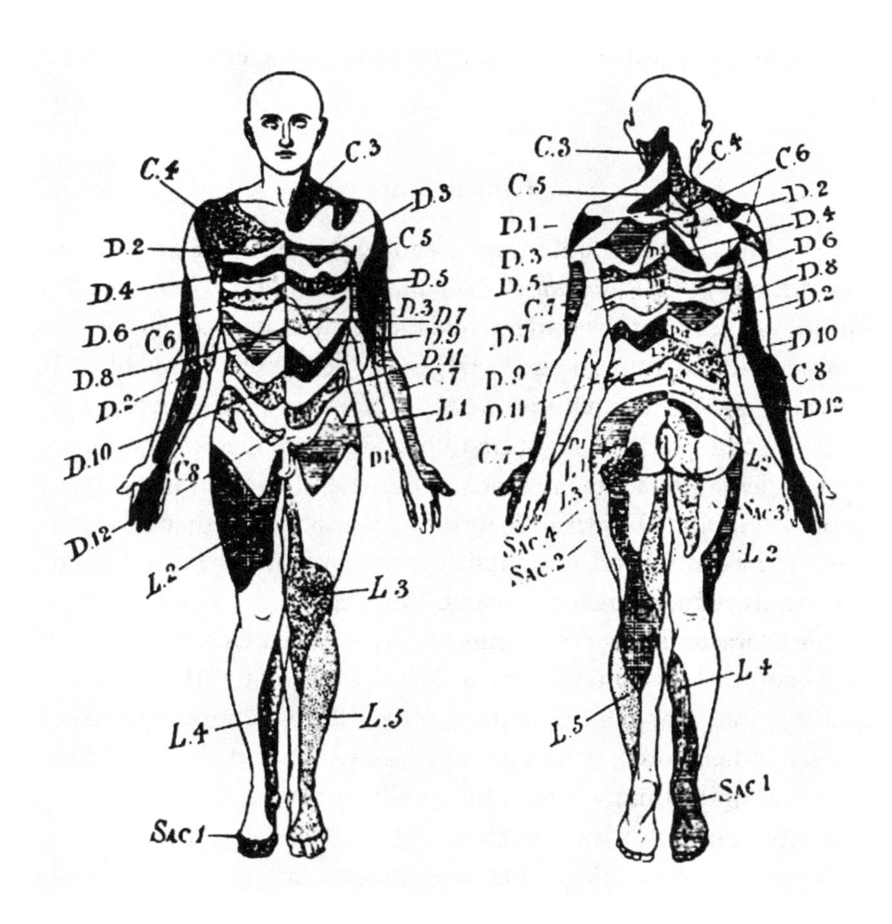

Figure 5.9. Reproducing Henry Head's zones, twentieth century. Unknown artist, illustration. Foerster's "Head's zones" were reproduced by Lu Gwei-djen and Joseph Needham. Although Lu was less inclined to compare meridians to nerves, Needham added that "the elaborate work of Henry Head published in the ensuing years became classical for its subtlety and originality" (207). From Lu and Needham, *Celestial Lancets*, 1980, 206, fig. 47

est, but its range of results both validated and undermined Lu's attempts to render Chinese medicine theoretically commensurate to biomedicine. The ontological similarities between meridian paths, embryological differentiation, and referred pain conflicted with the political binary of "East" and "West." The same political discourse that popularized acupuncture analgesia also drew boundaries around it, expanding its aesthetic appeal while limiting its ontological possibilities.

Conclusion

In 1980, Lu Gwei-djen and Joseph Needham published *Celestial Lancets: A History and Rationale of Acupuncture and Moxa*. It introduced a history of acupuncture and moxibustion as well as its contemporary reincarnation as acupuncture analgesia. They wrote: "This book is primarily about acupuncture, one of the oldest and most deeply characteristic of the techniques of traditional-Chinese medicine. . . . That the needles stimulate many kinds of nerve-endings and nociceptors is today undoubted, with all that this may imply of further repercussions in the central and autonomic nervous systems."[141] Lu and Needham portrayed acupuncture-moxibustion among "the oldest and most deeply characteristic" components of Chinese medicine, even though they knew that its national popular identity relied on early twentieth-century Cold War politics and modern technoscientific infrastructures. In other words, "Chinese" medicine existed between extremes. It was culturally unique and ontologically universal; it was medically innovative and historically unchanging.

These sets of paradoxes likewise trapped Lu and Needham between binaries of tradition and modernity. They marked Chinese medicine as an oriental object and a modern technique where needles could "stimulate many kinds of nerve endings and nociceptors." The ability of acupuncture and moxibustion to manipulate nerve endings and nociceptors served to "sketch a rationale of it in terms of modern knowledge."[142] This suggested that historically, physicians had always needled at nerves. Neuroanatomy rationalized acupuncture. It offered a logical framework while stripping it of its temporality. Despite their rhetorical confidence, Lu and Needham remained ambivalent towards the promise of nerves.

In *Celestial Lancets*, Lu Gwei-djen went on to discuss *jingluo* (經絡), which is spelled *ching-lo* in the Wade-Giles system. Lu emphasized that "well-defined points" facilitated the material movement of

Qi through hollow cavities.[143] She cited classical texts listing hundreds of bodily sites, cautioning that these existed "neither in the skin or the flesh, neither in the muscle nor the bones."[144] They were somewhere in between, somewhere else. Despite lying beyond muscle and bone, Lu emphasized that *ching-lo* were anchored to Yin and Yang areas of the body, which were themselves material objects. She described Yin as "palpable and material" and Yang as "formless and insubstantial," even though these dichotomies were not absolute.

Lu also included a modified illustration of the embryological sketch that she had made with Joseph Needham. The previously unarticulated bulge now looked amphibious (**Figure 5.10**). In this rendition, animal bodies universalized Yin and Yang. They modeled its embryological origins and mapped Yin and Yang areas not as relative spaces but as discrete areas on a body. For Lu, *ching* and *lo* ran underneath the surface of the skin like clusters of nerves and capillaries. She elaborated: "We have spoken of stars and railways but the really classical analogy was with earthly water-works. There is no doubt that in the *ching-lo* system we have to deal with a very ancient conception of a traffic nexus with a network of trunk and secondary channels and their smaller branches. From the beginning these were thought or in terms analogous to those of hydraulic engineering, involving rivers, tributaries, derivate canals, reservoirs, lakes, etc."[145]

Lu navigated between and translated among physiological structures through ontological resemblances. Meridians looked like nerves yet were not nerves. They acted like hydraulics and waterways yet were not pipes. They were like a city yet were not a city. These ontological resemblances enabled the temporal and technical variations of acupuncture analgesia to support new "gate control" theories of pain. Electric probes transformed acu-moxa sites into currents; controlling gates rendered meridian paths into neurotransmission; dermatomes translated meridian paths into spinal levels. These models conveniently correlated to the parallel paths that moved down the back of the bladder meridian man.

Still, these comparisons did not make sense. For Lu, meridian *tu* afforded more meaning than neurophysiological charts. In other

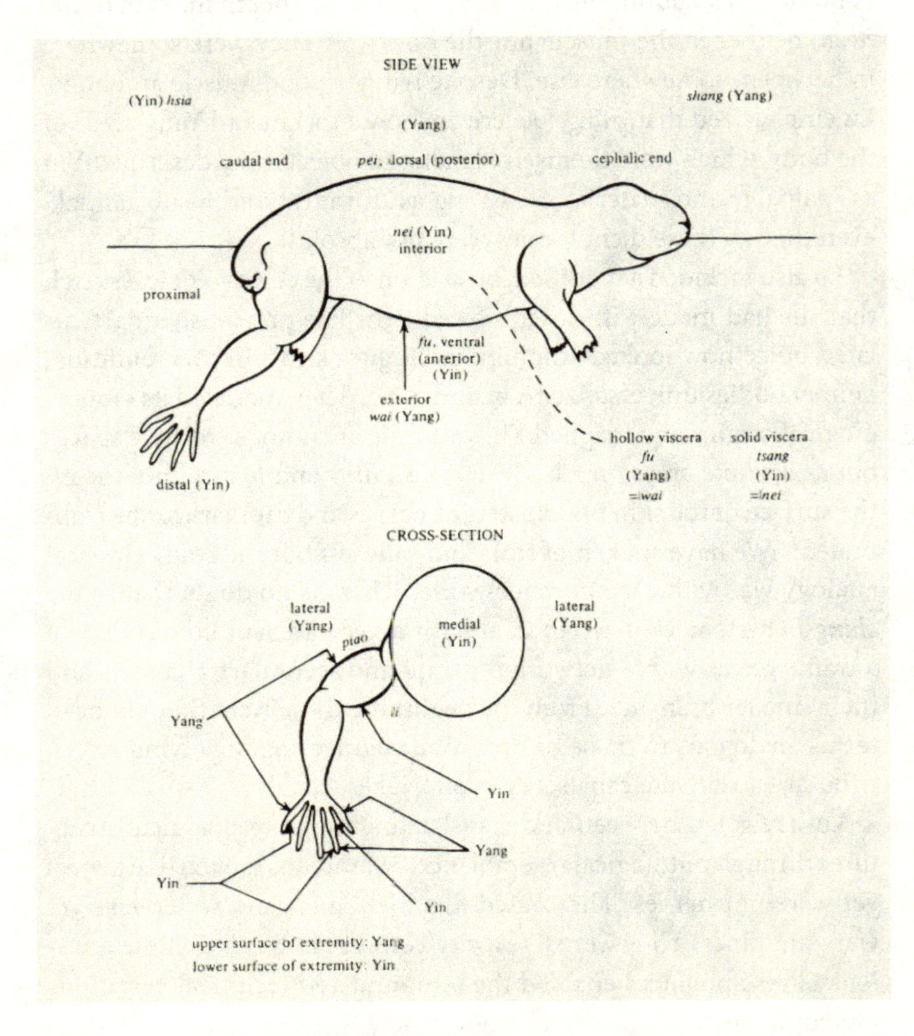

Figure 5.10. Animal orientations, twentieth century. Lu Gwei-djen (1904–1991) and Joseph Needham (1900–1995), illustration. To universalize Chinese medical orientations, Lu and Needham depicted them on an embryological blob. The original caption read, "Chinese usages of thought about Yin and Yang, inner and outer, above and below, etc., in relation to the mammalian body." From Lu and Needham, *Celestial Lancets*, 1980, 38, fig. 9

words, body maps were not qualitatively identical. They did not take the same form even when they referenced the same body parts, such as the head, hands, feet, and torso. Different areas of the body did different things. It was not that meridians could not be translated or explained but that they introduced new information; they presented new phenomena and new cosmologies. Lu's and Needham's messy labeling of Yin and Yang areas on the fingers, for instance, went beyond embryologic orientations of cephalic, caudal, distal, and proximal parts. Although the body remained the "same," with a head, hands, and feet, they could also give rise to Yin and Yang meridian pairs.

In this body, meridians as material *ching* and *lo* needed more conceptual unfolding. However, they became problematic when the political, discursive, and practical appeal of nervous stimulation and needling-induced numbness overshadowed historical epistemology. At stake for Lu was establishing Chinese medicine as Chinese, which backfired in the spectacular accounts of acupuncture analgesia. Lu had hoped to arouse a sense of urgency and revolution by integrating "all the true powers discovered in both China and Europe."[146] Yet, popular discourses that determined the "true" power of scientific practice allowed ontological uncertainty within some forms of knowledge but not others. Modern maps meant to facilitate diplomatic exchange could not resolve these differences.

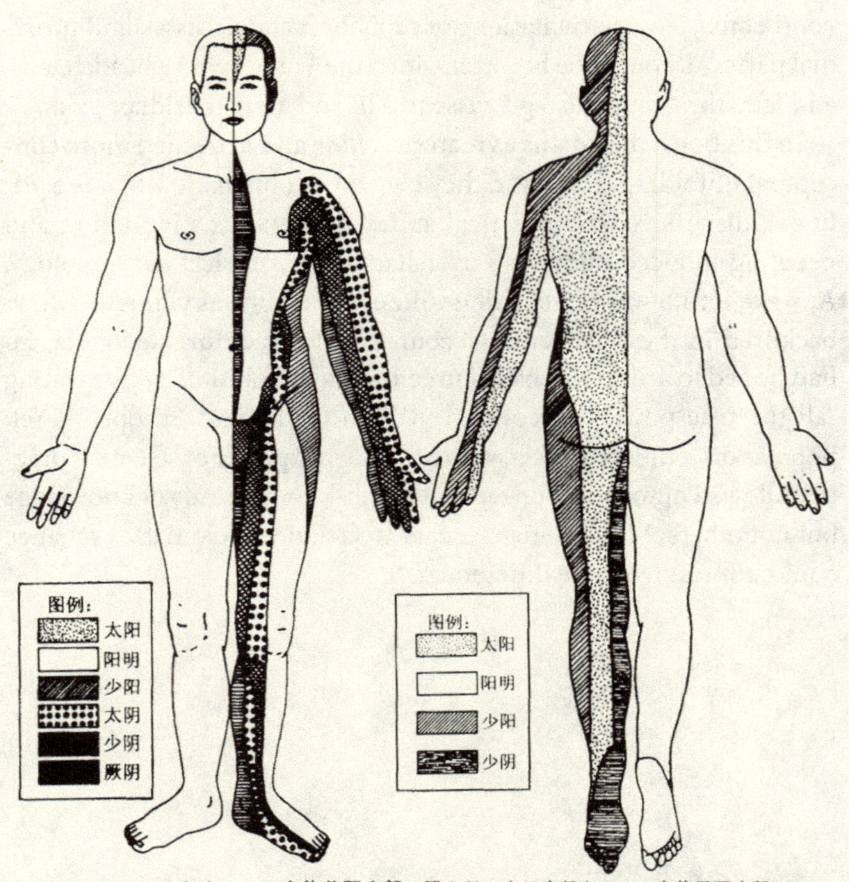

图 3-71　十二皮部之一——身体前面皮部　图 3-72　十二皮部之二——身体后面皮部

Epilogue

Shortly before his death, the illustrator Wang Xuetai edited a 2003 volume featuring over 600 figures and 60 color prints. Titled *Illustrated Guide to Chinese Acupuncture-Moxibustion*, the book involved 16 collaborating authors. Among the images in the book was Yin-Yang-territory man.[1] He floated at the center of the page, facing toward and away from the reader. The caption described the floating man showing 12 territories (*bu*) on the skin (*pi*). He was split in half, with one side left blank and the other side fully marked. Much like the split view of Henry Head's sensation man, the floating Yin-Yang man offered a double vision of the body. The blank half suggested an unencumbered surface. The shaded half suggested discrete expressive zones that were each as different as the varieties of lines and dots that defined them.

Figure 6.1. (opposite) Floating Yin-Yang man, twenty-first century. Wang Xuetai (1925–2008), ink on paper. The floating skin man stylistically referenced dermatome maps. The caption described 12 territories *bu* (部) on the skin *pi* (皮). The front of the body displayed territories defined by six kinds of Yin and Yang: *tai yang*, *yang ming*, *shao yang*, *tai yin*, *shao yin*, and *jue yin*. The back of the body displayed four kinds: *tai yang*, *yang ming*, *shao yang*, and *shao yin*. From Wang, *Zhonghua zhenjiu tijian*, 2003, 71–72

The floating Yin-Yang man made a direct stylistic reference to dermatome maps. His hands faced the viewer at the front; his feet hung from his body when turned away. Below the image, the caption explained that shaded areas represented 24 distinct territories (even though only 20 appeared in the legend). Like Henry Head's territories of affected sensation, these strips of skin had distinct, bounded, and fixed borders. Unlike Head's sensation man, the bands on floating Yin-Yang man did not correspond to spinal levels. They followed a different organizing principle. The front of the body displayed six types of Yin/Yang territories: *tai yang*, *yang ming*, *shao yang*, *tai yin*, *shao yin*, and *jue yin*. The back of the body displayed four types of Yin/Yang territories: *tai yang*, *yang ming*, *shao yang*, and *shao yin* (**Figure 6.1**).[2]

Wang's floating man was a strategic construction. He situated Yin and Yang paths within the visual grammar of neurophysiology. It extended beyond the embryological orientations that Lu Gwei-djen and Joseph Needham had emphasized in their writing. This image gave the body texture. The exterior composition defined interior spaces more clearly than Lu's and Needham's discursive distinction between Yin and Yang as different material qualities of *nei* (inner, but not inner) and *wai* (outer, but not outer). In this image, Wang's Yin and Yang sections were fixed on the page; they were discrete enough to give rise to Yin and Yang meridians. The fluid body that shifted beneath the skin appeared stable enough.

Wang Xuetai had spent over half a century drawing meridian maps.[3] He urged against relying on nerves as an explanatory mechanism for Asian forms of health and healing. Even if the therapeutic effects of needling and cauterization resembled viscerocutaneous reflexes, these reflex responses could not account for the range of diagnostic and therapeutic effects in historical and contemporary techniques. "[T]his is only a small portion of the larger practice," Wang explained in a recorded interview, adding, "many people are still confused about this."[4] Needling at nerves to treat chronic pain and induce anesthesia captured the popular imagination, while many other phenomena remained unaddressed.[5]

Like Wang, other physiologists have increasingly distanced themselves from relying on neurophysiology as the primary mechanism to explain needling therapies. For instance, research teams in the United States have examined the effects of needling in connective tissue.[6] Studies have examined how different types of needle rotation initiate different kinds of intracellular responses, such as activating cytoskeletal reorganization.[7] Rather than claiming alliance with an individual anatomical structure or physiological system, connective tissue has offered researchers an alternate space to conceptualize meridian paths and sites.

As physiologists have abandoned their efforts to conceptualize meridian paths as nervous structures, other inconsistencies abound. For instance, needling, heat, and electric stimulation are each unique forms of sensory input that invoke different kinds of cells and circuits. In other words, the presumed mechanisms for needling could not be generalized to heating; the cellular implications of heating could not be generalized to electric stimulation. Researchers have identified these generalizations as some of the "fundamental gaps" in understanding how different stimuli along meridian sites can register similar therapeutic effects.[8]

Neuroscientists have turned their attention to sensory organs beyond nerves in the last two decades.[9] For instance, researchers specializing in the science of touch have examined the role of keratinocytes, one of the main types of skin cells, in detecting physical and chemical stimuli.[10] At the innermost layer of the epidermis, Merkel cells have been identified as non-neuronal cells that play a role in distinguishing shape, form, and texture.[11]

Meanwhile, the ongoing search for biomarkers remains.[12] Laboratories supporting acupuncture (not always moxibustion) research have increasingly focused on identifying dynamic biomarkers that provide more sensitive assessments of physiological states.[13] Needling practices are not as remarkable as their physiological implications. In 2015, the Defense Advanced Research Projects Agency received a $5.4 million grant to research "bionodes," where electrically stimulating the vagus nerve on the neck induced an anti-inflammatory response in the body.[14] While researchers claim the mechanism to be completely unknown, acupuncture researchers have suggested that *jingluo* (經絡) may hold the answer.

Meridian *tu* (圖) operated as a technology for facilitating vision and evidence. It has persisted across long histories of translation that have accumulated numerous asymmetries and false equivalences between anatomical structures and physiological expressions. Across these long histories, improvising meridians on paper served as a way to navigate between textual traditions and their graphic counterparts. Improvisation through creating hand-drawn images served as a means for knowledge production. It allowed scholars like Lu Gwei-djen and illustrators like Wang Xuetai to create new images showing discrete Yin and Yang areas on the body. It allowed them to use visual representations of Yin and Yang territories to ground meridian paths. Yin and Yang side-stepped the promises of neuroanatomy and neurophysiology. It offered a fidelity to a seemingly indigenous East Asian body. Yin and Yang translated what biomedicine could not—if Yin and Yang could even be considered for facilitating translation.

Expressive structures introduced a range of moving targets. Body maps existed at the intersection of cultural, political, and social practices. Each of these maps manifested experience while also directing expectations for how to experience the body. Mapping served as a nebulous space of speculation and experimentation. Here, the body could not be fully known despite being fully depicted.[15] It was in the inscription of a line—a graphic line that lacked form, texture, and tangible quality—where ontology erupted. These lines took on different meanings when they were curved, straightened, thinned, thickened, broken, colored, or shaded.[16]

The tyranny of lines was not unique to representing meridians but applied to neurophysiology more broadly. Illustrations of the reflex arc as a single path from the moment of touch to the centers of cognition have spawned manifold interpretations. Concepts like the "labeled line" and "pattern" theory have extensively characterized neurophysiological pathways. Although less discursively popular than the reflex arc, "labeled line" and "pattern" theory has graphically underpinned nearly all small-scale neurophysiological images tracing neuronal circuits. Like images of the reflex arc, the labeled line served as a metaphor for mechanisms of neuronal coding. As much as neuroscientists have tried to follow the sequence of physiological movements inside the body, they still recognized that lines were "dirty" at best.[17]

Meridian *tu* and sensation maps encountered one another under a whirlwind of social and political instability. With acupuncture analgesia, practitioners black-boxed meridian paths using gate control theory as one potential explanatory mechanism without knowing how it actually worked. Patients who received needle-induced numbness could not escape from the language of nerves. Even now, acupuncture anesthesia remains a major neurophysiological event that eludes neurophysiological explanation. Researchers continue to track sites and tease apart sinews, publishing papers on how needling therapies can address neuritic and spasmodic pains, including lumbago, neuralgias, pain from amputation stumps, and dysmenorrhea, as well as allergies like asthma or eczema. Although anesthesia is not a total phenomenological mystery, its means of delivery are—especially through needling and heating.

These mysteries remain unresolved in part because nerves were historically hard to explain. Sensation man was by no means a stable representation of sensation. Henry Head's maps inspired a fascination with how sensation changed over time. The boundaries that Head inscribed on paper and his own body were highly volatile. He tried to develop a new theory of sensation through self-experimentation, but his conclusions remained open-ended. Surgeons and psychologists replicated Head's methods but failed to resolve the challenges of perceiving, describing, and inscribing sensation. Even early-career psychologists like Elsie Murray recognized that hierarchies of sensation more likely existed as "shades" on a spectrum. Rather than dealing with the shades of perception, neurosurgeons modified Head's 1893 sensation man and folded his work as part of the discovery of dermatomes. These discrete sections of the body corresponded to specific levels on the spine and provided a way to account for individual differences by generalizing variations in sensation observed during recovery. In effect, these modified maps offered an organizing principle that presented a static view of the body, fixed in time.

Interrogating body maps highlights the tensions among image, object, and epistemology. Interrogating body maps further highlights the role that they played in the political, social, and cultural impetus for creating a unified and a distinct form of "Chinese medicine" that rendered it exceptional. Similarly, the growing field of neuroscience has rendered neurons both timeless and essential for understanding the brain. Physicians and researchers navigating these systems have fixated on articulating how they work. Yet, whereas practitioners of a "Chinese" medicine await the revelation of new mechanisms to legitimate acupuncture and moxibustion, neuroscientists take on questions of mechanism as opportunities for new experimental design and discovery. These approaches are not parallel. As knowledge systems remain fluid with ever-porous boundaries, their practices stay diverse.

These splintering divergences remain relevant. In 2020, the World Health Organization established its Global Centre for Traditional Medicine and convened the inaugural Traditional Medicine Global Summit in 2023.[18] We have been here before, with large organizations pledging to incorporate evidence-based traditional medical knowledge into mainstream practice. Like most public health initiatives, these programs aim to expand access to care and improve health outcomes, especially for underserved communities. The stakes for global healthcare are high. Yet in these endeavors, familiar binaries emerge: East/West, South Asian/East Asian, rich/poor, science/tradition, and evidence/experience. Binaries bind us to difference. And difference further ossifies in public health discourse, compressing complex histories to make room for fantasies of innovation and newness.

As diverse and integrative practices continue to evolve, they will continue to rely on improvisatory practices to make sense of historical texts, translations, commentaries, and the numerous images accompanying them. Hand-drawn images will also persist. Improvisation as a way of knowing, as a mode of conceptual adornment, and as an aesthetic practice will continue to shape epistemic judgment.

Scholars will continue to attend to the social circumstances that determine the conditions of image-making, use, and interpretation. Moving forward, applying analytical frameworks attentive to critical feminist, post-/anti-/de-colonial studies can only serve to understand the sprawling implications of how practitioners create, select, and present images as evidence. It will better articulate how body maps and body *tu* manifest difference and reveal historical contingency. Rather than an account of stabilizing diversity, these images expose a history in motion that remains to be fully seen.

ACKNOWLEDGMENTS

I turned to the history of science and medicine in Asia as a way of connecting to my grandparents. My paternal grandfather, Li Xingbai, was an artist and art critic. My paternal grandmother, Liao Jialun, was a librarian and music teacher. Research trips to China allowed me to spend time with them as they aged and passed. I thank my parents, who deepened my appreciation for the relationship between music theory, material cosmology, and the history of science. Chi Li, my mother, and Guangming Li, my father, have been my collaborators, cheerleaders, and sympathizers. Their commitment to their craft inspires my pedagogy and research. I thank my maternal grandparents, Li Zhenxiang and Wang Zuixian, for their genuine surprise each time I visit. I thank my sister, Xuan Li, who reminds me that I am an artist (when she is the artist). I thank my extended family whose presence in my life reminds me to reach beyond academia.

I thank Dorothy Ko, Shigehisa Kuriyama, and Debbie Cohen for creating a foundation for me as a young scholar. Special thanks to my graduate school friends for their brilliance. Amy Johnson, Emily Harrison, Lisa Haushofer, Renee Blackburn, Mitali Thakor, Shreeharsh Kelkar, Nicole Labruto, Marie Burks, Canay Özden-Schilling, Tom Özden-Schilling, Clare Kim, and Jia-Hui Lee have profoundly shaped my intellectual world. I thank my graduate committee members, Emma J. Teng for her critical insight, David S. Jones for his close reading, Erica C. James for her attention to my sources, and Robin W. Scheffler for his consistent generosity that allowed me to survive the writing process. Special thanks to Harriet Ritvo, Christopher Leighton, Mike Fischer, Stefan Helmreich, Eugenia Lean, Danian Hu, and Zuoyue Wang, who commented on my research and offered guidance.

I thank Chris Walley and Chris Boebel for their course, DV Lab, which allowed me to connect my work to contemporary practitioners through film.

This project also benefited from the care and attention of colleagues and advisors whose abundance of knowledge about medicine, healing, and the body I have attempted to emulate. Conversations with Charlotte Furth, Vivienne Lo, Linda Barnes, Gao Xi, Ma Bo Ying, Bridie Andrews, Roberta Bivins, Yan Liu, He Bian, Sean Lei, David Luesink, Marta Hanson, Nicole Barnes, Max Moerman, Angela Leung, Daniel Trambaiolo, James Flowers, Claire Sable, Shan Jiang, and Ted Kaptchuk each left me thoroughly inspired. My sincere thanks to Andrew Macomber and Kevin Buckelew for thoughtfully engaging with this book and offering comments that have significantly improved it.

Lilla Vekerdy at the Smithsonian, Martin Heijdra at Princeton, Ma Xiao-he at Harvard, Katherine Fibiger and Martin Brennan at UCLA, Kelly Caldwell at Moody Medical Library, Philip Montgomery at the Texas Medical Center Library, Jet Prendeville at the Frondren Library, Chengzhi Wang at the Weatherhead East Asian Institute, and Michael Seminara at the Welch Medical Library generously offered their expertise when I had questions about anything from esoteric collections of classic medical texts to image use and copyright. At MIT, Michelle Baildon guided me through finding aids. Mabel Chin Sorett tolerated my many receipts. Karen Gardner was a powerhouse. A special note of gratitude goes to John Moffett at the Needham Research Institute, who played a tremendous role in shaping my research on Lu Gwei-djen.

Support for this work was made possible by the Paul & Daisy Soros Fellowship for New Americans, the Wellington & Irene Loh Grant, the NSF Graduate Research Fellowship, and the NSF Dissertation Research Grant. These generous sources of funding have allowed me to travel to archives that would have been beyond reach, including the China Academy of Medicine, China Center on Acupuncture Moxibustion Research, Needham Research Institute, and Wellcome Library, among others. I thank the Weatherhead Institute of East Asian Studies (Columbia), Heyman Center for the Humanities (Columbia), the Chao Center for Asian Studies (Rice), and the Human Research Center (Rice) for further funding that supported this work.

Research for this project has required trips to China, where I learned a great deal from new friends and colleagues. I am particularly grateful to Director Wang at the People's Liberation Army Second Artillery Corps General Hospital for facilitating introductions to physicians in Beijing. Madame Ye also generously offered her unconditional assistance in arranging meetings at the Institute of Acupuncture and Moxibustion, where He Wei and Li Liang significantly expanded my knowledge of the Museum on the History of Acupuncture Moxibustion in China. I am also grateful for the guidance of Huang Taiji at the Chengdu College of Chinese Medicine, Huang Longxiang at the China Institute of Chinese Medicine and Acupuncture Moxibustion Research, Li Xiuzhi at the Chengdu Second People's Hospital, Zhang Hui at the Chengdu Second People's Hospital and Yang Laoshi at the Hunan College of Chinese Medicine and Acupuncture Moxibustion. Finally, Cheng Baixian at Tongji University has generously supported this project ever since I first approached him in 2009.

Special thanks to Prakash Kumar, Projit Mukharji, and Amit Prasad for running the Decolonizing Science in Asia Workshop at Penn State, where I workshopped this preliminary work with feedback from Juno Salazar Parreñas, Dwai Banerjee, Bharat Venkat, Nicole Barnes, Lijing Jiang, Charu Singh, and Ran Zwigenberg. During my time as a Presidential Scholar in Society and Neuroscience at Columbia, I benefitted from generative conversations and mentorship from Kathryn Tabb, Eugenia Lean, and Kavita Sivaramakrishnan. Pamela Smith allowed me to attach myself to the Making and Knowing Lab and learn from its expansive community. Special thanks to Naomi Rosenkranz, Sophie Pitman, and Tillmann Taape for sharing their intellect and talent. Ellen Lumpkin welcomed me as a technician in her skin lab. Martin Picard opened me to the world of mitochondria. I learned endlessly from my colleagues, Kara Marshall, Ann-Sophie Barwich, Matteo Farinella, Nori Jacoby, Carmel Raz, Heidi Hausse, Andrew Goldman, and Federica Coppola. I thank Anthony Acciavatti for, in addition to being a model of ambitious creativity, inviting me to share early version of this work at the Princeton-Mellon Research Forum. I thank Debbie Cohen, who facilitated generous feedback from the History of Science & Medicine Colloquium at Yale.

I continued to work on this manuscript as a faculty at Rice. I thank Kathleen Canning for sponsoring a book workshop with comments from the illustrious Yi-Li Wu and Nancy Rose Hunt, as well as Eric Huntington and Farshid Emami, who provided invaluable insight into the manuscript. I am particularly indebted to Elizabeth Festa, who coached me through my writing with incredible patience and insight. I am also grateful to my colleagues in the History Department, including Caleb McDaniel, Alida Metcalf, Ussama Makdisi, Elora Shehabuddin, Lisa Balabanlilar, Carl Caldwell, Nate Citino, Daniel Cohen, Daniel Domingues Da Silva, Moramay López-Alonso, Paula A. Sanders, James Sidbury, Maya Soifer Irish, Lora Wildenthal, Fay Yarbrough, William Suárez-Potts, Sayuri Guthrie Shimizu, Kerry Ward, Aysha Pollnitz, Michael R. Maas, Randal L. Hall, Alexander Byrd, Beverly Konzem, and Marcie Newton, who welcomed me as a new faculty.

I am also deeply grateful to Kristen Ostherr, who built the medical humanities program at Rice and has been an unwavering supporter of my work. I thank Mary Horton and Ricardo Nuila for their life-changing conversations. At Rice, I started the Medicine Race Democracy Lab and thank Eddie Jackson, Sophia Peng, Taylor Philips, Jason Lee, Josh Harper, Katherine Wu, Summer Nguyen, and Brandon Ba for their early participation. Special thanks to Liz Petrick and Elizabeth Brake, who extended unimaginable generosity of friendship during COVID. Shoutout to Rodrigo Ferreira and Yael Loewenstein.

I thank Tani Barlow who has guided me to the ends of my wit and back. Izumi Nakayama has extended safety to me as an early-career scholar. Pierce Salguero has taught me the importance of self-realization. Projit Mukharji and Manjita Mukharji have inspired me with their compassion and love for their cats. I thank Alex Wragge-Morley for his uninhibited mind, Tianna Uchacz for her tremendous ambition, Bharat Venkat for his world-bending ideas, Lisa Haushofer for her indomitable integrity, Shireen Hamza for her wells of wisdom, and Victoria Massie for her shared attention to the stars.

At Hopkins, I am in great company. Jeremy Greene, Elizabeth O'Brien, Graham Mooney, Mary Fissell, Nathaniel Comfort, Sasha White, Yulia Frumer, Loren Ludwig, Lijing Jiang, and Tobie Meyer-Fong, thank you for your collegiality. None of my work would have been possible without the very capable Maggie Cogswell, Marian Robbins, Michael Seminara, and Danielle Stout. I am grateful to editors at the Johns Hopkins University Press, particularly Matt McAdams for ensuring that this book would be a beautiful object, Ahmed Ragab for facilitating reviews for the manuscript, Adriahna Conway for her care in the final edits, Robert Brown for expertly managing the master files, Deborah Bailey for her meticulous copyedits, and David Prout for compiling the index. I thank Elizabeth Brake, Bharat Venkat, Alex Wragge-Morley, Joe Vignone, and Anand Venkatkrishnan for offering valuable assistance with copyedits. I thank the many participants at the Hopkins History of Medicine Colloquium, who commented on a draft of chapter 4, and fellows at the Center for Black Brown and Queer Studies for commenting on a draft of the introduction. Special thanks to Fan Mang and Lauren Kim for assisting this book in its final stages. I thank Michael Stanley-Baker for generously reviewing the glossary and Farshid Emami for suggesting that I have one in the first place. I thank Laurel Crawford and the TOME grant committee for making this book open access and the Alexander Grass Humanities Institute (AGHI) and the American College of Acupuncture and Oriental Medicine (ACAOM) for providing additional book subvention funds. Shoutout to John Paul Liang, Gregory Sparkman, Thalia Micah, and Chris Hammond for helping me connect theory and practice. I am also grateful to Karen Darling at the University of Chicago Press for her preliminary interest in and support for the manuscript, as well as to the anonymous reviewers for their time and care in reading this manuscript.

Finally, I thank my cohabiting animals (Bunbun, Fez, Meowmi, and Coco), who have opened my heart. To other non-human material things: the six laptops that have sustained my research over 12 years, my Scrivener subscription, the dozens of hard drives that held my ideas, and Keynote and Photoshop for giving me a place to play. Thank you to the planes, buses, trains, and cars that have safely transported me abroad and delivered me back home. Thank you to the many cups of hot chocolate that comforted me when a chapter seemed impossible to complete. I thank the trees. I thank the blankets that protected me from winter drafts. And last but not least, I offer thanks to those who I have not listed specifically here. Beyond the tremendous support of those who have helped make this manuscript possible, any other lapses in gratitude or errors in judgment are evidence of a work in progress.

GLOSSARY A

Key Concepts

body map—A category of study in this book that refers specifically to hand-drawn anatomical atlases. These include drawings of meridian *tu* (圖) and illustrations of disturbed sensation. This idea draws directly from critical cartography to address body maps as constructed artifacts of visualization, empiricism, and imagination. Body maps actively construct rather than neutrally represent and express the imaginative distortions of their creators. They naturalize a body in constant fluctuation and mediate discursive, therapeutic, and epistemic practices.

generic images—A subcategory of genre that refers to recognizable, stereotyped images with established names, such as *mingtang tu* (明堂圖). As a mode of classification, generic images represent subjects and objects that are genealogically related, like the many versions of "meridian men" or "sensation men." They provoke debates around ontology through their stereotyped, named forms that direct perception and action. They are distinct from graphic genres.

genre—A mode of classification borrowed from literary theory and used in this book as an analytical tool to compare graphic images and their represented objects. Drawing from Carolyn Miller's work on genre as social action (1984) and Anis Bawarshi's expansion on genre function (2003), this book considers how images provoke responses and organize social action through their form. As ideological types that direct perception, genres are unstable categories enforced by how images are used, sorted, stored, and sold. In contrast to overly descriptive approaches, this book uses genre as a theoretical framework to examine the relationship between an image's form, action, and ontology.

graphic genre—Refers to the stylistic conventions and elements of *tu* (圖) and maps that mediate experiences and representations of the body. Reflecting particular logics of classification, graphic genres like anatomical diagrams present stereotypical, generic images that guide perception and action. The body maps examined in this book constitute graphic genres that frame social action through their form. Graphic genres operated as forms of life that structured historical modes of seeing, moving, and touching the body.

improvisation—A way of knowing characterized by the hand-drawn body maps examined in this book. Similar to stylized jazz performances, these images contained "expression marks" from intentional and accidental embellishments within established aesthetic constraints. This improvisatory subjectivity invoked by hand-drawn anatomical atlases was not free of invention but relied on routines, rituals, and practiced techniques. It represented a rigorous mode of knowledge production that evaded other technologies, such as analog and digital photography.

jing (經)—A shorthand referring to *jingluo* (經絡) and *jingmai* (經脈). The same character also appears in the titles of philosophical, medical, and religious texts, used to designate a "classic" or canonical work. The character *jing* is also used to denote philosophical treatises, medical corpora, and religious sutras.

jingluo (經絡)—The Pinyin romanization of 經絡 referring to corporeal and ephemeral objects that facilitated therapeutic practices for protecting and enhancing the body in East Asia. Appears as *chinglo* in Wade-Giles and pronounced *keiraku*, *kinh tuyến*, and *jaoseon* in Japanese, Vietnamese, and Korean. Etymologically, the characters *jing* (經) and *luo* (絡) refer to weaving. Structurally, *jingluo* comprised larger vessels (*luomai*) and smaller branches (*jingmai*), including divergent channels (*jingbie*). *Jingluo* were old objects referenced in second-century BCE manuscripts and tomb artifacts, portrayed as 11, 12, and 14 primary *jing*. Later texts described 8, 10, and 12 *jing* pairs. Accessible through the mind, skin, and gut, 12 primary *jing* were engaged by practitioners through palpating the wrist to reach 12 primary organs.

jingmai (經脈)—Therapeutic and diagnostic channels described as ancillary to *jingluo* (經絡), comprising a system of "auxiliary" paths not typically illustrated graphically. Referred to in this book as "flesh meridians." Some forms of *mai* (經) included a flesh radical (月) and *xue* (血), a substance resembling blood. Other forms included 永 to reference fluid but not blood. The word *mai* has been equated with "pulses," as seen in the title of a treatise on pulses, *Mai Jing* (脈經), and it appears as *kei* (經) and *myaku* (脈) in Japanese. Early texts did not distinguish between channels containing blood and channels containing Qi.

meridian *tu* (圖)—A generic reference to illustrations in East Asian medical texts depicting *jingluo* (經絡). As a graphic genre, meridian *tu* served as instructional diagrams that also determined particular ways of seeing and engaging an interior landscape. Examples include *mingtang tu* (明堂圖) and images associated with authors of the Northern Song dynasty like Wang Weiyi (987–1067) or with un-authored texts like *Huo ren shu* (活人書). In contrast to drawings on paper, paths cast on bronze sculptures are known as *tongrentu* (銅人圖), or "bronze body *tu*."

meridians—A metaphorical term used in this book to describe graphic representations of *jingluo* (經絡) and *jingmai* (經脈). Limiting my analysis to meridians as graphic representation acknowledges the simultaneous materiality and invisibility of these corporeal structures and highlights how visual genres shaped historical discussions of ontology and location. This term is not meant as a direct or literal rendering of *jingluo* and does not capture the full range of therapeutic paths described in texts. Historically, "meridian" has been a contested term among Sinologists. Other scholars have offered metaphors like "sinarteries" or "reticular conduits" to capture branching structures, which exist as a legacy of early modern translations of *jingluo* as arteries.

mingtang tu (明堂圖)—A prominent set of meridian illustrations depicting 12 paired paths in the torso and limbs and two special meridians. As a graphic genre, these images operate as cultural symbols and didactic diagrams, typically arranged in three to four columns showing the top, side, and back of the body. Occasionally accompanied by an image of the organs. With origins in the Sui and Tang dynasties, *mingtang tu* are associated with Sun Simiao (d. 682) and Wang Tao (670–755). As a graphic lineage, they have diversified across medical traditions while retaining recognizable features. *Mingtang* translates to "illuminated hall," referring to a room in imperial palaces where emperors convened advisors.

nei (内) and ***wai*** (外)—*Nei* (内) is often translated as "inner" and *wai* (外) as "outer." In this book, chapter 5 takes *nei* as an actor's category where Lu Gwei-djen (1904–1991) considered the duality of "inner" and "outer" to be more complex than denoting interior and exterior spaces. Lu found that these terms were not merely relational descriptions but also referred to material concepts (see **Yin** and **Yang**). For Lu, *nei* represented corporeal matter, whereas *wai* encompassed something beyond the corporeal. Lu privately maintained that *nei* and *wai* held deeper material implications in Chinese anatomy and grounded the corporeal aspects of structures like *jingluo* (經絡) and practices like cultivating longevity (*neidan* 内丹).

nerves—Historically, structures that registered humoral fluctuations and were animated by invisible animal spirits in Greco-Arab medicine. From Galenic taxonomy to Enlightenment mechanics, nerves persisted as speculative objects. Thirteenth- and fourteenth-century natural philosophers articulated nerves as material while struggling to localize mental pathologies. In early modern Europe, nerves offered a seat for the rational soul, defining the body-mind. Debates in the eighteenth century emerged as philosophers theorized nerves through hydraulic models. In Edo Japan, scholars selectively adopted terms resembling "nerves," such as *seibinseinon* (セイビンセイノン) and *zeinun* (ゼイヌン), to elaborate Asian healing traditions rather than a new conception of nerves.

Qi (氣, [illegible], 暣, 炁, 気, 餼)—A critical and complex entity in Chinese cosmology and physiology, associated with breath, vitality, primordial matter, food, stars, and more. Qi has over 35 definitions in modern dictionaries, over 230 compound words beginning with Qi, and countless more prefixing it. Historically, Qi is related to cosmic formation and primordial chaos in the fourth century BCE. It becomes part of medical cosmology in the third and second centuries BCE, and it becomes a coherent foundation to East Asian medicine in the first- and second-century CE corpora *Huangdi neijing*. It is associated with neo-Confucianist principles in the twelfth century; it becomes a material for transformation in the nineteenth century; and it is a highly debated entity in the twentieth century, with many medical textbooks in the 1950s simplifying Qi, emphasizing physiology over cosmology.

shinkei/shenjing (神經)—A term popularized by the Japanese physician Sugita Genpaku (1733–1817) to describe "nerves" in his 1774 text *Kaitai shinsho* (解体新書). *Shinkei/shenjing* consists of the composite words *shin/shen* (神) meaning an extra-human entity and *kei/jing* (經) denoting a network of pathways. This neologism rendered the ethereal nervous system responsible for animating the material body of flesh, skin, bone, and blood. Other Japanese scholars had previously offered terms for "nerves" such as *seibinseinon* (セイビンセイノン) used by Hōan Arasayama in 1683 and *zeinun* (ゼイヌン) by Sōtei Nakamura in 1684. However, Sugita's *shinkei/shenjing* has become a definitive standard term.

tu (圖)—An East Asian graphic genre translating loosely as "image-text" that engages the imagination. As a specialist term denoting a style of representation, *tu* encompassed maps, diagrams, and illustrations across domains including music, law, ritual, and medicine. As Zheng Qiao (1104–1162) noted, *tu* facilitated the meaning, power, and intention of the text. *Tu* offered templates for action and for communicating technical knowledge. Not equivalent to paintings (*hua*) or likenesses (*xiang*), *tu* were functional images expressing knowledge through their use. Following aesthetic principles as hand-drawn or printed genres, *tu* were made to be studied and enjoyed, embedded in rituals of looking and doing. *Tu* guided gesture, practice, and performance. In this book, meridians as graphic objects existed in *tu*.

Yin and **Yang**—Common binary motifs with textual origins preceding the Warring States period (481–221 BCE) in texts such as the *Book of Changes*, or *Yijing* (易經), that initially referred to movements of the sun and moon and extended to dichotomies like light/dark, male/female. Yin and Yang would, around the fifth and sixth centuries, become fundamental motifs that structured geographic and temporal space, including interior corporeal landscapes. They categorized blood, bone, flesh, meridians, organs, and diseases. Yin and Yang related the body to environment through patterns, like seasonal outbreaks of Yang diseases in winter and Yin diseases in summer. They are capitalized here to denote their role orienting meridian systems. For instance, Yin and Yang classified *jingluo* (經絡) paths and connected them to Yin and Yang organs which determined the names of the 12 meridian pairs illustrated in *mingtang tu* (明堂圖).

GLOSSARY B

Other Sinographic Terms

The entries below are Romanized words paired with Sinographic characters for terms, names, and titles mentioned in the book, including both Wade-Giles and Pinyin according to the historical context in which the terms appear in the book. A full translation of the titles of works is found in the bibliography.

Arashiyama Hōan 嵐山甫安
bagua 八卦
Bankoku chiryōhō ruiju 蕃国治方類聚
beibu 背部
Benjamin Hobson 合信
Bian Que 扁鵲
bu 部
bu naozhi 補腦汁
Chang Chi 張機
Chen Hanbiao 陈汉标
Chen Wang 陈旺
Chen Xiutang 陳修堂
Cheng Dan'an 承淡安
Cheng dan'an jianyi jiu zhi - dan fang zhiliao ji
承淡安簡易灸治 - 丹方治療集
chi 尺
chikaku 知覚
ching 精
Ching Tsong 景宗
ching-lo/keiraku/jingluo 經絡
chōhōki 重宝記

chongmai 衝脈
Chou Kung 周公
Chou Li 周禮
cun 寸
Cun zhen huan zhong tu 存真环中圖
daotian 道天
Dejima 出島
di qi (Earthly Qi) 地氣
dōmyaku 動脈
Dong Biwu 董必武
fei 肺
fen 份
fu (abdomen) 腹
fu (viscera)腑
fubu 腹部
fujie 腹結
furen 伏人
fushe 府舍
gan (liver) 肝
Gao Juefu 高觉敷
gaoji shenjing huodong 高级神經活动
go jingi 五神器
gokan 五官
Gotō Konzan 後藤艮山
guan/kan 管
Guan Maocai 管茂才
Guben jiaozhu, Shisi jing fahui 古本校注, 十四经发挥
gui 鬼
guishen 鬼神
Guo Moruo 郭沫若
Guo Yicen 郭一岑
Guo Yizhen 郭一岑
Han Yü 韓愈
Haogu 好古
hegu/hoku 合谷

henggu 横骨
hou (esophagus) 喉
Hu Se-Huei 忽思慧
Hua Boren 滑伯仁
Hua Shou 滑壽
Hua Tuo 華佗
Huainanzi 淮南子
Huang Chujiu 黄楚九
Huangdi neijing 黄帝内經
huaying 華英
hui 會
ishin 胃神
Ishinpō 醫心方
Itō Jinsai 伊藤仁斎
jianjing/katai 肩井
jiao 焦
jiaoqi/Chio Ch'i 腳氣
jimai 急脈
jing qi (Essential Qi) 精氣
jingluo buju ye 經絡不拘也
jingluo zhi qi 經絡之氣
jingshen 精神
jiuwei 鳩尾
Kagawa Shuan 香川周庵
Kaitai shinsho 解體新書
Kajiwara Shōzen 梶原性全
Katsuragawa Hoshū 桂川甫周
ke 刻
kei/jing 經
ketsumyaku 血脈
Kim Bong Han 김봉한
Kinsō no sho 金瘡書
kirikami 切紙
Kōmō hiden geka ryōjishū 紅毛秘伝外科療治集
Leung a-Fat 梁亞發

liang 量
liangshou liu mai 两手六脉
liu qi 六氣
Lu Gwei-djen 魯桂珍
luoyi bu jue 絡繹不絕
Maeno Ryōtaku 前野良沢
mai/myaku 脈
maiguan 脈管
Manase Dōsan 曲直瀨道三
mian mian 綿綿
monmyaku 門脈
Mubun 夢分
myaku/Miak 脈
nanban-jin (Southern Barbarians) 南蛮人
nao 腦
naoqi 腦氣
nei tan/nei dan 內丹
neiguan 內觀
neijing (inner canon) 內經
neijing (internal view of the body) 內景
neijing tu 內景圖
nishiki-e 錦絵
Odano Naotake 小田野直武
Ogata Kōan 緒方洪庵
Ogyū Sorai 荻生徂徠
oranda tsuji オランダ辻
Oranda zen mukuro naigai-bun aizu 和蘭全軀內外分合図
Oranda zenku naigai bungōzu 和蘭全軀內外分合図
Ozeki San'ei 小関三英
Pan Shicheng 潘仕成
Pan Shu 潘萩
pi (skin) 皮
pi (spleen) 脾
Pinyin 拼音
qi (spirits) 祇

Qian Xinzhong/Jin Xin-Zhong 钱信忠
Qianjin yifang 千金翼方
qijiu 氣僦
Quanti xinlun 全體新論
Ren Zuotian 任作田
renmai 任脈
renzheng mingling tu 正人明堂图
riyue 日月
san 散
sanjiao 三焦
Satake Yoshiatsu 佐竹 義敦
seibinseinon セイビンセイノン
seinon セイノン
sekitsui 脊椎
shangdi 上帝
shanghan 傷寒
Shanghan lun 傷寒論
Shanghan zabing lun 傷寒雜病論
shangqu 商曲
shen 腎
shen/shin 神
Shen Nung 神農
shenming 神明
shenqi 神祇
shenque 神闕
shensheng 神聖
shidou 食竇
shigeki 刺激
shimen 石門
shishin 視神
shishō 刺触
Shisijing fahui 十四經法會
shou 手
shou shao yin xinjing 手少陰心經
shou taiyin fei jing 手太陰肺經

shoujing 手經
shuifen 水分
Siegen K. Chou 周先庚
sishi qi 四時氣
Sōtei Nakamura 中村宗填
suan 酸
Sugita Genpaku 杉田玄白
Sugita Seikei 杉田成卿
sui hai zhi yin zhi zai tong wei di 髓海至陰之在通尾骶
suidō (waterways) 水道
suji 筋
Sun Simiao 孫思邈
Suwen (Simple Questions) 素問
Taixi renshen tushuo 泰西人身圖說
Takano Chōei 高野長英
Tanba no Yasuyori 丹波康頼
Tang Zonghai 唐宗海
tian 天
tianchi/Тянь-чи 天池
Tokudaiji Saneatsu 徳大寺実淳
Tongren shuxue zhenjiu tu jing 铜人腧穴针灸圖經
Ton'ishō 頓医抄
tou 头
uchibari 打針
Waitai miyao 外台秘要
Wang Bing 王冰
Wang Haogu 王好古
Wang Tao 王燾
Wang Weiyi 王惟一
Wang Xuetai 王雪苔
wei 胃
wei qi 衛氣
Wu Jianren 吳趼人
wu qi 五氣
Wu Youxing 吳有性

xi 系
xiang 相
xianghu douzhen 相互斗争
Xifang zi mingtang jiu jing 西方子明堂灸經
xin 心
xin qi 心氣
Xin zhenjiu xue 新针灸学
xiong 胸
xizhiluo (fine branched network) 細支絡
Xu Dachun 徐大椿
xue/chi 血
xueguan 血管
xun jing kao xue bian 循經考穴編
Yamawaki Tōyō 山脇東洋
Yang Jie 楊介
yang qi 陽氣
yangren 仰人
Yasuyori Tanba 丹波康頼
Yi Yin tang ye Zhongjing guang wei dafa 伊尹汤液仲景广为大法
ying qi 營氣
yingwei qi 營衛之氣
Yoshimasu Motoki 本木良意
Yoshimasu Tōdō 吉益東洞
Yu Yan 余巖
yuanqi 元氣
yunmen 雲門
zang 臟
zang qi 臟氣
zeinun ゼイヌン
Zengding zhongguo zhenjiu zhiliao xue 增訂中國鍼灸治療學
Zentai shinron 全体新論
zhang 丈
Zhang Panshi 张磐石
zhen qi (True Qi) 真氣
zheng qi (Ordering Qi) 正氣

Zhenjiu dacheng 针灸大成
Zhenjiu jiayi jing 针灸甲乙經
Zhenjiu jicheng 針灸集成
zhenjiu mazui 针灸麻醉
Zhenjiu xue shouce 针灸学手册
zhi-yue/Жи-юе, or *riyue* 日月
zhong (heavy) 重
zhongfa 中法
Zhonghua zhenjiu tujian 中华针灸图鉴
zhongxi 中西
zhongyi kexuehua 中醫科學化
zhongying 中英
Zhou Enlai 周恩来
Zhu De 朱德
Zhu Lian 朱琏
ziran zhishu 自然之數
zong qi 宗氣
Zōshi 蔵志
zu 足
zu tai yang pangguang jing 足太陽膀胱經
zu tai yin pangguang jing 足太阴膀胱经
zujing 足經
zusanli 足三里
zuyangming 足阳明

NOTES

Preface

1. In addition to the important texts on dissection and visualization by Katherine Park, Daniela Bleichmar, Projit Mukharji, Vivienne Lo, and Sean Lei, the literature on wax figures and dissection in Euro-American sources to which this passage refers includes the following (non-exhaustive list): *Courtiers' Anatomists* by Anita Guerrini, *Fabric of the Body* by K. B. Roberts and J. D. W. Tomlinson, *Dissection* by John Harley Warner and James M. Edmonson, *Ephemeral Bodies* edited by Roberta Panzanelli and Julius Schlosser, *Model Experts* by Anna Katharina Maerker, *Anatomy Museum* by Elizabeth Hallam, and *Malleable Anatomies* by Lucia Dacome.

2. The common refrain was *zhongyi meiyou jiepou* (中医没有解剖), although the history of dissection (or *jiepou* as dissection instead of "anatomy") can be considered a slightly different element of this story.

3. Here, the Pinyin romanization is not meant to be an accurate representation of how the characters are pronounced historically. Even now, languages related to the Sinographic sphere can pronounce *jingluo* (經絡) as *keiraku*, *kinh tuyến*, or *jaoseon*.

4. For an excellent review of the many types of images of channels, see Huang and Wang, "Acupuncture Illustrations," 2022.

5. Practitioners may recognize the international conventional nomenclature for *ermen* (耳門) as Sanjiao 21 (SJ21). Later in this book, I follow the transformation of site names in the twentieth century, where meridian paths were meant to mimic spinal levels, and reformers increasingly formalized numbering sites along the path.

6. A number of public lectures have attempted to explain the effects of acupuncture-moxibustion through nerves. See Zhu, "Essence of Acupuncture," 2015.

7. Molecular research on thermosensory and mechanosensory pathways discusses the ways in which intracellular pathways triggered by temperature or touch can be specialized or overlapping. On isolated and overlapping pathways, see Tan and Katsanis, "Thermosensory and Mechanosensory Perception," 2009; Ghitani et al., "Mechanosensory Nociceptors," 2017; McKemy, "Molecular and Cellular Basis," 2013.

8. Recent work in neuroscience has demonstrated an increasingly broad taxonomy of sensory cells, some of which are not typically considered neuronal cells such as Merkel cells and keratinocytes. These kinds of cells initiate a complex set of pathways that neuroscientists have attempted to map as individual paths or a network of paths. For a review of this field, see Lumpkin and Bautista, "Feeling the Pressure," 2005;

Lumpkin, Marshall, and Nelson, "Cell Biology of Touch," 2010; Xiao, Williams, and Brownell, "Merkel Cells and Touch Domes," 2014; Bouvier et al., "Merkel Cells Sense Cooling," 2018; Moehring et al., "Cells and Circuits of Touch," 2018; Sadler, Moehring, and Stucky, "Keratinocytes and Temperature Sensation," 2020.

9. Needles and moxa sticks are two different *kinds* of things that initiate two different *kinds* of sensations. Needling at the skin might be described as initiating mechanosensation; applying burning moxa on the skin might be described as initiating thermosensation. Considering these two kinds of input—of touch and temperature—generates a more complex picture.

10. In this book, I begin with meridian *tu* (圖) from the early modern period, attributed as a product of Song dynasty (~960–1279) printing practices.

11. I build on Kuriyama's seminal book *The Expressiveness of the Body* in spirit and in style. Other scholars such as Geoffrey Lloyd and Nathan Sivin have also contributed a great deal to thinking through deep comparative histories of medicine. I am also indebted to their scholarship. See Kuriyama, *Expressiveness of the Body*, 2002; Kuriyama, "Between Mind and Eye," 1992; Lloyd and Sivin, *Way and the Word*, 2002.

Introduction

1. Special thanks to Ingrid Lena Klein, who kindly facilitated my conversations with Nilza.

2. My interview with Nilza Domingues can be found in a documentary film, available at https://lan-a-li.com/Acupuntura-e-Moxabustao. I provide additional commentary on the interview in an essay forthcoming in *Asian Medicine* titled "*Acupuntura e Moxabustão*: Introducing Community Clinics in São Paolo."

3. For a survey of historical case studies of moxa, see Wilcox, *Power of Mugwort Fire*, 2008; Wilcox, *Moxibustion*, 2009.

4. Domingues, interview, 2013.

5. For a survey of acupuncture in South America, see Freidin, "Acupuncture in Argentina," 2013.

6. These concerns are related to my ongoing work as a documentary filmmaker from 2011 to 2021. Some interviews include the following: Beaton Starr, Harvard Medical School, August 6, 2012; Rolly Brown, via Zoom, July 30, 2021; Brendan Carney, Harvard Medical School, July 28, 2012; Baixian Cheng, Tong Ji University, August 15, 2009; Nilza Domingues, ATTA, São Paulo, Brazil, March 27, 2013; Carlos Eduardo, São≈Paulo, Brazil, March 27, 2013; Strambio Fabiana, São Paulo, Brazil, March 27, 2013; Jie Fan-Roche, Harvard Medical School, July 28, 2012; R. B. Gogate, IASTAM India, Mumbai, India, March 27, 2012; Caitlin Hosmer Kirby, Harvard Medical School, July 30, 2012; Hong Ji, Chengdu University of Traditional Chinese Medicine, July 30, 2014; Ted Kaptchuk, July 26, 2015; Matthew H. Kowalski, Harvard Medical School, November 8, 2012; Helene Langevin, Harvard Medical School, March 13, 2013; Donald B. Levy, Harvard Medical School, July 4, 2012; Li Cheng, Shanghai, China, August 18, 2009; Li Liang, Institute of Acupuncture and Moxibustion, April 24, 2015; Li, Shanghai, China, August 18, 2009; Li Xiuzhi, Chengdu Second People's Hospital, August 20, 2009; John

Paul Liang, Houston, Texas, August 4, 2021; Li Liang, Monterey Park, California, December 28, 2013; Shen Ping Liang, Houston, Texas, July 30, 2021; Mitchell Lynn, via Zoom, August 4, 2021; Bo-Ying Ma, Cambridge, United Kingdom, October 25, 2014 and March 1, 2015; Swati Mohite, IASTAM India, Mumbai, India, March 27, 2012; Randall Paulsen, Harvard Medical School, July 12, 2012; Fabiana Strambio, São Paulo, Brazil, March 27, 2013; Fang Wang, China People's Second Army Hospital, March 28, 2014; Zhiqing Wang, Monterey Park, California, December 27, 2013; Peter Wayne, Harvard Medical School, July 30, 2012; Jing Yang, Hunan University of Traditional Chinese Medicine, April 28, 2015; Hui Zhang, Chengdu Second People's Hospital, August 20, 2009.

7. This is a reversal of Bruno Latour's claim of having never been modern, which still undermines historical epistemology as being as un-modern as contemporary epistemic practices. Latour, *We Have Never Been Modern*, 1993.

8. Farquhar, "Chinese Medicine as Popular Knowledge," 2013.

9. See Scheid, "Remodeling the Arsenal," 2002; Scheid, *Chinese Medicine in Contemporary China*, 2002.

10. Although the ideas conveyed here are in conversation with an expanding "global" history of science, it does not represent a "global" history. Fa-ti Fan and Carla Nappi have urged scholars to critically assess the limitations and opportunities of "global" perspectives. For instance, Fa-ti Fan has pointed out how transmission histories follow themes set in motion by figures such as Joseph Needham and George Sarton. See Nappi, "Disengaging from 'Asia,'" 2012; Fan, "The Global Turn in the History of Science," 2012.

11. When considering the actual routes of the Silk Routes, Valerie Hansen's work has described a network of smaller routes primarily facilitating cultural exchange within Central and East Asia. Trade was largely centered around Chinese imperial wealth transfers to northwestern garrisons during the Tang Dynasty, rather than sustained long-distance commerce. See Valerie Hansen, *The Silk Road*, 2015.

12. Diplomacy did not always guarantee conversation. See Lightman, *Circulation of Knowledge*, 2013.

13. This work has been particularly well-researched with Lei, *Neither Donkey nor Horse*, 2014; Andrews, *Making of Modern Chinese Medicine*, 2015.

14. Here, I draw inspiration Pierce Salguero's multi-decade work that tracks a global history of Buddhist medicine. See Salguero, *A Global History of Buddhism and Medicine*, 2022.

15. In examining the transmission rates of medical knowledge, Ronit Yoeli-Tlalim's recent work provides an excellent account of medical histories in Central Asia. Her study primarily focuses on Tibetan medical manuscripts from Dunhuang and the Hebrew Book of Asaf. See Ronit Yoeli-Tlalim, *ReOrienting Histories of Medicine*, 2021.

16. Mei Zhan has famously borrowed from Pierre Bourdieu in her term, "worlding" Chinese medicine. See Zhan, *Other-Worldly*, 2009. For a fantastic social history of Chinese medical practitioners in the United States, see Venit Shelton, *Herbs and Roots*, 2019.

17. Hsu, "Tactility and the Body," 2005.

18. Pulse taking could deficiencies that manifested in symptoms like "wasting thirst" or hot flashes. For an analysis of case studies dealing with wasting thirst and hot flashes from Chunyu Yi's memoir, see Hsu, *Pulse Diagnosis*, 2010.

19. *Jing* (經) also appeared in the title of philosophical treatises, medical corpora, and religious sutras. It has been used to describe a "classical" text in addition to appearing in name of the treatise itself. Guangming Li has also pointed out that the famous set of Four Books and Five Classics were called the *sishu wujing* (四书五經), which provided standards for thought, morality, and behavior. Among the "five *jing*," only two included the word *jing* in their title. These classics included *Shi Jing* (诗經), *Shang Shu* (尚书), *Li Ji* (礼记), *Yi Jing* (易經), and *Chun Qiu* (春秋). Li, "Concerning the English Translation of 脈經 (*Mài Jīng*)," 2020.

20. Even early modern translations of longitude use the word *jing* (經). Historically, East Asian maps did not include longitude. See Yee, "Reinterpreting Traditional Chinese Geographical Maps," 1994. Recent scholarship has highlighted the metaphorical relationship between weaving and writing in early China, arguing that Han writers conceived of literary composition in weaving terms. This is reflected in key terminology emerging by the second century BCE, such as "classics" (*jing*), "weft-writings" (*weishu*), and "literature/texts" (*wen*), which draw on the language of warp threads and textile production. Archaeological evidence like weaving looms in the Tianhui tomb support the material origins of this metaphor. See Zürn, "The Han *Imaginaire* of Writing as Weaving," 2020.

21. Huang Longxiang and Wang Fang have translated *luomai* (絡脈) as "collaterals," which are related to *jingbie* (經別), translated as "divergent channels," located in the *pibu* (皮部), translated as the "muscle region." These paths appear in medical texts but are not illustrated in woodcuts and are rarely used in acupuncture and moxibustion practices. Huang and Wang, "Acupuncture Illustrations," 2022, 189.

22. This, of course, builds on the somewhat Kantian conception of beauty that Amitav Ghosh established in his seminal novel, *The Glass Palace*, 2001. To further situated Ghosh's work within postcolonial studies, see Su, "Amitav Ghosh and the Aesthetic Turn in Postcolonial Studies," 2012.

23. This is with the obvious exception of *Ishinpō* (醫心方) by Tanba Yasuyori (丹波康頼) (912–995). On a deep history of gender in medical iconography, see Wu, "The Gendered Medical Iconography of the Golden Mirror," 2008. We will see the ways in which some of these bodies were gendered and sexless later in the book. On gendered bodies in visual representation, see Park, *Secrets of Women*, 2010.

24. For a guide on reading different kinds of images, Elkins, *How to Use Your Eyes*, 2000.

25. Sources from antiquity that offered sophisticated articulations of meridians described the ways in which they were involved in cultivating longevity or treating illnesses. In particular, the *Maishu* (脈書) excavated from the Zhangjiashan (張家山) burial site offered the earliest and most comprehensive written description of piercing the skin to influence the movement of Qi (氣) in the body. See Lo, "The Influence of

Nurturing Life Culture on the Development of Western Han Acumoxa Therapy," 2001, 21.

26. Early silk and bamboo manuscripts portrayed 11 meridians, while corpora like the *Huangdi neijing* described 12 primary meridians. Other written sources described *jing* in 8, 10, or 12 pairs. Eventually, sources related to Daoist classical body practices described eight "extraordinary" or "auxiliary" paths; Buddhist treatises in Tibet described four paths. On the medical channels in tantric texts, see Gyatso, *Being Human in a Buddhist World*, 2015.

27. The 2023 special issue of *Asian Medicine*, titled "Classical Medicine: New Insights from Laoguanshan Cemetery" and edited by Vivienne Lo, Shelley Ochs, and Dolly Yang, presents groundbreaking articles offering detailed analyses of archaeological findings from the Tianhui and Laoguanshan tomb sites. The series provides a comprehensive overview of these excavations, which primarily comprised four Western Han tombs where archaeologists discovered relics and writings on bamboo slips. To give context, these studies complement existing research on medical documents and artifacts excavated from other significant sites, including Mawangdui (馬王堆) (tomb closed 168 BCE), Zhangjiashan (張家山) (tomb closed 186 BCE), and the first-century BCE and CE tomb sites at Mianyang (綿陽) and Wuwei (武威). For a detailed introduction to this research, see: Vivienne Lo, "Classical Medicine," 2024. Particular articles relevant to understanding the deep history of *jingluo*, see Lo, "Looms of Life," 2024, and Liu et al., "Bamboo Slip Medical Manuscripts Excavated from Tianhui Township, Sichuan," 2024.

28. The *Huangdi neijing* (黃帝內經) is best characterized as a compilation of theories and practices around the first century BCE. These classical themes were identified in the imperial bibliography as the "inner" and "outer" canon of the Yellow Emperor (*Huangdi neijing* and *Huangdi waijing*), although we will see in chapter 5 how translations of these titles were contested in the 1980s based on historians' insistence on the corporeal nature of the body in classical medical theory.

29. Although this book follows different iterations of images throughout time and geographic space, it also shows how images took on different meanings. This follows the idea that technologies and technological objects take on new meanings among different social contexts. See Arnold, *Everyday Technology*, 2013. On transmissions, see Secord, "Knowledge in Transit," 2004; Chambers and Gillespie, "Locality in the History of Science," 2001.

30. Technically, the 12 *jing* refer to meridians on one side of the body. Thus, in the *mingtang tu* (明堂圖), there are 24 primary meridians in the body, with two special meridians (*renmai* and the *dumai*) circling from the front to the back of the body. In the early modern period, the 12 *jing* were later canonized as 14 *jing* in texts like the fourteenth-century *Treatise on the Fourteen Meridians*, or *Shisi jing fahui* (十四經发挥) (1341, print 1528), which Po-Huei Hsieh, in his excellent master's thesis, has translated as *The Study of the Fourteen Cardinal Tracts*. Hsieh, "Paratextual Advantages of Versified Pulse Texts," 2019.

31. Historians of medicine have yet to work out how *mingtang tu* emerged as its

own graphic lineage, other than defining it based on what it is not. For instance, full-body meridian paths in *mingtang tu* stand in contrast to meridian paths illustrated on bronze sculptures and then transferred onto paper, images which are known as *tongren tu* (銅人圖), or "bronze body *tu*."

32. Here, *dao*, or "the Way," is also a polysemous category that historically extended from many different schools and practices, which are often conflated. For an example of this plurality, see Sivin, "On the Word 'Taoist' as a Source of Perplexity," 1978.

33. In its long history, *mingtang tu* (明堂圖) has become one of the most prominent set of meridian men, but still one of many other genealogies of meridian men that were attributed to many other authors, texts, and books. Huang Longxiang and Wang Fang introduce the wide variety of what they call "channel diagrams," including ten channel diagrams, point diagrams, bronze statues [*tongren tu* (銅人圖)], and images associated with individual authors such as Wang Weiyi (王惟一) (987–1067) of the Northern Song dynasty, Yang Jie (楊介) (1060–1113), Zhu Gong (朱肱) (1050–1125), or associated with unauthored books, such as *Huo ren shu* (活人書). See Huang and Wang, "Acupuncture Illustrations," 189.

34. To this end, practitioner-scholars like Volker Scheid have worked to identify the distinct genealogical networks of knowledge production. On these different genealogies, see Scheid, "The Globalisation of Chinese Medicine," 1999; Scheid, *Chinese Medicine in Contemporary China*, 2002; Scheid, "Traditional Chinese Medicine—What Are We Investigating?" 2007.

35. This is not to discount works that track the circulation of ideas, texts, material objects, and people among regions, which include the following: Elman, "Sinophiles and Sinophobes," 2008; Kim, "Specialized Knowledge," 2010.

36. In this book, I capitalize Yin and Yang (rather than italicize) to emphasize their role in orienting the meridians. The Romanization for Qi (氣) and Blood (血) are capitalized for the same reason. Evidence of the extensively developing ideas of Yin and Yang appear during the Warring States period (481–221 BCE). Yin and Yang referred to movements of the sun and moon and later extended to binaries including light and dark, male and female. It also came to organize geographic space and direction, and time. These kinds of naturalistic cosmologies came to reshape the spectrum of health and healing. See Lo, "The Influence of Nurturing Life," 19.

37. I take cosmology as meaning a framework for how individuals understood causal relationships. This draws on work by religion scholar Eric Huntington, who compares different cosmological images through different representations of the universe among regions related to the Himalayas. I, too, take the approach of comparing cosmologies from a humbler view of the human body. See Huntington, *Creating the Universe*, 2019.

38. This particular version has *mingtang tu* (明堂圖) dated to around 1301.

39. I will later describe my use of "genre," but the metaphor of class explicitly engages with discourses of genre studies. See Fishelov, *Metaphors of Genre*, 1993.

40. This was particularly the case for the 1682 set of meridian men in attributed to Michel Boym (1612–1659) and Andreas Cleyer (1634–1697?). If we look closely, all of

the men were most likely copied rather than traced, engraved, and facing the "wrong" way. See the original images via the Wellcome Collection referenced here: Boym, *Specimen Medicinae Sinicae*, 1682, available at https://wellcomecollection.org/works/hhnfmzea.

41. The claim that meridian man and sensation man were the "same" further falls into the trap of generality, which philosopher Ludwig Wittgenstein has critiqued in his proposal of "family resemblances." Whereas Wittgenstein's family resemblances refer to the "language games" of the polysemous nature of an "individual" word, this can also apply to images that are likewise embedded in what Wittgenstein calls "networks of similarity." See Wittgenstein, *Philosophical Investigations*, 2009; Griffin, "Wittgenstein, Universals and Family Resemblances," 1974; Arrington and Glock, *Wittgenstein's Philosophical Investigations*, 1991.

42. Zhu, "The Essence of Acupuncture Moxibustion," 2015.

43. These epistemic barriers led to ontological consequences. A notable example is the divergent conclusions drawn by neuroanatomists Camillo Golgi (1843–1926) and Santiago Ramón y Cajal (1852–1934) regarding nerve structure, despite their use of the same staining technique developed by Golgi. Whereas Golgi perceived an interconnected nerve network, Ramón y Cajal observed discrete individual nerves. Golgi's objection to Ramón y Cajal's interpretation stemmed from concern that it neglected the temporal distribution of nerves throughout the body. For more on Golgi and his debate with Ramón y Cajal, see Daston and Galison, *Objectivity*, 2010, 115.

44. Michael Polanyi's work on the tacit ways of knowing demonstrates how empiricism alone could not account for the scientific method. See Polanyi, *The Tacit Dimension*, 1966.

45. Mukharji, *Doctoring Traditions*, 2016.

46. Bynum, "Why All the Fuss about the Body?" 1995.

47. Of course, many scholars who focus on the body have built on Foucault's excavation of the body as a site of control, Bourdieu's description of a body as the locus of social practice, Scheper-Hughes' and Lock's formulation of the body as a somatic, social, and subjectified entity. On the chronic instability of bodies, see Vilaça, "Chronically Unstable Bodies," 2005.

48. For instance, I argue there that meridians were members of a flexible and dynamic body, in terms of Yin, Yan, Blood (*xue*), and the five phases or agents (*wuxing*), which use wood, fire, earth, water, and metal to articulate different states of health and disease. Furth and Ch'en "Chinese Medicine and the Anthropology of Menstruation in Contemporary Taiwan," 1992.

49. We will see in the following pages how embodied perception of objects is a deeply personal one, as the individual—whether the experimenter or the experimental subjects thoroughly inhabits the experience of being inscribed upon. Biehl and Locke, "Deleuze and the Anthropology of Becoming," 2010.

50. Latour, "The More Manipulations the Better," 2014, 347.

51. Cosgrove, "Introduction: Mapping and Meaning," 1999.

52. Monmonier, *How to Lie with Maps*, 1991.

53. See Rankin, "Cartography and the Reality of Boundaries," 2010; Corner, "The Agency of Mapping," 1999.

54. On graphic ontology in maps, see Carusi and Hoel, "Toward a New Ontology of Scientific Vision," 212; Fall, *Drawing the Line*, 2005.

55. Lynch and Woolgar, *Representation in Scientific Practice*, 1990. On histories of the role and value of visual representations and the writing practices associated with them, see Cambrosio, Jacobi, and Keating, "Ehrlich's 'Beautiful Pictures,'" 1993; Rotman, Becoming Beside Ourselves, 2008.

56. Kitchin, Gleeson, and Dodge, "Unfolding Mapping Practices," 2013.

57. Crampton, "Maps as Social Constructions," 2001.

58. Many scholars have written on the sovereign power of maps in the context of statecraft in South Asia and the Middle East. See Mishra, "Critical Cartography and India's Map Policy," 2015; Leuenberger, "Map-Making for Palestinian State-Making," 2013.

59. See, for instance, Fraley, "Images of Force," 2011; Schilling, "British Columbia Mapped," 2014.

60. On an example of the varying and contradictory control of maps in North America, see Oliver, "On Mapping and Its Afterlife," 2011; Dodge, Kitchin, and Perkins, eds., *Rethinking Maps*, 2009.

61. Cartographers critical of cartography furthermore describe mapping as unique in its variety of semiotics. Compared to spoken and written language, maps take on a particular set of verbal signs, graphic signs, and physical signs. There is no single cartographic sign or map language in the same way that there is no Platonic ideal. For more recent critiques, see Andrews, "Introduction: Meaning, Knowledge, and Power," 2002; Clarke, "What Is the World's Oldest Map?" 2013; Monmonier, "History, Jargon, Privacy and Multiple Vulnerabilities," 2013.

62. For a recent review of critical cartography, see Edney, *Cartography*, 2019.

63. Daston and Galison, *Objectivity*.

64. Modern medical pedagogy generated demands for clinical case studies, so that the patient's body was often in view and always in the mind. Already, topographic maps needed bodies to make sense. One's own body served as a point of reference for cartographic space. Things could be smaller than the body, larger than the body, surrounding the body, or completely enveloping the body. This, according to psychologist Daniel Montello, placed the body was at the center of different kinds of cartographic spaces. Making sense of maps required an active relationship between one's self, one's cartographic reference, and one's environment. Montello calls this "figural," "vista," "environmental," and "geographical." Montello, "Scale and Multiple Psychologies of Space," 1993.

65. See Merleau-Ponty, *The Primary of Perception*, 1964.

66. Sociologists have also engaged with the body as a site of vulnerability. Turner, *The Body and Society*, 1984.

67. Thomas Csordas has offered the term "somatic modes of attention" to describe the ways of attending to and with one's body in particular surroundings that include

the embodied presence of others. Csordas, *Body/Meaning/Healing*, 2002; Csordas, *Embodiment and Experience*, 1994.

68. Georges Métailié had described how often there were no explicit references to the images themselves in the text, as if their mere presence were enough. See Métailié, "The Representation of Plants," 2007.

69. Bray, "Introduction: The Powers of Tu," 2007, 35; Bray, Dorofeeva-Lichtmann, and Métailié, *Graphics and Text in the Production of Technical Knowledge*, 2007; Clunas, *Pictures and Visuality in Early Modern China*, 2012.

70. This of course is in reference to McCloud's classic graphic book that offers both a broad and specific definition of comics that relies on a particular visual vocabulary. To be sure, McCloud's consideration of comics is not a historical critique on the temporal and cultural transformations within the genre. See Scott McCloud, *Understanding Comics*, 1994.

71. Bray, Dorofeeva-Lichtmann, and Métailié, *Graphics and Text*, 2.

72. For instance, early modern scholars like the famous Zheng Qiao (1104–1162) recognized that *tu* (圖) facilitated the meaning, power, and intention of written words. *Tu* and text were entangled, or as Zheng described, *tu* were the stabilizing "warp" threads that secured the words, or "weft" threads. Words created patterns that *tu* secured in place. Together, they participated in the art of learning. In her introduction to *tu*, historian Francesca Bray cites Zheng Qiao in referencing the importance of *tu* in the "technique (or arts) of learning, *xueshu* 學術." Bray further explains that specialist knowledge also appeared in the style of a *tupu* (圖譜), an illustrated register that offered a "sequence of rubrics where for each item a graphic illustration was paired with an explanatory text." See Bray, "Introduction," 2007.

73. The distinction between scientific and aesthetic images is addressed in the following volume: Galison and Jones, eds., *Picturing Science*, 1998.

74. See Kusukawa, *Picturing the Book of Nature*, 2011.

75. It is also important to note that historically, these forms of expression were printed using shared technologies of production and on the same type of media, such as silk, paper, wood, and stone.

76. To understand the material and cultural significance of woodcut images, or *ukiyo-e* (浮世絵) in Japanese medical practice, Jiang Shan has examined how this genre of popular art during the Edo and Meiji periods often depicted moxibustion. Jiang notes that moxibustion became a common subject for *ukiyo-e* artists, particularly in scenes portraying *yūjo* (遊女), a specific class of women, self-administering moxa. See Shan Jiang, "The '*Ukiyo*' of Moxibustion Reflected in the *Ukiyo-e*," 2024.

77. These were also representative of Ian Hacking's concerns with the relationship between materialism and realism in practices of representation, intervention, and histories of ontology. See Hacking, *Historical Ontology*, 2002; Hacking, *Representing and Intervening*, 1983.

78. Billeter, *The Chinese Art of Writing*, 1990, 47.

79. This is not to suggest that lines can be universally understood. They, too, remain historically contingent. This claim is expressed within a particular modernist

lens of early psychological studies in America. Gibson, *The Ecological Approach to Visual Perception*, 1979.

80. Bill Rankin has demonstrated in his history of the GPS how coordinates shifted the logic of representational mapping where coordinate grids that overlay landscapes and took on a life of their own. Rankin, *After the Map*, 2016.

81. Tim Ingold's "ephemeral" lines offered direction for how one would "walk" through an unfamiliar terrain. Ingold, *Lines*, 2007, 84.

82. Here, Rheinberger has described lines as projecting time and space onto a two-dimensional surface to explore new ways of organizing data. These scribbles further become new resources and natural products that give research projects their own specific contour. Rheinberger, *An Epistemology of the Concrete*, 2010.

83. For instance, late-nineteenth-century French physiologists who used lines to track bodies at work—breathing, running, and pulsing—were particularly embedded within an industrial economy that prioritized energy and efficiency. See Brain, *The Pulse of Modernism*, 2016. In her study of these images, art historian Inge Hinterwaldner has argued that the expanding distance between icon and inscription generated a tension in the role of images to depict and facilitate. In particular, Hinterwaldner explores the work of Etienne-Jules Marey (1830–1904) and Friedrich Ahlborn (1858–1937). See Hinterwaldner, "Parallel Lines as Tools for Making Turbulence Visible," 2013.

84. Cosgrove, "Introduction: Mapping and Meaning," 1999.

85. Corner, "The Agency of Mapping," 1999.

86. Rankin, "Cartography and the Reality of Boundaries," 42.

87. Fall, *Drawing the Line*, 2005.

88. Lynch and Woolgar, *Representation in Scientific Practice*, 1990. On histories of the role and value of visual representations and the writing practices associated with them, see Cambrosio, Jacobi, and Keating, "Ehrlich's 'Beautiful Pictures,'" 1993; Rotman, *Becoming beside Ourselves*, 2008.

89. Chakrabarty, *Provincializing Europe*, 2007; Law and Lin, "Provincializing STS," 2017; Seth, "Putting Knowledge in Its Place," 2009; Seth, "Colonial History and Postcolonial Science Studies," 2017.

90. Even though I refer to "comparative" and "transnational" histories, I take seriously use my choice of frameworks that can critique and move beyond geopolitical binaries. With the orientalist underpinnings of "East" and "West," I only refer to "Chinese" and "European" as actors' categories and try to qualify them whenever possible. For instance, "Chinese medicine" is only "Chinese" in the context of state-centered narratives of medicine. Of course, this does not overlook the role that translation played across areas in closer proximity. Cook and Dupré, eds., *Translating Knowledge*, 2012.

91. Literary scholars have identified the unstable feature of genre, which in practice are enforced based on how images are used, sorted, stored, and sold. Postmodernists can treat genres like colors they mix to create a palette. See Bishop and Starkey, "Genre," 2006.

92. See Miller, "Genre as Social Action," 1984; Bawarshi, "The Genre Function," 2003.

93. Many philosophers have written about genre, critiquing the overly descriptive mode of genre. In contrast, scholars like Gérard Genette have claimed that genres establish a particular kind of logic and vice versa. See Efal, "Generic Classification and Habitual Subject Matter," 2014.

94. Here, I use "rendering" in reference to Lucia Allais's account of the transformation of rendering as a medium of historical imagination and the transformation. Specifically, Allais argues that in architectural drawing, computer rendering has become a proxy for human experience. Special thanks to Anthony Acciavatti for drawing my attention to this source. See Allais, "Rendering," 2020.

95. See Yood, "Writing the Discipline," 1968.

96. Bastian, "The Genre Effect," 2010; Bazerman, "The Life of Genre," 19.

97. Scholars of improvisation consider it as a unite of idiosyncrasy, or "inherently improvisational subjectivity." See Alperson, "A Topography of Improvisation," 2010.

98. Alperson, 276.

99. Butler, *Undoing Gender*, 2004, 1.

100. Scholars have also described improvisation as a form of cognition essential to interpreting any form of expression. Although I agree that improvisation scales to the unit of any mode of cognition, this scale renders improvisation too broad in this book. Instead, I draw boundaries around improvisation insofar as it grounds the expression and ontology of hand-drawn meridians and nerves in their particular visual genres. See Gould and Keaton, "The Essential Role of Improvisation," 2000. Science studies scholars will note that I do not invoke "improvisation" in the way that Julie Livingston has done to describe the biomedical modus operandi when facing volatile supply chains and scarce resources in African oncology wards. See Livingston, *Improvising Medicine*, 2012.

101. See Beaulieu, "Voxels in the Brain," 2001.

102. Goldman has further urged against conceptualizing improvisation based on novelty, spontaneity, and freedom, which all rely on the researcher's judgments of novelty and spontaneity. See Goldman, "Improvisation as a Way of Knowing," 2016.

103. See Hagberg, "Foreword: Improvisation in the Arts," 2000.

104. The impulse to localize was particularly problematic in the history of neuroscience. Katja Guenther has demonstrated how discourses of localization and connectivity competed with attempts to articulate sensory and motor pathways in central Europe. In Guenther's account, the mind was so difficult to localize that it eventually gave rise to psychoanalysis. Guenther, *Localization and Its Discontents*, 2015.

105. Early silk and bamboo manuscripts portrayed 11 meridians, whereas corpora like the *Huangdi neijing* described 12 primary meridians. In the early modern period, the 12 meridians were later canonized as 14 meridians in texts like *Treatise on the Fourteen Meridians*. The 12 primary meridians connected to internal viscera, whereas other sets of meridians did not link to an individual organ. To be sure, the *renmai* (任脈) and the *dumai* (督脈) are not paired paths, but individual paths, until the 12 meridians,

which refer to meridians on one side of the body. Thus, there are 24 primary meridians in the body with two special meridians, the *renmai* and the *dumai*, circling from the front to the back of the body. Other written sources paired meridians in 8, 10, or 12 paths. Sources related to Daoist classical body practices described eight "extraordinary" or "auxiliary" paths; Buddhist treatises in Tibet described 4 paths. On the medical channels in tantric texts, see Gyatso, *Being Human in a Buddhist World*, 2015, 193.

106. Here, I do not take for granted the many pitfalls of transliteration, given the evolution of spoken languages across medieval and early modern East Asia. For the purposes of this book, I use Pinyin or *romaji* when relating analytical categories. Pronunciation here is not part of my historical subject of study. Of course, given the book's engagement with conceptual and visual translations of meridians, there is certainly space to consider the role of spoken language. For a historical and methodological review of Chinese phonology, see Behr, "Discussion 6: G. Sampson, 'a Chinese Phonological Enigma,'" 2015.

107. This is not to reinforce the romantic notion that the Chinese writing system was an ocular mode of communication or ideographic in nature. Rather, I occasionally refer to script to reference texts within an image or as a rough guide for clarifying homophones for readers. An excellent critique of the debates and methodologies in the Chinese language can be found here: Gu, "Sinologism in Language Philosophy," 2014.

108. Historians of medicine unfamiliar with the history of music theory, itself a Eurocentric field, may not make this connection.

109. As we will see in the next chapter, they told time and fluctuated over time. Further, *jing* (經) referenced permanent objects, whereas *luo* (絡) referenced physical things. The earliest pulse diagnosis comes from the memoir of a Han dynasty physician named Chunyu Yi (淳于意), which later appeared in the 86 BCE text *Shijing* (史记) (*Records of the Historian*) by Sima Qian (司馬遷). See Hsu, *Pulse Diagnosis*, 3.

110. Uluğ Kuzuoğlu has offered a detailed study of the modern transformation of Chinese script. See Kuzuoğlu, "The Chinese Latin Alphabet," 2022; Kuzuoğlu, "Capital, Empire, Letter," 2021; Kuzuoğlu, "Telegraphy, Typography, and the Alphabet," 2020.

111. The World Health Organization (WHO) report documents the consensus of 12 experts on a standard international nomenclature for meridian paths and points to facilitate communication in practice and research while also addressing historical challenges in standardization. See WHO Scientific Group on International Acupuncture Nomenclature and World Health Organization, "A Proposed Standard International Acupuncture Nomenclature," 1991.

Chapter 1: Representing Meridians and the Mind

1. Although the distinction between early modern and modern China does not entirely follow the same social, political, legal, and cultural parameters as those in Europe, most political and cultural historians of modern China identify modern China as roughly beginning with the fall of the Ming dynasty. See Elliott, *The Manchu Way*, 2006; Brook, *The Troubled Empire*, 2013.

2. Symbolically, Han men secured their hair with a small scarf known as a *zicuo* (緇撮), which had been fashionable during the famous Song dynasty (960–1279). For a survey of headdresses during the Song, see Wang and Kong. "An Aesthetic Study of Song Dynasty Scholarly Dress," 2017. On the political utility of policing hair styles from the Tang to the Song, see Kuhn, *The Age of Confucian Rule*, 50.

3. For instance, one Qing dynasty version of the fourteen meridians showed a heart meridian man with chest hair. This was featured in Emperor Qian Long's collection of fine art, titled *Ling men chuanshou gang ren zhi xue* (凌门传授铜人指穴). See Huang, *Graphic History of Chinese Acupuncture-Moxibustion*, 2003, 301.

4. Many of these vernacular gestures—stylistically holding one's hair, sleeve, or beard—have been systematized as an operatic narrative device to express one's emotion, status, and behavior. On the use of symbolic gesture in Chinese opera, see Siu and Lovrick, "Using Costumes," 2014.

5. Other premodern texts such as the now famous *Shanghan lun* (傷寒論), attributed to Zhang Zhongjing (張仲景) (fl. 196–205), offered other configurations of meridian paths that focused on six primary meridians extending from the feet. These six meridians or *liujing* (六經) referred not only to meridians, but also described six stages of disease progression. Early modern texts that revived interest in the *Shanghan lun*, such as the twelfth-century text *Huorenshu* (活人书), followed this conceptual genealogy and began its sequence of meridians with the foot-greater-Yin (足太阳經) path. For an intellectual history of these texts, see Lu, "Development of Zhu Gong's Thought on Cold Pathogenic Diseases," 2011.

6. As Michael Stanley-Baker has shown in his excellent article on the history and emergence of many uses of the word "Qi," it first appeared in medical writing dated to excavated medical texts and figurines from the third to second centuries BCE, including the *Maishu* (脈書) and *Yinshu* (引書) from Zhangjiashan tombs dated 186 BCE. These sources also describe the role of physical sensations in determining the function of Qi before the *Huangdi neijing* (often translated as the *Yellow Emperor's Inner Canon*, ca. first century CE) systematized medical knowledge. These texts from Mawangdui, Mianyan, Shuangbaoshan, and Laoguanshan/Tianhui tombs show an increasingly refined language for coding bodily experience. These were not fully integrated into the *Huangdi neijing*, which defined Chinese medicine for the next 2,000 years. See Stanley-Baker, "Qi 氣," 2022.

7. The broad literature on Qi ranges from early excavated texts to contemporary studies, examining Qi across history from paleographic sources and early medical texts to Neo-Confucian thought, Buddhist and Daoist cultivation practices, and modern phenomenological perspectives. Major works include Kuroda, *Research on Ki*, 1977; Onozawa, Fukunaga, and Yamanoi, *Thoughts on Qi*, 1978; Schäfer, *10,000 Things*, 2015; Kubny, "Qi," 2002; Sakade, *Taoism, Medicine and Qi in China and Japan*, 2007; Jiang and Zhao, *Acupuncture and Qi*, 2018; Stanley-Baker, "Qi 氣," 2022.

8. This is in reference to the "aesthesis" of morbid anatomical bodies in Hendriksen, *Elegant Anatomy*, 2014.

9. Nutton, "Physiologia from Galen to Jacob Bording," 2012.

10. Anatomy only came to overtake physiology in the mid-sixteenth century, beginning with a simple editorial fluke. French physician Jean François Fernel (1497–1558) was credited for publishing the first book on physiology in the 1540s. However, Fernel's book was originally focused on anatomy before his editor changed the title to *Physiologia* to boost sales. The original title was "On the Natural Part of Medicine." Nutton, "*Physiologia* from Galen to Jacob Bording," 2012.

11. Wright, "Ventricular Localization in Late Antiquity," 2018.

12. Lloyd, "Pneuma between Body and Soul," 2007; Martin, "Paul's Pneumatological Statements and Ancient Medical Texts," 2006; van der Eijk, "Nemesius of Emesa and Early Brain Mapping," 2008.

13. Martin, *The Corinthian Body*, 1999.

14. Vidal, "Brainhood," 2009.

15. This was central to Greek philosophical and medical discourses from the fifth century BCE onward. Wright, "Ventricular Localization," 7.

16. This was a general consensus of an idea from the second to the seventh centuries. Wright, 6.

17. Galen and Singer, *Selected Works*, 1997, 189.

18. Galen and Singer, 189.

19. Galen and Singer, 230.

20. Galen and Singer, 192.

21. Galen and Singer, 197.

22. See, for instance chapters by Gavrylenko, Taub, and Orland, in *Blood, Sweat and Tears*, 2012.

23. Detailing this variety of objects like the soul further builds on Caroline Bynum's insistence that engaging with what medieval theologians and philosophers discuss as soul (*anima*) and body (*corpus*) is not meant to translate easily into a Cartesian mind/body problem, which we will encounter later in chapter 3. For a review of Bynum's concerns with discussions about "the body," see Bynum, "Why All the Fuss about the Body?" 1995.

24. In the early modern period, the materiality of these states was more deeply informed by Christ in the Eucharist, where the body could shift into blood and bread. Moreover, the fact that these transformations occurred at all overpowered the need to explain how they happened. See Browe, *Die Verehrung der Eucharistie im Mittelalter*, 1967.

25. See Bynum, *Wonderful Blood*, 2007.

26. Although I sometimes refer to 14 *jingluo* (經絡) as a single set, it includes two main *jingluo* that looped from the front and back of the body. Meanwhile, the remaining 12 meridians accounted for paths only one side of the body. The number of full-body meridians is 26 if you double one side and include the two special meridians.

27. See Hsu, "Tactility and the Body in Early Chinese Medicine," 2005.

28. Medical historians Vivienne Lo and Catherine Despeux translate *wuzang* (五臟) as "five viscera." For a discussion of the translation of *wuzang* (五臟), see the translation and compilation by Major et al., eds., *Huainanzi*, 2010.

29. Early archaeological manuscripts reveal intricate face reading practices in early China. Over 80 facial sites were inspected for color and complexion to determine health and disease prognosis. The earliest references to this practice appear on 1500 BCE silk manuscripts excavated from the Dunhuang caves. See Despeux, "From Prognosis to Diagnosis of Illness in Tang China," 2005.

30. See Despeux, "The Body Revealed," 2007, 635.

31. Certain paths were associated with musical modes and sounds in early China. Musical modes later became associated with colors. Among the Yin organs and meridians, the heart and heart-membrane meridians were associated with the *zheng* (徵) mode and the color red (赤). The spleen meridian was linked to the *gong* (宫) mode and yellow (黄). The lung meridian corresponded to the *shang* (商) mode and white (白). The kidney meridian aligned with the *yu* (羽) mode and black (黑). The liver meridian matched the *jiao* (角) mode and teal *cang* (苍).

32. These animals were known as the *bai shou* (白獸) (white beast) in the lungs, the *zhu que* (朱雀) (vermilion bird) in the heart, the *long* (龍) (dragon) in the liver, the *feng* (風) (phoenix) in the spleen, and the *lu* (鹿) (deer) in the kidneys. See Zhang and Dear, "Embodying Animal Spirits in the Vital Organs," 2018.

33. These dates are contested given that historians suggest that Yang compiled the images in 1113 and served as Emperor Huizong's physician (1082–1135), well after Yang's presumed death. See Miyasita, "A Link in the Westward Transmission," 1967.

34. This is not to say that Yang Jie was the only individual producing images. For a long history of medical images that began during the Song dynasty, see Despeux, "Picturing the Body in Chinese Medical and Daoist Texts," 2018.

35. Scholar-practitioner Volker Scheid takes Yang Jie's images as evidence of critical medical innovation. Scheid writes that the focus on the organs demonstrates "critical textual research, empirical observation, and the ongoing effort to improve clinical practice and reflected a shift of attention in the investigation of things away from cosmological resonance and introspection toward more directly observable relationships." See Scheid, "Transmitting Chinese Medicine," 2013, 323.

36. Huang, *Graphic History of Chinese Acupuncture-Moxibustion*, 363.

37. More on this later in the chapter. The original Chinese is "咽吞物爲扼要嗌也." See Wang, *Yi Yin's Grand Method of Decoctions Expanded by Zhongjing*, 1234, 6.

38. For a detailed overview of the five phases in medical geography, see Marta Hanson, *Speaking of Epidemics in Chinese Medicine: Disease and the Geographic Imagination in Late Imperial China*, 2011, 25–34.

39. The organs also guarded the body. Yin-oriented organ systems were responsible for Yin-related body parts. They were arranged in a kind of physiological hierarchy, joining together different depths of the body. As Paul Unschuld translated from the *Su Wen*, "The lung rules the body's skin and body hair. The heart rules the body's blood and vessels. The liver rules the body's sinews and membranes. The spleen rules the body's muscles and flesh. The kidneys rule the body's bones and marrow." Unschuld and Tessenow, *Huang Di Nei Jing Su Wen*, 213.

40. The character *zang* (臟) explicitly featured the flesh radical 月. For instance,

among the five primary *zang*, four included the flesh radical—spleen (脾), kidney (腎), stomach (胃), and liver (肝)—each had fleshy references. For a discussion on the materiality of flesh among the internal viscera, see Li, "Medical Poetics," 2020.

41. These are also known as *suihai* (髓海), *xuehai* (血海), *qihai* (氣海), and *shuigu zhihai* (水穀之海). See Unschuld, *Huang Di Nei Jing Ling Shu*, 2016, 361.

42. The ocean itself, though inscribed in the head was more importantly linked to the stomach, and organ/orb/storehouse/agent, whose character 胃 was also tied to flesh. The ocean of bone marrow was then linked to the stomach, the breast and a special meridian known as the *chongmai* (冲脈). The *chongmai* was not part of the 12 primary meridians, but the eight extraordinary or auxiliary meridians known as the *qijing bamai* (奇經八脈). The original Chinese in the *Lingshu* on the Four Oceans, reads, "胃者， 水穀之海， 其輸上在氣街， 下至三裏； 衝脈者， 爲十二經之海， 其輸上在於大杼， 下出於巨虛之上下廉； 膻中者，爲氣之海， 其輸上在於柱骨之上下， 前在於人迎； 腦爲髓之海， 其輸上在於蓋， 下在風府." Unschuld, 362.

43. In the original inscription: "髓海 至陰之在 通尾骶." The spacing of the text varies among different versions of the woodcut. Early modern texts that related to medicine and body practices offered many different names for substances in the head. This included *suihai* (隨海), *niwan* (泥丸), and *tiangu* (天谷), to name a few.

44. The lung, heart, liver, spleen, and kidneys are connected to the hair, arteries, blood, membranes, muscles, sinew, flesh, bone, and bone marrow. Yin organs that expressed Yin-specific disease symptoms registered specific signs. For instance, problems in the Yin-organs manifested on the face, specifically, the left cheek, right cheek, forehead, nose, and chin. Redness in the left cheek suggested heat problems in the liver; redness in the right cheek heat problems in the lung; redness in the forehead suggested heat disease in the heart; redness around the nose suggested a heat disease in the spleen.

45. Scholar-practitioners have offered literary analyses of the many uses of the character 氣 in classical texts such as the *Huangdi neijing*. See Jiang and Zhao, *Acupuncture and Qi*, 2018.

46. For more an excellent and deep survey on the etymology of 氣, see Stanley-Baker, "Qi 氣," 2022.

47. Quoted in Tiquia, "The Qi That Got Lost in Translation," 2011, 39.

48. This is not to say that Qi (氣) exists in the mind through imagination. Rather, here I aim to emphasize the multiplicity of Qi (氣), which often appears in body cultivation practices, or *yangsheng* (養生) that directly manifests as feelings and fluids in the body. For more on the visual representation of body cultivation practices, see Wang and Fuentes, "Chinese Medical Illustration," 2018.

49. Tiquia, "The Qi That Got Lost in Translation," 40.

50. For more on these distinctions, see Ge, "On the Concept and Classification of Qi," 2008.

51. These body parts marked sites vulnerable to disfigured Qi (氣) that knocked

disease into the bones. It could break out disease in the blood; it could burst into the flesh. Qi (氣), in other words, was also a menace.

52. For instance, Yang diseases broke out in the winter, while Yin diseases broke out in the summer. Liu et al., "The Functioning of Yin and Yang," 2019, 122.

53. For a linguistic analysis of 心 as heart-mind, see Yu, *The Chinese HEART in a Cognitive Perspective*, 2009.

54. For reflections on understanding heart-mind-brain as an example of medical poetics, see Li. "Medical Poetics."

55. As the full description goes: the Qi (氣) of the kidneys reacts to extreme joy; the Qi of the liver reacts to extreme anger; the Qi of the lungs reacts to extreme sadness; the Qi of the spleen reacts to extreme fear; the Qi of the heart reacts to extreme anxiety. From engaging with the emotions in the organs, Qi then extended from the organs and into the bones. This is my translation of the text from the *Su Wen*; for a comparison of this translation, see footnote 132 in Unschuld and Tessenow, *Huang Di Nei Jing Su Wen*, 117.

56. Daoist texts also describe individual animals inhabiting each organ, which historians have cleverly translated as "animal spirits" not to be confused with early modern European animal spirits. See Zhang and Dear, "Embodying Animal Spirits in the Vital Organs."

57. The Yin organs included the lungs (肺), heart (心), heart membrane (心包) (pericardium), liver (肝), spleen (脾), and kidneys (肾). The Yang organs included: the large intestine (大肠), small intestine (小肠), triple burner (三焦), stomach (胃), bladder (膀胱), and gallbladder (胆).

58. Although published in 1341, the earliest remaining copy of the *Treatise on the Fourteen Meridians* is a Japanese print from 1625. See Chen, "A Survey of Hua Shou's Life and Writings," 2004.

59. Hua Boren had been primarily practicing medicine along the southeast coast of China around Jiangsu and Zhejiang Provinces. He had been prolific with nineteen published books, although few of his books survive. For a review of Hua's work, see Niu, "Spreading of Hua Shou's Medical Books in Japan," 1988.

60. A mode is a way of organizing musical notes. In the context of lydian/hypolydian modes, scales are composed of seven notes defined by a prescribed tonic. Similarly, musical modes are also relevant to meridian paths, with particular differences in tones and scales. To learn about the relationship between musical modes assigned to meridian paths, see "Modes," 2022, which can be accessed here: https://pulses.blogs.rice.edu/2021/05/21/modes/.

61. See Messer, "Hot/Cold Classifications and Balancing," 2013.

62. For a long history of the humors and their later relationship to contemporary forms of neurasthenia, see Arikha, *Passions and Temper*, 2008.

63. Galen and Singer, *Selected Works*, 246.

64. Galen and Singer, 246.

65. Galen and Singer, 258.

66. Galen and Singer, 315.

67. "Apples and pears, too, cause less harm when cooked; if they are raw," wrote Galen, "those which are most suitable for preservation are less bad. But I digress." Galen and Singer, 306.

68. For instance, blood was particularly omniscient in criminal cases where blood would flow out of the victim's body if the killer were nearby. Bildhauer, ed., "Medieval European Conceptions of Blood," 2013.

69. Blood had the ability to compose and transform. For instance, in the *Book of Nature*, a white snail born from putrid grass turned into blood when sprinkle with salt. Bildhauer, 66.

70. These causes could be read and predicted if they manifested viscerally and visibly. Arnold then attempted to illustrate what Galen described as the three discrete sections of the brain (phantasy, intellect, and memory), offering one of the earliest illustrations of the triple partitioned or tripartite brain. Yet, the image itself did not dictate the more practical concerns of mental pathologies. It was hard to localize disease states. When Arnold listed the five forms of alienation—folly, mania, melancholia, lovesickness, and irrational hatred—he struggled to pin these states onto the page. See MacLehose, "The Pathological and the Normal," 2018.

71. Here, I do not assume empiricism to be based on "objective facts" when the objectification of knowledge is instead the goal of empirical practice. Instead, I refer to empiricism as a practice in the ways that Schaffer and Shapin have discussed: empirical practices rely on assortments of social and material technologies that reify and distribute information to convince readers of "facts"—to extend the experience of witnessing. See Shapin and Schaffer, *Leviathan and the Air-Pump*, 2011.

72. See Wragge-Morley, "Imagining the Soul," 2018.

73. Willis, *Cerebri anatome*, 1663, 435.

74. The original text being: *Nervi brachiales per processus nerveos transversos et se mutuo intersecantes, invicem communicant*. Willis, 435–437.

75. This is a critical focus point to emphasize that meridians are material structures manifested within the body. These were interior structures that engaged with interior ecologies. What is also interesting here is that, unlike other texts that focused on sets of six meridians that focused on the feet, these six meridians focused on the hands.

76. Among public library entries, Wang Haogu's active dates are listed as 1298–1308. Chinese language Wiki and Baidu searches also list his birth and death dates as 1162—1249 or 1279–1368. For the purposes of this book, I will maintain the dates of his birth and death as described in scholarly biographies while also recognizing that these dates remain contested.

77. Early lists of the many types of Qi (氣) show both an empirical basis in observing Qi from different sources circulating inside and outside the body, as well as a numerological structuring using significant numbers to organize and memorize observational data, not just to construct micro-macrocosmic correlations; for example, the number five structured multiple unrelated Qi typologies like calendrical progressions, facial complexion, external vapors, and *zang-fu* organ Qi. Numerology provided a mnemonic

device to systematize sensations of the material world within and outside the body. See Stanley-Baker, "Qi 氣," 34.

78. The first mass woodblock printing of texts occurred during the Tang dynasty, which reprinted numerous Buddhist scripts. State publishing of medical texts peaked in the eleventh century under the Bureau for Editing Medical Treatises, where the Song dynasty's central government compiled sixteen authoritative editions of medical classics along with eighteen new medical treatises. For a list of these books published by the Bureau for Editing Medical Treatises, see Hinrichs, "The Song and Jin Periods," 2013, 106.

79. The original caption reads "*liangshou liu mai da yao hui* (兩手六脈大要會)," suggesting a "concise" or "important" sketch of the six *jingluo* (經絡) on two hands.

80. The text here is inconsistent where some descriptions of the six *mai* indicate the head rather than the stomach.

81. The full caption reads "天地有自然之氣 自然之數 人稟天地而生 氣數與天地等 修真之士 窮造化之源 知升降之路 安神定息 一念不生 湛然無慾 真氣周流 自然造化 老子曰綿綿若存 用之不勤 太素曰出入廢則神機化滅 升降息則氣立孤危." I have added spaces between the text to suggest how it might be read in segments.

82. The original text read, *xiuzhen zhi shi* (修真之士).

83. The original text read, *zhen qi zhou liu ziran zaohua* (真氣周流自然造化).

84. The original text read, *shengjiang xi ze qi li gu wei* (升降息則氣立孤危).

85. For instance, the status of medical personnel was also shifting as groups of physicians were trained specifically to treat imperial family members.

86. Hanson, *Speaking of Epidemics in Chinese Medicine*, 35.

87. These new scholar-officials concentrated in urban centers that shaped powerful economic forces, ideological, literary, artistic, and technological achievements. See Kuhn, *The Age of Confucian Rule*, 2011.

88. During the Song dynasty, the idea of the five circulatory phases and the six influences, or the *wuyun liuqi* (五運六氣), gradually took hold over every aspect of medicine, including prognosis, diagnosis, prevention, pharmacotherapy, and acupuncture-moxibustion practices. The imperial court eventually recognized this as such a central aspect of medicine that it was included in the official examination system. See Despeux, "The System of the Five Circulatory Phases," 2001.

89. Although most scholars place Wang Haogu in the Southern Song (1127–1279) and Yuan dynasties (1271–1368), records of his movements between Kaifeng and Shanxi suggest that he spent part of his adult life in the Jin empire instead (1115–1234), although his early life and childhood remains largely unknown.

90. Wang's style name was Haicang (海藏) and courtesy name was Jinzhi (进之). Members of the literate elite, who were often male, were known by many names.

91. The record of 33 epidemics was stark in comparison to only one case that was recorded in the eleventh century. Qiao and Su, *Wang Haogu*, 8.

92. *Shanghan lun* (傷寒論), often translated as *On Cold Damages*, is an amalgamation of writings attributed to Zhang Zhongjing (張仲景) (fl. 196–205). At its inception, *Cold Damages* had articulated and defined cold diseases and their etiologies, but

it was only one of many medical discourses and approaches. Zhang described cold diseases as distinct from other diseases known as *za bing* (雜病). The legacy of *Cold Damages* was later popularized in the Northern Song, the Ming, and the Qing dynasty. See Hanson, "The 'Golden Mirror,'" 118.

93. These more popular disease models were attributed to legendary physicians like Hua Tuo (華佗) (c. 140–208). On Hua Tuo, see Cao, "The Historical Significance of Hua Tuo's 'Six Divisions and Three Methods,'" 2002.

94. Jin forces had taken control of the city from the Song dynasty nearly a century earlier, in 1127, but established its capital elsewhere.

95. For a military history of this period, see Johnson, *Women of the Conquest Dynasties*, 2011. And though by no means definitive, a helpful sequence of events can be found in Waterson, *Defending Heaven*, 2013.

96. Wang may have traveled over 400 miles to Shanxi, although his exact destination in this period is unclear.

97. Qiao and Su, *Wang Haogu*, 11.

98. The exact date of Wang's death remains speculative as the preface for one of his final books, *Cishi nanzhi* (此事難知), or *Subjects of Great Difficulty*, is dated 1308 when he was 108 years old. Qiao and Su, 12.

99. In his words, "嗟乎遊魂行屍," and because commas and exclamation marks were introduced much later in China, the punctuation given here is my own interpretation. Wang, *Yi yin tang ye zhong jing guang wei dafa*, 4.

100. To a similar degree, Wang had said, "不必重樓幽闕, 明堂絳宮" or translated more literally, "No need to climb to the highest peaks/palaces/shrines, to seek the heart of the empire." 4.

101. For instance, Wang was particularly inspired by Zhang Ji's (张机) second-century text, *Shanghan zabing lun* (傷寒雜病論), one of the most important medical classics restored in this period. Wang admitted that he did not fully understand all of the contents of the *Shanghan zabing lun*, but still incorporated its recipes and comments on materia medica to produce an improved, comprehensive view of the body. Hanson, *Speaking of Epidemics*, 35–36.

102. Wang would introduce the work of Hua Tuo in one of his final books, *Cishi nanzhi* (此事難知), or *Subjects of Great Difficulty*.

103. Although the *cun* (寸) is often translated as an "inch" and *chi* (尺) often translated as a "foot," the ratios between these three units could also compare *cun* to a "centimeter," a *chi* as a "decimeter," and a *zhang* (丈) as a "meter."

104. For a survey of this history, see Yee, "Taking the World's Measure," 1994.

105. This translation is taken from Yee, 97.

106. Sun had furthermore taken on the task of regulating meridian sites and meridian names by matching needling sites with corresponding symptoms and therapeutic functions. Sun also developed the *a-shi* point, or "touch point," that could be directly applied to an active meridian site rather than stimulating the area at a distance along the same meridian path. See Ka-wai, "The Period of Division and the Tang Period," 2013, 94.

107. As historian Marta Hanson has described, the hand's graphic representations were important in medical and cosmological practices. Representations of the hand and palm performed symbolic mediation. They carried abundant information and featured mnemonics that encoded technical, medical knowledge. They helped physicians remember the names of needling and heating sides, track the ingredients for therapeutic recipes, and complete complex calculations for making diagnoses and determining treatments. See Hanson, "Hand Mnemonics in Classical Chinese Medicine," 2008.

108. Wang also stabilized the body through reiterating absolute measurements given by the fourth-century physician Bian Que (扁鵲). In Wang's estimation, Bian Que had offered the depth, width, and weight of different body parts:

> The mouth is three *cun* wide.
> The lips are nine *fen* long.
> The distance from behind the teeth to the throat is 3.5 *cun*, it has a total of five *he*.
> The tongue weighs 11 *liang*, measures 7 *cun* long, 2 *cun* wide.
> The esophagus entrance weighs 11 *liang*, measures 2 *cun* wide, and 1 *chi* 6 *cun* long.
> The throat weighs 12 *liang*, measures 2 *cun* wide, and 1 *chi* 2 *cun* long divided in nine sections.

Wang, *Comprehensive Theories*, 32.

109. For instance, Wang counted 21 vertebrae, representing three sections of seven bones, although later treatises described 24 vertebrae based on 24 solar periods in a year. Meanwhile, texts such as *Taixi renshen tushuo* (泰西人身圖說) (*Explications of the Illustrations of the Human Body by Westerners*), enumerated the vertebrae sections as follows: cervical—7 vertebrae; dorsal (thoracic)—12 vertebrae; lumbar—5 vertebrae; sacral—6 vertebrae; and coccygeal (tailbone)—4 vertebrae. The total count results in 24 vertebrae, aligning with the 24 divisions of the solar calendar. Wang, 652.

110. Historian Catharine Despeux has emphasized that the range in these numbers demonstrated that they were meant to be symbolic representations of the body. For instance, even the number of bones correlated with the number of days in the year, which were sometimes described as 360, 365, or 366. Furthermore, when considering body "sections" based on bone divisions, these numbers also reflected what counted as bony material, which included nails, teeth, cartilage, and varying standards in bone division. See Despeux, "The Body Revealed," 648.

111. I have presented a preliminary translation of Wang's text in as part of the Forum for the History of Health, Medicine. See Li, "Measure Your Meridians," 2023. The full video can be accessed here: https://www.youtube.com/watch?v=D5go9q4wJgQ.

112. The numbers among meridian lengths, lengths of time, and individual breaths were all connected. For instance, 1,620 *cun* ÷ 6 *cun* = 270 breaths (inhale, exhale); 270 × 50 cycles = 13,500 breaths (inhale, exhale).

113. Later in the text, Wang suggests that people's bodies have become increasingly stagnant over time, such that it takes longer for Qi to travel along the meridians with

each breath. Specifically, whereas previously each full inhale-exhale cycle took the equivalent of two "steps" (*ke*) on the water clocks, Wang asserts that contemporary people move at four "steps" per breath cycle. This means each inhale and exhale takes one hour, rather than the previous 30 minutes.

114. Kuriyama, "The Imagination of the Body and the History of Embodied Experience," 2001, 25.

115. This was a famous text. It had circulated between China and Japan through new trade routes that emerged at the beginning of the thirteenth century. Buddhist priests who had contributed to the printing revolution during the Song also facilitated intellectual exchange and the reprinting and distribution of treatises like *Fourteen Meridians*. See Gole, "Song Printed Medical Works and Medieval Japanese Medicine," 2013.

116. Measurements of human body proportions continued to emerge in the sixteenth century, such as *Xun jing kao xue bian* (循經考穴編) (*On Locating Meridian Sites*) that appeared in 1575, the images of which were reproduced in Lu Gwei-djen's magnum opus *Celestial Lancets* in 1980.

117. In the sixteenth-century treatise *Xun jing kao xue bian* (循經考穴編) (*On Locating Meridian Sites*), meridian sites took the form of a schematic *tu* that transformed the body into a system of plots. Two grids displayed the front and back of the body, or the "belly area," *fubu* (腹部) and "back area," *beibu* (背部).

118. This inscription appears on the right arm as *ren you daxiao changduan bu deng wei tong shen cun keyi qu zhi* (人有大小長短不等惟同身寸可以取之).

119. This inscription appears on the left arm as *ren zhang ze cun chang ren duan ze cun duan ying ru lao you jie ran* (人長則寸長人短則寸段嬰儒老友皆然).

120. For instance, Hua's images, more importantly, identified the location between meridian sites based on bony protrusions. Physicians had often examined bony protrusions at the surface of the body as a way to access internal visceral. See Despeux, "The Body Revealed," 638.

121. Dürer's proportion studies were translated into French (1557) and Latin (1532). However, the Italian version, reprinted in 1594, had the greatest impact. It expanded the availability of anatomical information in Italy and remained the most cited version in later Baroque treatises.

122. The full book title reads: *Hierinn sind begriffen vier Bucher von menschlicher Proportion durch Albrechten Durer von Nurerberg erfunden und beschuben zu nutz allen denen so zu diser kunst lieb tragen*, or "Here are conceived four books of human proportions invented by Albrecht Durer von Nurerberg [*sic*] And for the benefit of all those who love art."

123. Dürer and Price, *De symmetria*, 16.

124. Dürer had also been long interested in the proportions of the horse. See Cuneo, "The Artist, His Horse, a Print, and Its Audience." 2011.

125. Dürer and Price, *De symmetria*, B1v.

126. Camerarius had translated *Human Proportions* along with Dürer's *Underweysung der Messung* (1525), and synthesized the internal structure, principles, and

precision of Euclid's *Elements*. For a close reading of *Underweysung*, see Andrews, "Albrecht Dürer's Personal Underweysung Der Messung," 2016.

127. Dürer, *De symmetria*, 10.

128. Dürer, 17, view 10: B2r.

Chapter 2: Early Modern Metaphors as Translation

1. Lu and Needham, *Celestial Lancets*, 1980, 272.

2. Japanese scholars such as Seiichi Iwao had been writing about the "Dutch Doctor in Old Japan" in the 1960s, but Lu never cited Iwao in her own work.

3. Historian Wei Yu Wayne Tan has further argued that Ten Rhijne visited Edo during a significant transformation across medical practices distinct from medical lineages in the Qing empire. These changes included developments among the secretive Irie (入江) lineage and the Mubun (夢分) lineage that specialized in the *uchibari* (打針) technique, or hammering needles into the stomach. See Tan, "Rediscovering Willem Ten Rhijne's *De Acupunctura*," 2020; Vigouroux, "Reception of the Circulation Channels," 2015.

4. In particular, see Bivins, *Acupuncture, Expertise and Cross-Cultural Medicine*, 2000; Bivins, "Imagining Acupuncture," 2012.

5. Lu's observations conceptually moved across Wittgenstein's "family resemblances," speaking directly to comparative circumstances. Per Wittgenstein, faces with eyes, noses, and mouths do not make family resemblance—particular combinations of features determine relationships. See Wittgenstein, *Philosophical Investigations*, 2009. For a lucid account of resemblances that are qualitatively "determinant" but not qualitatively "identical," see Khatchadourian, "Common Names and 'Family Resemblances,'" 1958.

6. Furthermore, Ten Rhijne arrived just as Dutch East India Company (VOC) trade patterns shifted, deprioritizing direct China trade for the Indian Ocean, ceding more China routes to the British. See Blussé, "No Boats to China," 1996.

7. On physician kidnappings, see Nakajima, "Sixteenth-Century and Seventeenth-Century East Asian Seas and Migration of Chinese Intellectuals," 2004.

8. For a comprehensive and excellent history of this period, see Macé, *Médecins et Médecines dans l'Histoire du Japon*, 2013, 92–130.

9. For an analysis of literature from this period focusing on Arashiyama Hōan's 嵐山甫安 (1633–1693) use of *seibinseinon* (セイビンセイノン) in *Bankoku chiryōhō ruiju* (蕃国治方類聚) (1683) and Sōtei Nakamura's 中村宗填 *zeinun* (ゼイヌン) in *Kōmō hiden geka ryōjishū* (紅毛秘伝外科療治集) (1684) to articulate Qi-Blood (気血) movement, and Yoshimasu Motoki's 本木良意 (1628–1697) distinction between *suji* (筋) and *kei* (経) in *Oranda zenku naigai bungōzu* (和蘭全躯内外分合図) (1690), see Matsumura, "The Japanese Conception of 'Nerve,'" 1998. Special thanks to my manuscript reviewer for recommending this text.

10. In this chapter, I maintain the common translation of *shin/shen* (神) as a material "spirit" in the same way that Qi (氣) remains invisible but material. This decision speaks to the combination of analytical methods I use throughout the book

that build on and diverge from prominent sinologists. For instance, Paul Unschuld's seminal translation of *Huangdi neijing ling shu*, defines *shen* as "spirits" or something other-worldly. As he put it, "It was, along with 'heaven,' an abstract, supernatural power and authority that decided human fates. The demons, ancestors, other spirits, the deities—all these numinous powers were subsumed under the term *shen*. It was, in league with 'heaven,' the expression of the human being's existential heteronomy." To be sure, Unschuld approaches this translation as a sinologist rather than as a historian, which speaks to the limits of the methods necessary for deep histories of medicine. See Unschuld, *Huang Di Nei Jing Ling Shu*, 2016, 9. In a similar vein, Unschuld has defined *shenqi* (神氣) as "spirit Qi" that relates to Yang orientations. He explains that Chinese commentators often offer several interpretations of *shenqi*, all of which are "entirely convincing," 39.

11. For instance, Mathias Vigouroux compellingly argues that some of Ten Rhijne images may have portrayed acupuncture mannequins rather than meridian illustrations. Made of bronze, wood, and paper, mannequins offered tangible learning aids, conveying knowledge distinct from speculative maps. Vigouroux, "What Knowledge to Transmit?" 2019.

12. Mathias Vigouroux (155–192) offers an excellent analysis of this image in the context of Japanese image production and medical theory.

13. The Northern Renaissance, which included artists like Albrecht Dürer, refers to art from Northern Europe in the fifteenth and sixteenth centuries, including Germany, France, England, and the Netherlands. It retains more Gothic elements than Italian art. See Smith, *The Northern Renaissance*, 2004.

14. This is not to say that the portrait of the goatee man had no place in the early modern period. Perhaps it aligned more with Netherlandish art that emerged from the Low Countries, including modern day Belgium, the Netherlands, and Luxembourg, which featured secular works and portraits. See Pacht, *Van Eyck and the Founders of Early Netherlandish Painting*, 1999.

15. Mathias Vigouroux has described how wood and paper acupuncture mannequins were also key in assimilating and disseminating vessel theory in seventeenth century Japan. See Vigouroux, "What Knowledge to Transmit?" 181.

16. These were particularly used for examination. Lu and Needham, *Celestial Lancets*, 133.

17. In the twentieth century, bronze model replicas began to feature male genitals. For examples, see Huang, *Graphic History of Chinese Acupuncture-Moxibustion*, 2003, 222–234.

18. On the history of three-dimensional meridian men in Japan, see Nagano, *The Museum of Acupuncture and Moxibustion*, 2001.

19. "*Foramina indicabat interpres; fortè quod æneæ illæ machinæ in locis acupungendis & inurendis foraminibus pertusæ fuerint.*" Ten Rhijne, *Dissertatio de Arthritide*, 1683, 160.

20. The Latin is "*Donyn figuram ex ære.*" In Latin, *aeneus* could mean either bronze or copper.

21. A note on capitalization and Willem ten Rhijne's name. In Dutch, "ten Rhijne," it likely means "by the Rhine" or "at the Rhine," suggesting that his family's origins were near the Rhine River or in an area named after it. I follow historians like Harold Cook and Wei Yu Wayne Tan in capitalizing the preposition "ten" when writing "Ten Rhijne."

22. Ten Rhijne, *Dissertatio de Arthritide*, 190–191.

23. The *uchibari* (打針) technique uniquely focused on skilfully hammering the needle into the abdomen, an area that in some schools of acu-moxa was considered as forbidden. Specifically, Mubun (夢分) emphasized that weaker individuals needed to be needled in the abdomen whereas stronger bodies required puncturing in the back. See Tan, "Rediscovering Willem ten Rhijne's *De Acupunctura*," 2020, 125.

24. Vigouroux has argued that these images were not intended for the general public, and more likely produced to communicate technical knowledge between established acupuncturists or to help apprentice acupuncturists who could benefit from a master's explanations to educate their gaze. Vigouroux, "What Knowledge to Transmit?" 187.

25. For a close study of these sources, see Goble, *Confluences of Medicine in Medieval Japan*, 2011.

26. For instance, Federico Marcon showed images in earlier *materia medica* books were decorative, not informative. Marcon, *The Knowledge of Nature*, 2015, 232–234.

27. Vigouroux, "What Knowledge to Transmit?" 156.

28. For an excellent study of this period, see Trambaiolo, "Diplomatic Journeys and Medical Brush Talks," 2014.

29. Joshi and Kumar, "The Dutch Physicians at Dejima," 1062.

30. This was known as the Shimabara Rebellion in 1637–1638. Hidetada (r. 1605–1623) and Iemitsu (r. 1623–1651) eventually restricted strictly commercial exchange with China and Holland, and commercial and diplomatic exchange with Korea and Kyukyu, which Japan also considered as vassal countries. On these restrictions, see Kazui and Videen, "Foreign Relations during the Edo Period," 1982; Toby, *State and Diplomacy in Early Modern Japan*, 1991.

31. For instance, small groups of scholars edited a Dutch-Japanese dictionary called *Zufu Haruma*, which was edited by Carl Peter Thunberg (1743–1828), Isaac Titsingh (1745–1812), and Henrik Doeff (1777–1835).

32. *Kasuparu-ryū geka*, referred more specifically to the Casper School of medicine.

33. Joshi and Kumar, "The Dutch Physicians at Dejima," 2002, 1064.

34. Siebold enjoyed collecting maps. His daughter, Oine, remained in Japan and became the first female physician. Joshi and Kumar, 1066.

35. For instance, Tokugawa Yoshimune (1716–1745) was famous for recruiting scholars who were in astronomy, medicine, *materia medica*, and math. Sugimoto and Swain, *Science and Culture in Traditional Japan*, 1978.

36. Interpreters trained in Nagasaki until they were called upon to meet with new arrivals in Dejima or representatives in Edo. These trips to trading posts or the capital continued regularly until 1790, when it was reduced to once every four years. See Horiuchi, "When Science Develops Outside State Patronage," 2003, 150.

37. For instance, when polymath Otsui Gentaku (1757–1827) engaged with foreign guests, he relied on numerous interpreters, such as Motoki Ryōei (1735–1794), Yoshio Kōzaemon, Ishii Shōsuke, Shizuki Tadao, and Baba Sajūrō. See Macé, *Physicians and Medicines in the History of Japan*, 2013, 91–92.

38. For a comprehensive history of medical exchange in this period, see Michel-Zaitsu, "Exploring the 'Inner Landscapes,'" 2018, 10.

39. This was particularly the case of the second-century text, the *Shanghan lun*. On formal textual transmission, see Elman, "Sinophiles and Sinophobes in Tokugawa Japan," 2008.

40. In some of the oldest medical books, there were 390 cases of heating compared to only six of needling. Physicians in the Heian and Kamakura period also considered moxa as more effective than needling. Vigouroux, "Book Trade and Diplomacy," 2013, 109.

41. Another of Dōsan's disciples named Oze Hoan (1564–1640) also published *Fourteen Meridians*, which was reprinted 21 times. Vigouroux, 122.

42. On the transmission of Chinese medicine treatises, see Mayanagi Makoto's many publications and talks, such as "Transmission of Medical Texts," 1990; Mayanagi, "Import and Reprinting," 1998; Mayanagi, "Japaneseization," 2008; Mayanagi, "Japanese, Korean and Vietnamese Medicine," 2010.

43. Mathias Vigouroux has tracked how editions of *The Fourteen Meridians* [*Shisijing fahui* (十四經法會)] were imported in 1641, 1706, 1712, 1714, 1724, 1763, 1842, and 1843, and *The Great Compendium of Acupuncture Moxibustion* [*Zhenjiu dacheng* (针灸大成)] was imported in 1638, 1710, 1719, 1722, 1725, 1735, 1759, 1763, 1837, 1839, 1844, 1845, and 1849. Vigouroux, "Book Trade and Diplomacy," 112.

44. Although Japan imports also included Korean texts, these were rarely reissued. Vigouroux, 124.

45. Silver was such a valuable commodity that the Tokugawa government eventually limited the number of VOC boats that could enter the docks in the Nagasaki harbor to limit the amount of copper and silver that left with Dutch traders. On the variety of trading materials throughout the VOC, see Prakash, *The Dutch East India Company and the Economy of Bengal*, 1985; Iwao, "Cultural and Commercial Relations between Japan and the Netherlands," 1969.

46. Next to the needle was the image of a mallet that Ten Rhijne noted was sometimes made of ivory, ebony, or other kinds of hard wood. "*Acus fit longa (q) acuta, (r) rotunda; cochleatum habeat manubrium; (s) conficiatur ex auro, rarius ex argento, nunquam ex alio metallo.*" Ten Rhijne, *Dissertatio de Arthritide*, 183.

47. Otori, "The Acceptance of Western Medicine in Japan," 1964, 255.

48. Historian Mathias Vigouroux has offered a brief transcription of one of these encounters, which had taken place throughout 1636 and 1764. See Vigouroux, "Book Trade and Diplomacy," 133.

49. On the Franco-Dutch War, see Pritchard, *In Search of Empire*, 2004, 265–295.

50. Ten Rhijne studied medicine at Leiden University, where he produced a thesis

titled *De Dolore Intestinorum a Flatu* on internal viscera and earned his doctorate at the age of 21. Iwao, "A Dutch Doctor in Old Japan," 1961, 172.

51. On the social history of Dejima, see Leonard Blussé, Willem Remmelink, and Ivo Smits, *Bridging the Divide: 400 Years, The Netherlands-Japan*.

52. Iwao, "A Dutch Doctor in Old Japan," 173.

53. For a history of Ten Rhijne, see van Dorssen, "Willem Ten Rhijne," 1911; Otsuka, "Willem Ten Rhyne in Japan," 1971; Michel, "Willem ten Rhijne und die Japanische Medizin (I)," 1989; Michel, "Willem ten Rhijne und die Japanische Medizin (II)," 1990; Cook, "Medical Communication in the First Global Age," 2004.

54. These translations draw on Carrubba, Bowers, and Ten Rhijne, "The Western World's First Detailed Treatise," 1974, 375.

55. This is a curious detail in the text given that Japanese models during the Edo period made of paper and wood were not necessarily filled with water. In China, models were entirely covered in wax and filled with water or mercury. When the student correctly inserted the needle, water would flow out of the acupuncture point. Mathias Vigouroux has described that a Song dynasty replica was transmitted to Japan as early as the fourteenth century by the physician Takeda Shokei (1338–1380), but this type of mannequin had no influence until Japanese acupuncturists assimilated vessel theory at the beginning of the Edo period. Vigouroux, "What Knowledge to Transmit?" 181.

56. On the history of Dutch trade routes and knowledge networks, see Huigen, Jong, and Kolfin, *The Dutch Trading Companies as Knowledge Networks*, 2010; Jackson, *Network of Knowledge*, 2016.

57. On early printers to the Royal Society, see Rivington, "Early Printers to the Royal Society 1663–1708," 1984.

58. Paranavitana, "Medical Establishment in Sri Lanka During the Dutch Period (1640–1796)," 1988.

59. Ten Rhijne *Dissertatio de Arthritide*, 147. In Latin, "*Qvomodo in patentissimo macrocosmi alveo appellendum deteget locum ratis rector, nisi latitudinis gradu pernoto cursum, quem circino in mappis interstinxerat, firmare, syrtes & scopulos providè effugere, & de navis progressu ex secundis vel adversis, fluctibus, etiam inconspicuis, citiore vel tardiore, probabiliter conjectare nonverit?*" Sincere thanks to Dr. Tillmann Taape for his assistance for improving on the translation offered by Bowers. Unless otherwise notes, Latin translations in this chapter have been reviewed by Dr. Taape.

60. In this body, the *soma* operated beyond the limitations of vision so that it could only be accessed through *tekhnē*, or the reconstruction of meaning through reasoning. Intervention rested on logic. The body was a strange object where daemonic energies haunted the *psukhē*, or psyche. See Holmes, *The Symptom and the Subject*, 2014.

61. "*Quo nomine nervos, venas & arterias vulgo comprehendunt Sinæ*," in the Latin. Ten Rhijne, *Dissertatio de Arthritide*, 159.

62. "*Quæ lingitudinis in arteriis, cum humidi radicalis tum calidi nativi, differentia ex arteriarum magis minusque contortis mæendris & locorum distantia procedit*," in Latin. Ten Rhijne, 165.

63. "*Arterias voco, quoniam hæ nomen à pulsu sortiuntur: Miak enim Japonibus pulsum notat.*" Ten Rhijne, 167.

64. For instance, blood was particularly omniscient in criminal cases where blood would flow out of the victim's body if the killer were nearby. See Bildhauer, "Medieval European Conceptions of Blood," 2013, 58–59.

65. In the early modern period, the materiality of these states was more deeply informed by Christ in the Eucharist, where the body could shift into blood and bread. And the fact that these transformations occurred at all overpowered the need to explain how they happened. Browe, *The Veneration of the Eucharist*, 1967.

66. A "natural spirit" defined the heart, whereas a "vital spirit" defined the liver. Blood could compose and transform. For instance, in the *Book of Nature*, a white snail born from putrid grass turned into blood when sprinkled with salt. Bildhauer "Medieval European Conceptions of Blood," 66.

67. Both experts and non-experts understood this knowledge about the volatility of humoral fluids as expressions of unseen events in the body. See Orland, "White Blood and Red Milk," 2012.

68. In the Qing dynasty, physicians explained that women's and men's health were the same except for menstruation, pregnancy, and postpartum. Wu, *Reproducing Women*, 2010, 89.

69. This was made famous in the classical phrase from the *Huangdi neijing*, "men are ruled by *qi*, and women by *Blood*." Wu, 28.

70. Historian Yi-Li Wu has carefully demonstrated how over time, *xue/chi* (血) no longer referred to a metaphorically female body but actual female bodies. In an attempt to define gender difference, scholar-physicians in the seventeenth century suggested that female bodies carried more *xue/chi* than male bodies. New forms of early modern female alchemy identified *xue/chi* as the source of female vitality and the equivalent of semen in men. Although *xue/chi* defined gender difference, it was not exactly a gendered object. Wu, 53.

71. Blood as *xue/chi* (血) was also linked to fire and disorders related to nongendered organs such as the spleen, stomach, heart, and liver. According to the Confucian scholar Zhu Zhenheng, the body had an innate inclination toward excess fire that was generated in the heart, which was filled with Yang. Meanwhile, the spleen and stomach were involved in turning food and water into Qi, which then needed to be regulated to ensure Qi and Blood's health. See Zhu, 42.

72. The meridian terms *mai*, *luo*, and *jing* reflected this through names like *baomai* ("womb *mai*") and *baoluo* ("womb *luo*"). Zhu, 90–91.

73. "*[P]romiscuè enim sumunt Sinæ ut sæpius dictum.*" Ten Rhijne, *Dissertatio de Arthritide*, 166.

74. The *Oxford English Dictionary* identifies antiquated definitions of "arteries" from the fourteenth century as denoting anatomical structures: "The trachea or windpipe (occasionally including the larynx); (in early use also, in plural)." In contrast, early definitions of "vein" appeared as "A tract of ground or water, a mineral deposit, and a

small natural channel or fissure within the earth, through which water trickles or flows; a flow of water through such a channel or fissure."

75. Structures named arteries compared to structures named veins became more distinct the closer they were to the heart, which propelled a force of movement sustained by the thickly coated walls of the arteries. Numerous histories of Harvey's contributions to the circulatory system are abundant from the late nineteenth century to the present. Although these histories identify a fracture with humoral medicine in favor of a mechanical body, there is evidence for continuity across these histories, especially in neo-Galenic physicians who drew on humoral medicine to characterize nerve juices. Willis, *William Harvey*, 1878; Harvey, *On the Motion of the Heart and Blood in Animals*, 1889; Shackelford, *William Harvey and the Mechanics of the Heart*, 2003; Wright, *William Harvey* 2013.

76. Ten Rhijne, *Dissertatio de Arthritide*, 148.

77. In Latin, Ten Rhijne uses the word "*vorago*," which also referenced a deep hole, chasm, watery hollow, or vortex.

78. Matsumura, "The Japanese Conception of 'Nerve.'"

79. This was in reference to the 1683 treatise *Bankoku chiryōhō ruiju* (蕃国治方類聚).

80. This was in reference to the 1684 treatise in *Kōmō hiden geka ryōjishū* (紅毛秘伝外科療治集).

81. This was in reference to the 1690 treatise *Oranda zenku naigai bungōzu* (和蘭全躯内外分合図).

82. The famed physician Manase Dōsan (曲直瀬 道三) (c. 1507–1594) had founded Japan's first major medical school for Kampo medicine that trained over 800 disciples and focused teaching on clinical experience. On the evolution of prescribing practices in Kampo medicine from the Edo period into the modern era, see Suzuki and Endo, "Schools of Japanese Kampo Medicine," 2011.

83. Dōsan Manase had been investigating acupuncture through studying classical texts like the *Huangdi neijing* and through clinical experience and his masters' teachings. See Amano "On Koteimeidokyukyo Fushinshosho," 2015. For an examination of Manase's flexible dosing practices, see Yakazu, "On the Medicine Established by Dōsan Manase," 1991.

84. By situating key texts like Tōyō's *Zōshi* (蔵志) (1759) and Tōdō's *Yakuchō* (薬徵) (1785) within the context of Confucian evidential scholarship and urban medical commercialization, this represented a pivotal epistemological synthesis that paved the way for engaging with Western science. See Trambaiolo, "Ancient Texts and New Medical Ideas in Eighteenth-Century Japan," 2015.

85. For an overview of the empiricist dissenting trends in eighteenth-century Japanese medicine and the stimuli for investigating anatomy, see Rosner, *Medizingeschichte Japans*, 1988, 58.

86. Gotō conducted novel animal dissections to directly observe bodies, which helped introduce Western notions of blood circulated through capillaries. Macé, *Physicians and Medicines in the History of Japan*, 97.

87. Special thanks to my anonymous reviewer for pointing out *ketsumyaku* as the appropriate reading of 血脈 instead of *kechimyaku*.

88. The original text reading: 門脈血門二脈之支。共從此入肝○其附屬而在其內者。小血道也。裏其表者。 Sugita et al., *Kaitai shinsho*, 1774, 2.

89. The original text reading: 血脈之細絡多焉。.

90. Gabor Lukacs has found that *Kaitai shinsho* inspired more marginalia than any pre-modern Japanese medical work based on his study of its composition, production, distribution, and reception. See Lukacs, *Kaitai Shinsho*, 2008. In his analysis of Lukacs's study of *Kaitai shinsho*, Grégoire Espesset cautions that outdated Orientalist value judgments persist, as evidenced by Lukacs leaving the significant Chinese medical phrase *nei dui dantian* (內對丹田) untranslated. He advocates nuanced analysis of non-Western knowledge on its own terms. See Espesset, "Traditional Chinese Knowledge," 2014.

91. Historian Annick Horiuchi has argued that Sugita acted as a government agent to officially appropriate Dutch knowledge, or *rangaku*, as an object of state control. Horiuchi, "When Science Develops Outside State Patronage," 151.

92. The preface to Sugita autobiography, *Dawn of Western Science in Japan*, explained that he worked with the surgeon Gentetsu Nishi and the Confucian scholar Saburoemon Miyase. Sugita, *Dawn of Western Science in Japan*, 1969.

93. Sugita, 33.

94. Sugita, 34.

95. Naganari produced medical texts like *Geka soden* (外科宗伝) (1706) incorporating some Dutch knowledge, and Yoshio Kögyü (?–1713) came from an interpreter family that later translated important Dutch medical works. These interpreter lineages transmitted knowledge across epistemological boundaries, laying foundations for the Dutch studies movement. See Rosner, *Medizingeschichte Japan*, 58.

96. Sugita, *Dawn of Western Science in Japan*, 3.

97. Horiuchi, "When Science Develops Outside State Patronage," 167.

98. Sugita, *Dawn of Western Science in Japan*, 31

99. This has already been argued through detailed analysis in Lukacs, *Kaitai Shinsho.*

100. Alpers, *The Art of Describing*, 1984, 133.

101. Quoted in Alpers, 157.

102. Alpers, 157.

103. Iwao, "Nagasaki and Western-Style Paintings," 1969.

104. Screech, *The Lens within the Heart*, 2002, 140.

105. Screech, 161.

106. Kuriyama, "Between Mind and Eye," 1992.

107. Hall, *Tanuma Okitsugu*, 1982.

108. In the late seventeenth century, Neo-Confucianists such as Itō Jinsai (1627–1705) and Ogyū Sorai (1666–1728) were emphasizing the study of original Confucian texts and not interpretations, and this trend of studying classical sources also continued into the eighteenth century. Vigouroux, "Book Trade and Diplomacy," 116.

109. Sugita, *Dawn of Western Science*, 38.

110. The image was a mirror version of Vesalius's 1543 *Fabrica* before being copied by the Spanish painter Gaspar Becerra (1520–1570) and traced by the French engraver Nicolas Beatrizet.

111. See Chen, *Investigation on Oracle Bone Inscriptions*, 2001.

112. See Lei, *Investigation on "Shi Bu" in Shuowen*, 2000.

113. Sterckx, "Searching for Spirit," 23.

114. Winslett, "Deities and the Extrahuman in Pre-Qin China," 2014, 940.

115. *Shen* can also be a kind of stative verb, such as "to be divine" or "to be numinous," and it is used to modify superhuman intelligence. These creates terms like "divine sages" or "divine men," and it can also imply a kind of "uncanny sage." Winslett, 942.

116. Sterckx, "Searching for Spirit," 25.

117. In this period, Matteo Ricci (1552–1610) had considered Confucianist natural philosophy as an imperfect ethical theism that could be ultimately compatible with Christianity, although Dominicans and Franciscans later disagreed, being anti-literati and anti-Confucian. These debates continued into the nineteenth century as Roman Catholics, British and American Protestants, British Anglicans, and Scottish Presbyterians created new terms for God as other missionary projects extended into Korea. See Oak, "Competing Chinese Names for God," 2012, 91.

118. Sugita, *Dawn of Western Science*, 36.

119. Boerhaave actively participated in academic medical discourse, and his English and Dutch contemporaries credited him as the founder of physiology. However, Boerhaave's ideas were so commonplace that his French contemporaries took little notice of his work. Cook, "Boerhaave and the Flight from Reason in Medicine," 2000.

120. Knoeff, "Herman Boerhaave's Neurology and the Unchanging Nature of Physiology," 2012, 201.

121. As a dedicated Calvinist, Boerhaave was deeply invested in predestination, which manifested in his belief on teleological processes in the body. For instance, if the body was full of pores, then water and oil that penetrated the pores held the body together. Or, if he took particular interest in fire, then fire was the primary source of life and motion. See Knoeff, *Herman Boerhaave*, 2003, 193.

122. In cases of stagnation, Boerhaave suggested a kind of "shock" therapy to stimulate the internal space from an external input. For instance, epilepsy patients demonstrated a disturbance of the entire sensorium commune.

123. While "grafting" is used here metaphorically, Nana Osei Quarshie has deftly demonstrated in his forthcoming book *An African Pharmakon: Psychiatry and the Mind Politic of Modern Ghana* the ways in which "grafting" appeared as an actor's category in appropriating and accommodating diagnostic and therapeutic approaches to mental distress in West Africa. For one example of the legal, social, and economic dimensions of what Quarshie describes as the "African Pharmakon," see Quarshie, "Spiritual Pawning," 2023.

124. To be sure, Ten Rhijne was not alone in his struggle. One of his supervisors in

Batavia, Andreas Cleyer (1634–1697), also published an image of meridians in a 1682 edited volume titled *Specimen Medicinae Sinicae*.

125. Historian Roberta Bivins has described how hostile responses to body maps extended from conflicting representational practices. Bivins, "Imagining Acupuncture," 2012.

126. Wotton, *Reflections upon Ancient and Modern Learning*, 1694, 152. Similarly, the German physician Engelbert Kaempfer (1651–1716) who joined the Dutch East India Company and set out for Japan in 1690, also critiqued the practice of locating sites of burning moxa, writing that "sound reasoning does not permit us to testify in defense of all of them." Bowers and Carrubba, "The Doctoral Thesis of Engelbert Kaempfer on Tropical Diseases," 1970, 310.

127. These included physicians such as Takano Chōei (高野長英) (1804–1850), Ozeki San'ei (小関三英) (1787–1839), Ogata Kōan (緒方洪庵) (1810–1863), and Sugita Seikei (杉田成卿). For a close analysis of this history, see Macé, *Médecins et Médecines dans l'Histoire du Japon*, 2013.

128. For instance, in 1832, Takano Chōei proposed the neologism *shigeki* (刺激) for "stimulus," which combined characters for "pierce/excite" and "intensity." Ozeki San'ei instead suggested *shishō* (刺触), replacing the second character with one meaning "burst," which dominated until the 1870s, when Takano's original *shigeki* became standardized. Takano also dubbed the five senses *goshinki* (五神器), inserting *shin* (神) when most doctors described the five senses as *gokan* (五官). He also deployed 神 in coining terms like *shishin* (視神) for vision and used *ishin* (胃神) or "stomach spirit" for gastric sensitivity, rather than *chikaku* (知覚) or "feeling."

129. The contradictions that underlay Lu Gwei-djen's scholarly work ran parallel to those that underlay her social and political position, which will be explored in chapter 5.

130. Carrubba, Bowers, and Ten Rhijne, "The Western World's First Detailed Treatise on Acupuncture," 396.

131. Iwanaga Soko (1634–1705) was one of four physicians sent to Deshima to serve as an interpreter for Dutch merchants. In particular, he was tasked with producing a list of 165 questions for Willem ten Rhijne to answer. The questions and responses were compiled in the first volume of a series titled "*Zen seishitsu iwa*" and involved the work of four additional translators. On this collection, see Otsuka, "Willem ten Rhyne in Japan," 257.

Chapter 3: The Limits of Anatomy through Tu (圖)

1. To provide context, cameras first arrived at Shanghai's docks in 1873, bringing models from Britain, America, Belgium, and Japan. By the late 1920s, the dominant personal camera was the German Agfa brand, frequently seen in the municipal newspaper *Shenbao*. Cheng Dan'an likely used this model to take these photos and his particular use of photography in this manner is notable. Wang, *The History of Photography in Shanghai*, 14.

2. Special thanks to Alex Wragge-Morley for reviewing this chapter and clarifying my early modern perspectives of the mind and body.

3. These debates engaged with vitalist discourses like that of Margaret Cavendish (1623–1673), who developed an early vital matter theory. Vitalism emerged in the eighteenth century, and held that a vital force distinguished living from non-living matter. Although vitalism declined by the mid-nineteenth century with the rise of organic chemistry and cell theory in biology, it was empirically grounded and integral to the development of modern biology. Historians now recognize the diversity of vitalist thought, which ranged from the vital matter theories of the late seventeenth century to the medical vitalism of Paul Joseph Barthez (1734–1806) and Théophile de Bordeu (1722–1776). Rather than rejecting mechanism, vitalism offered an alternative molecular and dynamic vision of matter, life, and physiology. For an introduction to vitalism, see Reill, *Vitalizing Nature in the Enlightenment*, 2005.

4. This is not to say Cartesian ideas can be used to generalize conceptions of the mind at the time. For instance, the English natural philosopher Kenelm Digby (1603–1665) viewed matter as inherently vital and prone to self-organization, unlike Cartesian mechanical philosophy.

5. Descartes made the famous distinction between the human-specific rational soul and the relatively unremarkable animal spirits in *L'Homme* (*Treatise of Man*). There were certainly finer distinctions between the two upon which many Descartes' contemporaries elaborated, including Tommaso Cornelio (1614–1684), Nicolas Steno (1638–1686), and Claude Clerselier (1614–1684). For an excellent series of essays on the reception of *Treatise of Man*, see Antoine-Mahut and Gaukroger, *Descartes' Treatise on Man and Its Reception*, 2017.

6. Wragge-Morley, *Aesthetic Science*, 2020, 63.

7. Wragge-Morley has also pointed out that Willis's description of the animal soul increasingly began to appear similar to that of the rational soul.

8. Wragge-Morley has recounted how Willis's contemporaries, like Nicolas Steno, were profoundly skeptical of Willis's narrative about the brain. Willis had primarily relied on metaphors of objects like the bain-marie to dream up the functions of the brain's dura mater.

9. Salisbury and Shail, *Neurology and Modernity*, 2010.

10. The concept of "sympathetic" nerves arose from the belief that similar-looking nerve structures resonated with each other. These resonant qualities defined the "sympathetic" nervous system, wherein form dictated function: structure preceded purpose. This fixation on physical form welcomed metaphors likening the body to a musical instrument, with sympathetic vibrations between related parts. Salisbury, *Neurology and Modernity*, 4.

11. House, "Beyond the Brain," 41.

12. Salisbury, *Neurology and Modernity*, 15–16.

13. Salisbury, 19.

14. Benická, "Xin as a 'Qualitatively Equal' Co-constituent of Phenomena in Chinese Mahayana Buddhism," 2006.

15. Ng, "An Early Qing Critique of the Philosophy of Mind-Heart (*Xin*)," 89.

16. For more on translation works and institutions during this period, see Tsien,

“Western Impact on China Through Translation,” 1954; Wright, “The Translation of Modern Western Science,” 1998.

17. To offer some context, Hobson was one of many physician-translators working with acupuncture-moxibustion texts in this period including Jean-Baptiste Sarlandière (1787–1838), Jules Cloquet (1790–1883), and Louis Berlioz (1776–1848). Notably, German physician Franz Philip von Siebold (1796–1866) had been intimately involved with Japanese acu-moxa, and French anatomist/physiologist Jean-Baptiste Sarlandière (1787–1838) published on “electroacupuncture” (*l'électro-puncture*). See Vigouroux, “The Surgeon's Acupuncturist,” 2017; Macé, “The Medicine of Shizaka Sōtetsu 石坂宗哲 (1770–1841),” 1995; Sarlandière, *Mémoires sur l'électro-puncture*, 1825.

18. This idea, also known as *qiongli* (窮理), was first espoused centuries before by the early modern scholar Lu Jiuyuan (陸九淵) (1139–1192). In Lu's words, “Should there arise a sage in the Sea of the East, or that of the West, or North, or South, it would be the same mind, and the same principle.” Quoted in Chan, “Sinicizing Western Science,” 553.

19. Hong Kong had just become a British concession following the Treaty of Nanjing. See Chan, “Sinicizing Western Science,” 534.

20. Andrews, “Tuberculosis and the Assimilation of Germ Theory in China,” 1997, 122.

21. For a linguistic analysis of these terms, see Li, “Investigation of Medical Terms in *Quanti Xinlun*,” 2018.

22. Other publications attributed to Hobson include *Annual Report of the Missionary Hospital at Canton* (惠愛醫館年紀) (1850), *Treatise on Physiology* (全體新論) (1851), *Theological Evidences* (上帝辯證) (1852), *Commentary on the Gospel of John* (約翰真經解釋) (1853), *Forms of Prayer* (祈禱式文) (1854), *A New Treatise on Natural Philosophy and Natural History* (博物新編) (1855), *Explanation of Faith* (信德之解) (1855), *Summary of Christian Principles* (問答良言) (1855), *Selections from the Holy Scripture* (聖書擇錦) (1856), *Critical Ancient Excerpts* (古訓撮要) (1856), and *Advent of Christ* (基督降世傳) (1856).

23. The first edition of *Treatise on Physiology* had an initial print run of 12,000 copies. Elman, *On Their Own Terms*, 292.

24. In the same year, Daniel Jerome Macgowan (1814–1893) also published a book on animal magnetism, nerves, electricity, and electrotherapy titled *Bowu tongshu* (博物通书) (1851).

25. The 10-volume compendium *Zhenjiu dacheng* (针灸大成) published in 1601 was attributed to physician Yang Jizhou (杨继洲) (1522–1620). It sourced entries from over 20 preexisting medical treatises, some of which also included content from the canonical *Huangdi neijing*.

26. These examples refer to *zhu guan* (竹管), *ji ling guan* (鸡翎管), and *mao guan* (笔管), which may have been used to describe tubes of different materials and sizes to blow air in the ears. These references are found in Books 1, 3, and 5 of the *Zhenjiu dacheng* (针灸大成). See Zheng and Liu, eds., *Zhenjiu dacheng* (针灸大成), 1936.

27. Chen relied on the ambiguity of language to crystalize the ontological certainty

of nerves. For instance, words like *hua* (化) could simultaneously have singular meanings like transform, grow, generate, or mean all three. Similarly, words like *sheng* (生) could individually mean generate, maintain life, grow, or all three.

28. Instead, the word *nao* (脑) appears in the word *naokong* (脑空) to reference the empty space in the skull, or the shell of the head. This appears in Books 5 and 7 of *Zhenjiu dacheng* (针灸大成).

29. Other translations of medical texts by John G. Kerr (1824–1890), John Dudgeon (1837–1901), and John Fryer (1839–1928), eventually overshadowed *Quanti xinlun* (全體新論) (1851–1857).

30. The original text reads: *jiu dui jisui zuoyou gongsheng sanshiyi dui* (九對脊髓左右共生三十一對).

31. Pan was involved in updating an 1852 anthology of 56 works by Chinese and Western authors who had written about topics in science and medicine. Elman, *On Their Own Terms*, 293

32. Letter to Dr. Tedman, Canton, August 21, 1852, Council for World Mission Archives, South China, Incoming Letters, box 5, 1851–1852, microfiche, no. 92. Quoted in Li, "An Early Version of the Composer's Selection," 208.

33. For more on biology and modernism, see Pauly, "Modernist Practice in American Biology," 1994.

34. As historian Robert Brain has recounted, Huxley became known as one of the "prophets" of the protoplasm. See Brain, *The Pulse of Modernism*, 37.

35. Head, "On Disturbances of Sensation," 1893, 10.

36. Head, 63–69.

37. While early modern microscopic inspection filled anatomists with wonder at discoveries like the pancreatic duct, kidney structures, uterine follicles, salivary and tear glands, and finer nervous structures, observation alone had significant limitations. Cook, "Medicine," 2016.

38. For a nuanced history of this period, see Guenther, *Localization and Its Discontents*, 2015.

39. Head and Sherren, "The Consequences of Injury to the Peripheral Nerves in Man," 117.

40. Prior to this period, the London Hospital's location in the infamously rough East End docks area had hampered its ability to become a renowned research institute. Its relocation to the more affluent West End shifted its patient demographic, allowing Head to collect more diverse clinical cases. Morris, *A History of the London Hospital*, 1910.

41. Anon, "London Hospital Pathological Institute," 1901.

42. The two nerves that Sherren severed were the radial and "external cutaneous" nerves.

43. Head and Sherren "The Consequences of Injury to the Peripheral Nerves in Man."

44. Jacyna, *Medicine and Modernism*, 2008, 125.

45. Head and Rivers, "A Human Experiment in Nerve Division," 343.

46. Herzig, *Suffering for Science*, 2005.N.J.: Rutgers University Press, 2005.

47. Head, "Correspondence, Henry Head to Ruth Mayhew," November 6, 1903.

48. Head, "Correspondence, Henry Head to Ruth Mayhew," November 27, 1903.

49. Head, "Correspondence, Henry Head to Ruth Mayhew," May 15, 1903.

50. Rivers and Head, "A Human Experiment in Nerve Division," 332.

51. Rivers and Head, 332.

52. In 1889, Gaskell posited that nerves and nerve cell groups in the cerebrospinal axis were arranged not in a continuous chain but "metamerically," in a linear series. He argued this evidence showed vertebrate nervous tissue originated in a distinctly segmented animal. See Gaskell, "On the Relation between the Structure, Function, Distribution and Origin of the Cranial Nerves," 1889.

53. "Deep" sensations extended from motor nerves and operated independently from sensory nerves in the skin. These deeper sensations responding to pain produced by excessive pressure or joint injuries. Meanwhile, "protopathic" and "epicritic" sensations functioned closer to the surface of the skin. "Protopathic" responses were initiated by painful stimuli and extreme temperatures and acted as part of the reflex system with a "widely diffused response" without any appreciation of discrete, localized stimuli. Meanwhile, "epicritic" responses distinguished more discrete sensations, such as the location multiple stimuli and finer grades of temperature. Where protopathic represented a basic process for extreme stimuli, epicritic served as a more sophisticated process for finer stimuli. Head, Rivers, and Sherren, "The Afferent Nervous System from a New Aspect," 1905, 111.

54. Rivers and Head, "A Human Experiment in Nerve Division," 389.

55. Head, "On Disturbances of Sensation," 390.

56. Yeh, *Shanghai Splendor*, 2007.

57. These manifested as numerous terms: "Chinese-Western," or *zhongxi* (中西), "Chinese-French," or *zhongfa* (中法), and "Chinese-English," or *zhongying* (中英) and *huaying* (華英). See Chang, "From Heart to Brain," 6.

58. Huang sold his brain tonic by promoting ingredients he knew could feed the brain after studying American products that used phosphorous, cod liver, iron, and vitamins A and D. Chang, 17.

59. For the original advertisement, see Anon., "*Bu naozhi qianjin yi xiao shuo*," 1904.

60. Historian Hugh Shapiro argues neurasthenia shifted from an elite disorder of the kidneys to a disorder of the laboring body in the early Republican era, with weakness and fatigue now stemming from atrophied nerves. A key turning point was the transformation of *shèn jīng shuāiruò*, previously kidney weakness, into *shénjīng shuāiruò*, or neurasthenia, a weakness of the nerves. See Shapiro, "The Puzzle of Spermatorrhea," 1998.

61. Huang had dropped out of school to sell his mother's eye drops while also peddling aphrodisiacs. Chang, "From Heart to Brain," 7.

62. See Zhu, "Body, Soul, and God," 1996.

63. Lei, *Neither Donkey nor Horse*, 75.

64. Interpreting *jingluo* (經絡) as blood vessels did not introduce a new idea of blood vessels, as medical commentators in the eighth-century used the observation of arterioles, neuronal branches, and capillary vessels to describe the branching meridian channels. Lu and Needham, *Celestial Lancets*, 18.

65. Tang had been reading Wang Qingren's 1830 critique of classic anatomy, *Correcting the Errors of Medicine*, and John Dudgeon's 1886 translation of *Gray's Anatomy*. The full title of Tang's book is *Zhongxi huitong: Yijing jingyi* (中西匯通醫經精義) [*The Essential Meanings of the Medical Canons: (Approached) through the Convergence and Assimilation of Chinese and Western Medicine*]. Lei, "Qi-Transformation and the Steam Engine," 320.

66. Tang Zonghai's famous 1892 treatise *Essence of Medical Canons* had combined numerous sources, including critical reflections and recent translations of *Gray's Anatomy*.

67. Here, Tang reinterpreted descriptions of Qi-transformation, or *qihua*, from eighth- and tenth-century texts. Lei, *Neither Donkey nor Horse*, 72.

68. Tang's ideas greatly influenced Zhang Xichun (1860–1933), a scholar and physician celebrated as a pioneer in integrating Chinese and Western medicine in early-twentieth-century China. See Andrews, "From Case Records to Case Histories," 333.

69. As historian Hans van de Ven has pointed out, the Republican central governments were less financially endowed than the provincial governments, which led to the decentralization of power. See van de Ven, "Public Finance and the Rise of Warlordism," 1996.

70. Medicine in China at this point still ranged on a broad spectrum from itinerate healers to scholar-physicians. The debates on meridian maps here extend from scholar elites (Tang Zonghai) to former political exiles (Zhu Lian). To learn more about the spectrum of Chinese healing practices, see Andrews, *Modern Chinese Medicine*, 2015, 26–50.

71. Lei, "How Did Chinese Medicine Become Experiential?" 2002.

72. To further blur the distinction between experience and experimentation, a number of historians of science and of medicine have contributed longer histories of scientific and everyday experiences of observation. See Daston and Lunbeck, *Histories of Scientific Observation*, 2011.

73. In the 1920s, China did not have a universal system for medical licensing. For more on professional societies in China, see Xu, *Chinese Professionals and the Republican State*, 2001.

74. For more about these sentiments, see Bu, "Social Darwinism, Public Health, and Modernization in China, 1895–1925," 2009, 96–98.

75. In doing so, medical reformers produced a new dichotomy between Chinese medicine and biomedicine that relied on different ontologies. The former was described as experiential and subordinate to the latter, which was experimental. See Lei, "How Did Chinese Medicine Become Experiential?"

76. Bridie Andrews notes that students in the first wave studying medicine in Japan in the early twentieth century came mostly from northern China and learned German-

style medicine. Chinese students sent abroad via medical missionaries were typically southern and trained in Britain / North America. See Andrews, *Modern Chinese Medicine*, 146.

77. Lei, *Neither Donkey nor Horse*, 12." Meanwhile, the Ministry of Health declined to regulate Western medicine when the National Medical Association proposed an intermediary council to set standards for it. Andrews, *Modern Chinese Medicine*, 153.

78. Quoted in Andrews, 337.

79. Ironically, Yu Yan had avidly supported these new reforms and was so excited by the Movement that he could not sleep soundly for days. Lei, *Neither Donkey nor Horse*, 146. Based on Yu's writings in Yu, "My Opinion on the Proposal for Sorting out Chinese Medicine," 1936.

80. Lei, *Neither Donkey nor Horse*, 78.

81. Communist Party members later cast Yu Yan as the primary opponent of Chinese medicine. Traditional practitioners even rewrote the title of his proposal from "Abolishing Old-Style Medicine in Order to Clear Away the Obstacles to Medicine and Public Health" to "Original Text of the Proposal to Abolish the Practice of Chinese Medicine, Discussed and Decided by the Central Board of Health." See Lei, 86.

82. Hinrichs and Barnes, *Chinese Medicine and Healing*, 2013.

83. Cheng's grandfather specialized in pediatrics, and his father specialized in pediatrics and surgery.

84. Nanjing Medical College Archives. Cheng Dan'an Papers. Uncatalogued.

85. Andrews, *Modern Chinese Medicine*, 201.

86. Mayanagi Makoto (真柳誠) has examined the history of acupuncture and moxibustion between Japan and China, detailing Cheng Dan'an's activities and arguing that the preservation of classical scholarship the Edo period enabled its later revival in twentieth-century China. Special thanks to my reviewer for this reference. Mayanagi, "The Contribution of Japan," 2006.

87. Cheng would turn the institute into a school that trained students in Mandarin and Japanese. It has since matriculated over 3,000 students. Just as Cheng started expanding his school's library, fighter jets encroached along the coast.

88. Peattie, Drea, and van de Ven, eds., *The Battle for China*, 2013.

89. Radical artist Liu Haisu (1896–1994) famously introduced nude human models for art education a decade earlier. For more about this period, see Danzker et al., *Shanghai Modern*, 2004.

90. Wang, *The History of Photography in Shanghai*, 13.

91. Cheng, *Revised Approaches to Studying Chinese Acupuncture Moxibustion Therapy*, 1933, 35.

92. Cheng, 34.

93. Cheng, 34.

94. Andrews, *The Making of Modern Chinese Medicine*, 127.

95. Numerous histories have described foreign presence representing foreign domination, encouraging reform attempts during this period. For an overview of movements like the "Self-Strengthening Movement" and "Foreign Affairs Movement"

from 1860 to 1895 setting the stage for conceiving Chinese medicine, see Andrews, *Modern Chinese Medicine*, 14–24.

96. These sensations indicated different textures, densities, and speeds rather than individual moments. Although classical texts often described these sensations as characteristics of palpating along the *jingluo*, Cheng insisted that these would be sensations that the patient needed to feel and self-report. Cheng, *Revised Approaches*, 36.

97. Cheng, 271. This dynamic sensation of heaviness, sourness, or heavy-sourness from the patient and heavy-tightness from the physician has also interested acu-moxa researchers like Park et al., "Does Deqi (Needle Sensation) Exist?," 2002, 45–50; Hui et al., "Perception of Deqi," 2011, 2; Hui et al., "Perception of Deqi by Chinese and American Acupuncturists: A Pilot Survey," 2.

98. Cheng, *Revised Approaches*, 36.

99. Cheng, 50.

100. Cheng writes, "已根本動搖 . . . 則亦不能 成立矣," 51.

101. Cheng, 35.

102. This, he later claimed, was the sign of a matured scholar-practitioner. Hai-shu Sun et al., "Analysis on Cheng Dan-An's Educational Thought in His Book Chinese Acupuncturology," 39.

103. Nagahama, *A Study of Meridians*, 1.

104. For instance, Cheng explained, "To treat asthma, locate the site by wrapping a string around both feet. Then fold the string in half around the neck and [see] where the two ends meet at the center of the back. Apply fifteen [stubs of] moxa a day; continue for seven days, rest for seven days, then treat for another seven days. Repeat [this sequence] for a few months and [the patient] will feel better regardless of age." Cheng, *Cheng Danan's Simple Moxibustion Method*, 31.

105. Andrews, *Modern Chinese Medicine*, 204.

106. Murray, "A Qualitative Analysis of Tickling," 322.

107. Murray, 293.

108. Murray, 296.

109. Murray, 301.

110. Murray (305) appeared to use "contact pressure" and "contact-brightness" interchangeably, demonstrating the blurry distinctions of pain, pressure, and contact.

111. For instance, when Rivers plucked a hair from Head's arm, he would say, "You are touching me, you tickle me," conflating "touch" with basic "contact." Murray, 298.

112. Murray, 298.

113. Trotter was a British surgeon at East London Hospital. Ten years older than Davies, Trotter mentored Davies in surgery at University College Hospital. Davies' obituary describes Trotter and Davies experimenting on each other, eventually disagreeing with Head's epicritic and protopathic sensation theories, though their published paper suggests otherwise. C. P. T., "In Memoriam," 1965.

114. Trotter and Davies, "Experimental Studies of the Innervation of the Skin," 1909, 136.

115. Trotter and Davies, 138.

116. They explained that their choice of nerves allowed them to compare larger areas supplied by one nerve and changes in response to anesthesia.

117. Trotter later served as Sergeant Surgeon to three successive English kings: George V, Edward VIII, and George VI. For more on Trotter, see Elliott, "Wilfred Batten Lewis Trotter," 1941, 325.

118. Trotter and Davies, "Experimental Studies of the Innervation of the Skin," 138.

119. Despite his fixation on precision, Trotter's Royal Society obituary described him as speculative yet experimental. Elliott, "Wilfred Batten Lewis Trotter," 326.

120. Trotter and Davies, "Experimental Studies of the Innervation of the Skin," 207.

121. Boring pursued experimental psychology after abandoning electrical engineering training. He had been significantly influenced by German structuralism during his time at Cornell. See Boring, "Oral History Interviews," July 5, 1961

122. Boring, "Cutaneous Sensation after Nerve-Division," 1961, 80.

123. Boring, 27.

124. Nichols, "Correspondence," June 26, 1925.

125. The physical quality of the nervous system, both in the body and in the mind overwhelmed the soldiers that Head treated and "produced more individual and corporate misery than cancer." See Anon., "National Council for Mental Hygiene," 766.

126. Quoted in Jacyna, *Medicine and Modernism*, 98.

127. See Crary, *Techniques of the Observer*, 1990; Hughes, *Consciousness and Society*, 1958; Ryan, *The Vanishing Subject*, 1991.

128. Walshe, "The Anatomy and Physiology," 50.

129. Henson, "Henry Head's Work on Sensation," 537.

130. During this period, other figures in Japan, Germany, and France this period continued to publish on theories of acupuncture and anatomy. For instance, historian Mathias Vigouroux has written about early-nineteenth-century collaboration between Japanese acupuncturist Ishizaka Sōtetsu (石坂宗哲) (1770–1841) and German surgeon Franz Philipp von Siebold. Vigouroux argues that Siebold was particularly interested in Ishizaka's innovative assimilation of Chinese nutritive Qi [*ei ki* (栄気)] into arteries [*dōmyaku* (動脈)] and protective Qi [*e ki* (衛気)] into veins [*seimyaku* (静脈)]. See Vigouroux, "The Surgeon's Acupuncturist," 2017.

131. Cheng, *Cheng Dan'an's Simple Moxibustion Method*, 3.

Chapter 4: Generic Maps and the Failure of Standardization

1. For an abbreviated version of these materialist debates, see Li, "Communist Materialism and Illustrating Medical Textbooks," 2022.

2. As I have noted in the introduction, rather than translating "*zhenjiu*" as "acupuncture," I draw on the dual nature of *zhenjiu* as understood by my historical actors who wrote about "*zhen*" (acupuncture) and "*jiu*" (moxibustion) as two separate, but complementary, therapeutic practices.

3. Zhu, *New Approaches*, 1954, 12.

4. Kim Taylor has described Zhu's organization of the body as reflecting military metaphors. In this analysis, I focus on anatomy as Zhu's way into subverting classic

theories of acupuncture-moxibustion. For a political history of Chinese medicine in early Communist China, see Taylor, *Chinese Medicine in Early Communist China*, 2005.

5. Special thanks to Jet Prendeville for her assistance in studying the material construction of the book.

6. Wang's uses of these marks suggests that he was working with a Japanese *mingtang tu*, which included similar symbols, but for different sites.

7. In the third edition of *New Acupuncture-Moxibustion* published in 1980, editors removed ⊗ and replaced it with ○ to indicate points that only prohibited needling. This shows how acupuncture-moxibustion fluctuated as a practice.

8. *Fen* (分) is approximately one-tenth of a *cun* (寸), here 33.33 millimeters, although *cun* was also a relative measurement using middle finger phalanges. In chapter 1, I discussed how *cun* referenced the "same body-inch" from the seventh century. By the twentieth century, *cun* was also the East Asian "inch" standard, longer than the British unit.

9. Zhu, *New Approaches*, 1954, 136–145.

10. For a comprehensive biography of Zhu Lian's life, see Zhang, *Zhu Lian Yu Zhenjiu*, 2015.

11. Other biographical details can be found in Li, *Biographical Dictionary of Chinese Medicine*, 1988, 260–261.

12. For a collection of Zhu Lian's photographs, see Zhang, *Zhu Lian yu zhenjiu*, 2015.

13. A commemorative essay on Zhu Lian can be found at Wei, Pan, and Mo, "Our Teacher Zhu Lian," 2014.

14. Schmalzer, *Red Revolution, Green Revolution*, 2016.

15. The phrase being "*zhongyi kexuehua, xiyi dazhonghua.*" Although the effort to render Chinese medicine "scientific" in the National Medicine Movement during the 1930s extended from Chinese medical practitioners' effort to demonstrate their relevance to the state, the project of "scientification" in 1944 came from the Communist Party to incorporate Chinese medicine as a part of their state identity. Taylor, *Chinese Medicine*, 17.

16. Lu, "Postcript," 277.

17. Lu, 278.

18. Wei, Pan, and Mo, "Our Teacher Zhu Lian."

19. Zhang, "My Encounter with Acupuncture-Moxibustion," 1954, 7.

20. Zhang, 7–10.

21. Zhu, *New Approaches*, 1954, 1.

22. Zhu, *New Approaches*, 1980.

23. Pavlov, who experienced four Russian revolutions first-hand, maintained a cautious approach to ideological affiliations with political parties. Nevertheless, he was not apolitical. For example, prior to the 1905 revolution, he actively supported the movement for broader democratic rights and even ran for a seat in the Duma, representing the moderate conservative Octobrist party, though unsuccessfully. Todes, *Ivan Pavlov*, 2015, 3.

24. Fan, "Pavlovian Theory and the Scientification of Acupuncture in 1950s China," 2013, 143.

25. This urge connected to Pavlov's early training as a priest where he had first wrestled with metaphysical debates on the materiality of the soul. Todes, *Ivan Pavlov*, 12. The Cartesian split suggested that the rational soul was immaterial whereas the lower animal spirits were material. In contrast, early modern neurophysiologists like Thomas Willis rendered the rational soul material by placing it in the brain. See chapter 2.

26. To this end, Pavlov and his many collaborators did not engage with speculating on ultimate causes. They resigned themselves to studying proximal reflexes where behaviours, whether intended or unintended, manifested as physiological movements.

27. For instance, Ivan Sechenov's *Reflexes of the Brain* also claimed that the "psychical activity of the brain" manifested in function, physiology, and personality. Sechenov, *Reflexes of the Brain*, 1863, 2.

28. In contrast, George Lewes's *Physiology of Common Life* identified digestion as the most basic and observable kind of physiological function. Lewes began his treatise with observations relating to hunger and thirst, food and drink, digestion, and indigestion, before moving to blood circulation, respiration and suffocation, and finally the mind and the brain. See Lewes, *The Physiology of Common Life*, 1859.

29. Some dogs better tolerated a fistula in their stomach, others better tolerated a hole in their pancreas. This meant that some were better suited for collecting juices from gastric glands, whereas others were better suited for collecting juices from the pancreas. For more on Pavlov's dogs, see Todes, *Ivan Pavlov*, 250–251.

30. Sechenov had studied physics and mathematics and was inspired by the second law of thermodynamics in conceptualizing different types of forces in acting on a discrete system. His early education furthermore introduced him to a German electromagnetic machine, which he used to derive more metaphors for functions in the body. Sechenov, *Reflexes of the Brain*, 119–120.

31. Sechenov, 20.

32. Todes, *Ivan Pavlov*, 248.

33. As Todes notes, Pavlov understood canine physiology to map directly onto human bodies. In other words, people were only more complex dogs (248).

34. Todes, 197.

35. Even though Pavlov insisted on the visibility of objects pressing on animals' sensory organs—inhibiting and exciting nerves—multiple input forms did not consistently act on the body. Different experimental inputs competed. Multiple signals seemed to generate interference. For instance, a steady metronome sound paired with food induced six saliva drops in a dog. But a metronome with a buzzer and food only produced three drops. Pavlov explained the metronome had an excitatory effect by increasing saliva, whereas the buzzer had an inhibitory effect by decreasing it. Yet both were auditory inputs working through the same system with different bodily effects. Perhaps nerve clusters could be simultaneously excited and inhibited to produce a general effect.

36. In his private papers, Pavlov speculated that reflex centers possibly acted more like gravitational attractions. The kind of "attraction" that Pavlov and his team described potentially explained how the strength of signals diffused in the brain. Different centers in the body simply tugged at each other. Dan Todes cites these sources from Pavlov's research notes during 1911–1913.

37. Pavlov, *Lectures on Conditioned Reflexes*, 1941, 162.

38. Pavlov, 102.

39. Todes, *Ivan Pavlov*, 492.

40. Pavlov, *Lectures on Conditioned Reflexes*, 355.

41. Engels famously created his own laws of dialectical materialism for the natural sciences. For a historical genealogy of dialectical materialism, see Pang, "Dialectical Materialism," 2019.

42. For a review of Mao's 1937 speech on contradiction, see Rojas, "Contradiction," 2019, 43–48.

43. Rather than accessing original publications in Russian, physiologists first encountered Pavlov in English before translating his essays into Chinese.

44. Zhao, "A Ten-Year Survey of Pavlovian Theory," 1959, 475.

45. In 1929, Pan Shu and Gao Zhaoyi worked together to translate V. M. Borovski's "Psychology of the Soviet Federation," and "Psychology in the Soviet Union." Li and Yan, "Chou Siegen K.," 2014, 333.

46. Chou had been studying psychology at Stanford University since 1925 before graduating in 1930.

47. When Chou returned to Beijing, he bought a British photo-polygraph called a Darrow machine to directly measure electric potential at the skin. If reflexes could be excited and inhibited, then they potentially exhibited a magnetic pull that translated into electric currents. See Li and Yan, "Chou Siegen K.," 334.

48. In the process of translating Pavlov from Russian, Chou realized that the English translation of "conditioned reflex" was misleading. Chou corrected "conditioned" to "conditional," recognizing that behaviors were shaped by environmental factors.

49. Zhang, "Review of Soviet Union Research," 2019.

50. Guo, "Remembering Pavlov," 1936.

51. Zhu, "Second Preface," 1954, 24.

52. Zhu, 24.

53. Zhu, 23.

54. Zhu did not elaborate on her position or speculate on how the nerves connected to the meridian points. She later explained that her many administrative obligations prevented her from writing more extensively on her own take on the meridian points.

55. Ma, "Anatomical Cìjī Chart," 1952.

56. Wang, "Xuetai's Theory on Acupuncture-Moxibustion," 2008, 309.

57. The Soviet Union had sent medical experts to Beijing to study acu-moxa, which later led to a collaborative effort of translating Zhu's book. The second edition of the book completely sold out, but for various reasons, the third edition would be

published posthumously and wouldn't appear until 1980, two years after Zhu Lian's death. Zhu, *New Approaches*, 1954, 18.

58. Needling and heating texts had apparently been introduced by P. Charukovsky (П. Чаруковский) to St. Petersburg in 1828. Qian, *Kitayskaya Narodnaya Meditsina*, 1959, 7.

59. Qian, *Kitayskaya Narodnaya Meditsina*, 9.

60. These Romanization reform attempts had first started with the Ministry of Education in 1912. Simmons, "Whence Came Mandarin?" 2017.

61. Chappell, "The Romanization Debate," 1980.

62. It is possible that China was more eager for the split with Russia because of Mao's pursuit of various policies that would undermine the relationship and lead to economic isolation. But it is also possible that the United States may have also played a role in this division.

63. The standard English translation of Engels' quote in *Anti-Dühring* is from Lawrence and Wishart, 1987, 112. It appears in Wang's abstract. See Wang, "Embodying the Law of Contradiction," 1959.

64. Wang, 7.

65. Zhu, *New Approaches*, 1954, 24.

66. Wang, *Handbook on Acupuncture and Moxibustion*, 33.

67. Wang, 33.

68. In 1956, Mao's Hundred Flowers Movement encouraged contending ideas, but critics were branded "reactionary idealists" and fields like genetics were stunted. See Schneider, *Biology and Revolution in Twentieth-Century China*, 2003.

69. There are over 33 styles of drawing the 14 meridians, commonly organized into three main groups based on the sources on which they are based. The texts that separate the groups include *Shisi jing fahui*, *Zhenjiu juying*, and *Lei jing tu yi*. See Huang, *Graphic History of Chinese Acupuncture-Moxibustion*, vol. 1, 2003, 293.

70. To Wang, any sources before the Republican era were considered as "*gudai yixue*," or "ancient medicine."

71. Wang selected this quote by Li Ting from his book *Yixue Rumen*, or *Introduction to Medicine* from the Ming dynasty. Wang, *The Handbook on Acupuncture and Moxibustion*, 55.

72. Wang, 55.

73. Wang, 55.

74. Wang, 56.

75. Wang first wrote about this type of movement in a series of articles published on a seminar he led on acupuncture moxibustion. Wang, "Acupuncture-Moxibustion Seminar (continued)," 1960.

76. This refers to an image by Manase Dōsan that historian Mathias Vigouroux has analyzed in detail. Vigouroux focuses on two of Manase Dōsan's 1571 *kirigami* (切紙) notes that features an image titled "Diagram of the Circulation of the Nutritive and Defensive [qi] Following or Against the Flow of the Twelve Channels" (十二經脈栄衛流注迎隨逆之圖), which showed the six hand and six foot channels along with their

entry/exit points. A thin line linked them to convey nutritive and defensive Qi pervading the entire body. See Vigouroux, "The Reception of the Circulation Channels Theory in Japan (1500–1800)," 2015.

77. There were 14 primary *jingluo*. The 12 that are illustrated here do not include the *renmai* and the *dumai* meridians, which are often described as operating independently from the other 12 *jingluo*.

78. This idea of a disabled body follows anthropologist Zoë Wool's history of Walter Penfield's disabled homunculus man. Wool, "Homunculus Revolts," 2019.

79. Wang took a formal position at the Chinese Medical Research Institute (which Zhu Lian helped establish), though this did not entirely protect him from Cultural Revolution violence. Zhu Lian would take on more administrative duties, eventually becoming a political prisoner condemned by her own Party during the Cultural Revolution. Wang, "Xuetai's Theory on Acupuncture-Moxibustion," 310.

80. Wang, "Perspectives on Improving Acupuncture Moxibustion Research," 1984, 28.

Chapter 5: Modern Mediations in Difference and Diplomacy

1. For instance, the Michigan-based embryologist Russell Woodburne had commissioned the artist J. C. Berger to illustrate one of the first three-dimensional cross-sections of skin in the early 1960s. Woodburne was praised for his novel approach to introducing human anatomy in his book, which began with organizing the human body based on embryological orientations. See Woodburne, *Essentials of Human Anatomy*, 1965.

2. See the image caption for a reproduction of Lu's original description. Lu and Needham, *Celestial Lancets*, 1980, 17.

3. Primary sources in this chapter primarily use Wade-Giles instead of Pinyin. For instance, *lo* and *pieh* in Wade-Giles would be *luo* (絡) and *bie* (别) in Pinyin. For assistance converting Wade-Giles to Pinyin, see https://www.chineseconverter.com/en/convert/wade-giles-to-chinese.

4. For an overview of the temporal and material variety of acupuncture analgesia in this period, see Li, "Pinpricks," 2018.

5. Lu shunned fame and prioritized privacy; until her death, few Chinese newspapers documented her contributions. Reports of Needham's visits to China from the 1950s to the 1980s describe Lu as his *zhushou*, or "assistant."

6. Lu and Needham, *Celestial Lancets*, 16.

7. Roy Porkert proposed the term "sinarteries," which Lu and Needham rejected on the grounds that the tracts were never conceptualized as tubes—a point of visual contention centered on the distinction between tube-like and tract-like lines lacking shading. Lu and Needham emphasized the concept of branching meridians, suggesting that translations such as "reticular conduit" were only marginally adequate (17).

8. Some of these comparisons, in particular Felix Mann's early books introducing meridians published in 1963, can be found in my article: Li, "Invisible Bodies," 2018.

9. Liu, Karl, and Ko, *The Birth of Chinese Feminism*, 2013.

10. Lei, *Neither Donkey nor Horse*, 2014; Andrews, *Modern Chinese Medicine*, 2015; Scheid, *Chinese Medicine in Contemporary China*, 2002; Taylor, *Chinese Medicine*, 2005.

11. Sean Lei observes that during this process of reinterpretation, some reformers reduced the concept of *jingluo* to mere blood vessels. This idea had been a subject of debate since Tang Zonghai first expounded on his interpretation of the material properties of *jingluo*. Lei, *Neither Donkey nor Horse*, 155.

12. In the 1920s, China did not have a universal system for medical licensing. For more on professional societies in China, see Xu, *Chinese Professionals and the Republican State*, 2001.

13. For more about these sentiments, see Bu, "Social Darwinism, Public Health, and Modernization in China, 1895–1925," 2009.

14. Winchester, *The Man Who Loved China*, 2008, 6.

15. Wang, *Lu Gwei-Djen and Joseph Needham*, 1999, 232–235.

16. The first big student strike was in 1919, which led to the formation of Ginling's Student Union.

17. The Henry Lester Institute was established as a private research institute in 1932 and held a number of specialists in chemistry, physiology, pathology, entomology, surgery, and medicine collaborated. For more on the Lester Institute in historical context, see Fu, "Houses of Experiment," 2016.

18. On the displacement of scientists and research institutions during the second Sino-Japanese war, see Fu, "Houses of Experiment."

19. Lu, "The First Half-Life of Joseph Needham," 1982, 4.

20. Lu, 7.

21. This disease state and its diagnosis involved a range of symptoms and differing explanations. For a comprehensive history of the shifting history of foot *qi* in East Asia, see Smith, *Forgotten Disease*, 2017.

22. Lu, "Pyruvic Acid and Muscle Metabolism," 1939, 2.

23. In contrast to Siegen Chou at Stanford, who focused on translating Pavlov's works into Chinese, Lu Gwei-djen devoted her efforts to photographing, replicating, and translating select plant entries from a revised fourteenth-century *materia medica* text. These plant illustrations, printed on glossy pages, were prominently featured in her thesis, standing in stark contrast to a brief list of structural forms of vitamin B1 also included in her work.

24. At the University of California, Berkeley, Lu continued her nutrition research before transferring to Columbia University, a move precipitated by her severe allergic reactions to acacia. Wang, *Lu Gwei-Djen and Joseph Needham*, 232.

25. Winchester, *The Man Who Loved China*, 53.

26. Lu, "The First Half-Life of Joseph Needham," 6.

27. Wang, *Lu Gwei-Djen and Joseph Needham*, 73.

28. See Emma Teng's 2017 exhibit on Chinese scientists in the US: http://china comestomit.org.

29. Wang, *Lu Gwei-Djen and Joseph Needham*, 67.

30. Her sources included records from court officials and Buddhist alchemists. Needham, "Wellcome Trust Reports," October 1, 1958, 6.

31. Needham, "Wellcome Trust Reports," January 10, 1958a, 3.

32. Needham, 3.

33. Nathan Sivin remarked that compared with her peers, Lu uniquely mastered hundreds of Chinese and English texts. Sivin, interview by author, 2011.

34. Lei, *Neither Donkey nor Horse*, 75.

35. The interpretation of *jingluo* as blood vessels did not introduce a novel concept of vascular structures. Medical commentators in the eighth century had already employed observations of arterioles, neuronal branches, and capillary vessels to describe the branching nature of meridian channels. Lu and Needham, *Celestial Lancets*, 18.

36. Cheng, *Revised Approaches*, 1933, 51.

37. Lu and Needham, "Dictionary of Medical Terms," 1964, 141.

38. Lu and Needham, "Notes for Revue Bibliographique de Sinologie," 1971.

39. Lu and Needham, "Index Termini Technologici," 1964.

40. Lu and Needham, "Index Termini Technologici."

41. Lu and Needham, "Notes for Revue Bibliographique de Sinologie."

42. While *nei* is often translated as "inner" and *wai* as "outer," Lu's classical study complicated this duality. For Lu, *nei* denoted corporeal matter and *wai* exceeded corporeality.

43. The concept of ontological projection bears similarity to mapping projections in linear algebra, where two entities are asymmetrically correlated. An illustrative example is the projection of a three-dimensional object onto a two-dimensional surface as a shadow: all points from the object map onto the shadow, but the reverse mapping is not possible. In this context, physiological orientations asymmetrically map onto Yin/Yang categories. Special thanks to Dr. Andrew Goldman for elaborating this analogy.

44. Lu, Needham, and Sivin, "Correspondence," 1972.

45. Lu, "The Inner Elixir (Nei Tan)," 1971, 3.

46. Lu, 5.

47. Lu, 5.

48. For a detailed history of late-nineteenth-century encounters of the occult, see Winter, *Mesmerized*, 1998.

49. Recent revisionist historiography of alchemy recontextualizes the discipline as an integral part of early modern European knowledge production and aesthetic practice, rather than relegating it to the realm of the occult. Numerous scholars have elaborated on these histories, in particular, Smith, *Body of the Artisan*, 2006; Newman, *Atoms and Alchemy*, 2006; Smith, *Business of Alchemy*, 2016; Wragge-Morley, *Aesthetic Science*, 2020.

50. Before the twentieth century, *weisheng* represented Chinese cosmology for protecting life and conquering disease. See Rogaski, *Hygienic Modernity*, 2004.

51. Lu, "Inner Elixir," 5.

52. Lu, 25. Italics added by the author to represent sections modified in Lu's draft.

53. Lu, 8.

54. Lu, 3.

55. For a more detailed history of this practice and background on select case studies, see Li, "Pinpricks."

56. Li, "Pinpricks," 217.

57. Some sources cite techniques of using acupuncture to relieve pain (migraines, toothaches, etc.) in the eleventh century. Still, Lu identified these techniques being described even earlier, in around the first century. Lu and Needham, *Celestial Lancets*, 115.

58. Han, "Scientific Evidence of Acupuncture," 1964, 8.

59. Han, 8.

60. For instance, bladder meridian man appeared on the cover of Felix Mann's *Acupuncture: The Ancient Chinese Art of Healing.*

61. As discussed in preceding chapters, each of the 14 meridian men was depicted wearing a distinct head ornament known as a *guanjin* (綸巾/纶巾). These ornaments varied in design, ranging from simple square handkerchiefs to elaborate panelled headpieces.

62. Melzack, "Shutting the Gate on Pain," 1975, 58.

63. Melzack, 60.

64. Melzack and Wall, "Pain Mechanisms," 1965, 971–979.

65. Melzack and Wall, 971.

66. These inconsistencies are similar to the ones that inspired Henry Head to experiment on his own body to develop new theories of pain.

67. Melzack and Wall, "Pain Mechanisms," 977.

68. Melzack, "Shutting the Gate on Pain," 65–66.

69. Katz and Rosenbloom, "Golden Anniversary," 2015, 285–86.

70. To be clear, this is with the exception of *dumai* (督脈), which was the only path that ran directly on the spine.

71. Illustrators depicted certain meridians, such as the foot-Yang-ming meridian (足陽明), as bifurcating into distinct tracts. For example, Wang Xuetai's illustrations portray this meridian branching into two separate paths on the face and in the lower calves. Notably, these divergent paths do not extend to or reference the back of the body.

72. An illustration of the minor-Yin-bladder-foot meridian man, bearing closer resemblance to classical depictions from the *Treatise of the Fourteen Meridians*, appeared on the cover of Felix Mann's 1962 publication, *Acupuncture: The Chinese Art of Healing*. This image is analyzed in greater detail later in the chapter.

73. Jeans was instrumental in establishing one of Canada's first interdisciplinary pain clinics at the Montreal General Hospital. Her career was marked by numerous administrative leadership positions, including Executive Director of the Montreal General Hospital pain clinic, Head of McGill University's School of Nursing, and leadership roles in the Canadian Nurses Association and the Academy of Canadian Executive Nurses.

74. Melzack and Jeans, "Acupuncture Analgesia," 1974, 162.

75. Melzack and Jeans, 163.

76. Katz and Rosenbloom, 2015.

77. Melzack, "Shutting the Gate on Pain," 1975, 67.

78. For instance, another famous case beyond China is the obstetrician Anton Jayasuriya (1930–2005) in Colombo performing cesarean sections with his needling techniques.

79. Reston, "Now, about My Operation," 1971.

80. Reston, "Reston Tells of Surgery," 1971; "Reston Has Appendectomy," 1971.

81. Dimond, "Acupuncture Anesthesia," 1971, 1563.

82. Dimond, 1563.

83. Schmalzer, *The People's Peking Man*, 2008.

84. On the representation of medical and revolutionary expertise during this period, see Li, "The Edge of Expertise," 2015.

85. Lee, "Acupuncture Anaesthesia—Its Theory and Practice," 1971, 3.

86. Lee, 6.

87. Needham, *Mao and the Dark Aspects*, 1970, 54.

88. Needham, 54.

89. Joseph Needham had previously forfeited a professorship promotion at Cambridge University after exposing US military plans to develop bacterial weapons for use against China during the Korean War. Given this history, he was unwilling to further jeopardize his position by criticizing the country to which he had dedicated his life's work. See Li, Zhang, and Cao, *Explorations in the History of Science and Technology in China*, 1982, 40.

90. Lu and Needham, *Celestial Lancets*, 1980, 219.

91. In his notes, Needham does not indicate the name of the operation.

92. Lu and Needham, 281.

93. Lu and Needham, 281.

94. In classical Chinese medical texts, the term for placenta exhibited semantic flexibility, potentially meaning either the uterus or the placenta, with the specific meaning determined by contextual factors. On the history of the female anatomy, see Raphals, "Treatment of Women," 2013, 42; Wu, *Reproducing Women*, 2010.

95. Needham, "Joseph Needham 1972 Notes," 1972, 118.

96. The surgeons in Guangzhou reported that acupuncture anesthesia did not work in around eight percent of their operations, after which they resorted to chemical anesthesia. Needham, "Notes on China," 1972, 137.

97. Needham, 139.

98. Needham, 179.

99. Needham's notes read, "In abdom. operations chem. anaesth. still nec. hence it might be that acup. has little effect on autonomic and sympathetic" (137).

100. Needham, "Notes on China," 118.

101. Lu and Needham visited around 20 different hospitals and research institutions during their trip. These excerpts come from separate interviews conducted at hospitals in Beijing and Shanghai (121).

102. Needham, "Notes on China," 141.

103. Needham, 165.

104. Needham, 121.

105. Needham, 141.

106. Needham, 165.

107. Needham, 141, 167.

108. See Henry Head on "referred pain" that describes a particular relationship among neuronal sensation.

109. Needham, "Notes on China," 213.

110. Needham noted that this point was called the *san li* point that traveled along a foot meridian called *zú yīn míng jīng* (足陰明經), which is now commonly known as ST 36, stomach 36 (211).

111. Needham, "Notes on China," 211.

112. Needham, 215.

113. Needham, 143.

114. Needham, 207.

115. This is mostly inferred from the fact that Gwei-djen would research and provide notes for the *Science and Civilization* series while Joseph did all the writing. See Ho, *Reminiscence of a Roving Scholar*, 2005, 102.

116. Chiang et al., "Studies on Spinal Ascending Pathway," 1975; Man and Baragar, "Local Skin Sensory Changes," 1973; Chang, "Integrative Action of Thalamus," 1972.

117. Pilla, "Mechanisms of Electrochemical Phenomena," 1973.

118. Brown, Ulett, and Stern, "Acupuncture Loci," 1974, 67.

119. Mann, *Scientific Aspects of Acupuncture*, 1983, 80.

120. All of the books that Mann published early in his career introduced acupuncture to a popular audience. For instance, see Mann, *Acupuncture*, 1962; Mann, *The Meridians of Acupuncture*, 1964.

121. Mann, *Scientific Aspects of Acupuncture*, 80.

122. Lu, "Notes on Acupuncture," 1975.

123. Mann et al., "Treatment of Intractable Pain," 1973.

124. Mann, "Acupuncture Analgesia," 1974.

125. Mann, 363.

126. In his personal correspondence, Mann disclosed that only approximately 5 percent of his experimental cases yielded a satisfactory anesthetic effect. Mann, "Correspondence," 1975.

127. Mann, "Acupuncture Analgesia," 364.

128. Mann, *Acupuncture*, 1971, 5.

129. Joseph Needham received an English translation of Kim Bong Han's article from Poul Bonnevie, who at the time served as president of the Danish Society for Protection of Scientific Work and was affiliated with the World Federation of Scientific Workers. Kim, "On the Kyungrak System," 1963.

130. Mann, *Acupuncture*, 10.

131. The original maps appeared in Keegan and Garrett, "The Segmental Distribution of the Cutaneous Nerves," 1948.

132. The practice of numbering points along meridian paths was reminiscent of Cheng Dan'an's photographic representations from the 1930s. While some practitioners attribute the initial numbering of meridian paths to George Soulié de Morant (1878–1955), the visual correlation established between meridian points and dermatome levels held new ontological implications. For examples of de Morant's imagery, refer to the reprinted illustrations in George Soulié de Morant's *Acupuncture Chinoise Atlas*, 2018. For recent scholarship on George Soulié de Morant, see Candelise and Guill, "Chinese Medicine Outside of China," 2011; Dubois, "Revisiting the Medical Work," 2019.

133. In particular, Mann pointed to the following experiment as evidence for this phenomenon: Ray, Hinsey, and Geohegan, "Observations on the Distribution," 1943.

134. Mann, *Acupuncture*, 12.

135. Mann, *Scientific Aspects of Acupuncture*, 16.

136. Mann, 16.

137. In particular, Mann references Charles Sherrington, one of Henry Head's contemporaries. See Sherrington, *The Integrative Action of the Nervous System*, 1961.

138. Needham, "Notes on Meeting with Felix Mann," 1973.

139. Lu and Needham, *Celestial Lancets*, 207.

140. Mann, *Scientific Aspects of Acupuncture*, 89–90.

141. Lu, *Celestial Lancets*, ix.

142. Lu, ix.

143. "*Xue*" (穴) or meridian sites, translates to "hole," "minute cavity," or "crevice." Lu, 13.

144. The *Huangdi neijing* identified 160 sites by name, described 295 unique locations, and correlated 365 sites with the 365 degrees of the celestial sphere. In the latter half of the twentieth century, meridian sites were standardized to approximately 670 individual locations. Of these, 450 are explicitly recognized in practice, while 40 to 50 are commonly used. Lu, 13–15.

145. Lu, 22.

146. Lu, ix.

Epilogue

1. The idea of Yin-Yang man used here is distinguished from the language used for transgender surgeries as where Yin-Yang represented female-male binaries. For more on the history of transgender and intersex politics in Asia, see Chiang, "Christine Goes to China," 2017; Chiang, *Transtopia in the Sinophone Pacific*, 2021.

2. These can be translated to "Greater Yin," "Yang Ming," "Lesser Yang," and "Lesser Yin."

3. Throughout his extended career, Wang had served as the senior researcher at the Chinese Medical Research Institute, the senior adviser to the Chinese Acupuncture-Moxibustion Society, the honorary life-term president of the World Federation of

Acupuncture-Moxibustion, and the research director of the Institute of Acupuncture-Moxibustion, among other positions.

4. Wang, interview, "*Ruhe jianbie zhenjiu de liaoxiao*," 2006.

5. For instance, herbal remedies also operate through meridian paths and sites.

6. Helene Langevin's work has contributed the most to studying the effects of manipulating meridian sites on connective tissue. See Langevin et al., "Evidence of Connective Tissue Involvement in Acupuncture," 2002; Langevin et al., "Dynamic Fibroblast Cytoskeletal Response to Subcutaneous Tissue Stretch Ex Vivo and in Vivo," 2005; Langevin et al., "Connective Tissue Fibroblast Response to Acupuncture: Dose-Dependent Effect of Bidirectional Needle Rotation," 2007; Langevin et al., "Tissue Stretch Induces Nuclear Remodeling in Connective Tissue Fibroblasts," 2010; Engell et al., "Differential Displacement of Soft Tissue Layers from Manual Therapy Loading," 2016.

7. See Langevin et al., "Connective Tissue Fibroblast Response to Acupuncture," 3.

8. Langevin et al., "Manual and Electrical Needle Stimulation," 2015.

9. Lumpkin, Marshall, and Nelson, "The Cell Biology of Touch," 2010.

10. Zylka, Rice, and Anderson, "Topographically Distinct Epidermal Nociceptive Circuits," 2005.

11. Halata, Grim, and Bauman, "Friedrich Sigmund Merkel and His 'Merkel Cell,'" 2003.

12. Of course, researchers have continued to search for a neuroanatomical explanation for the therapeutic effects of electroacupuncture in the leg. For instance, while searching for sensory neurons in the leg, recent publications have identified PROKR2Cre-marked sensory neurons as linked to activating an anti-inflammatory response when stimulated by electroacupuncture. See Liu et al., "A Neuroanatomical Basis for Electroacupuncture to Drive the Vagal–Adrenal Axis," 2021.

13. Some examples include researching noncoding regions of the DNA and analyzing mosaic patterns between coding and non-coding regions through different algorithms to establish meaningful statistical correlations. This early work by Chung-Kang Peng led to the founding of the Center for Dynamical Biomarkers at Harvard. See Peng et al., "Mosaic Organization of DNA Nucleotides," 1994.

14. Purdue News Service. "Purdue DARPA (ElectRx) Project Focuses on Developing Implantable, Nerve-Stimulating 'Bionode' to Treat Inflammation," 2016.

15. Keller, *Making Sense of Life*, 2002.

16. Although this book has focused on hand-drawn inscriptions on the body and on paper, there remains a vast trove of three-dimensional models, from bronze sculptures to ceramic figurines to plastic dolls that can offer further insight into the material culture of presenting and re-presenting the body. The production, translation, transformation, and circulation of these pedagogical artifacts may follow a different path than maps of meridians and affected sensation.

17. These comments about the "dirty" aspects of labeled line and pattern theory were especially prominent during the 65th annual Montagna Symposium on the Biology of the Skin, titled "The Skin: Our Sensory Organ for Itch, Pain, Touch and

Pleasure," which I attended with the Lumpkin Lab during October 20–24, 2016. For a review of the conference, see Clary et al., "Montagna Symposium 2016," 1401.

18. Specifically, both the Centre and the summit took place in India. World Health Organization, "WHO Convenes First Summit on Traditional Medicine," 2023.

BIBLIOGRAPHY

Primary Sources

Needham Research Institute

Anon. *Lu Gwei-Djen: A Commemoration*. Edinburgh: Pentland, 1993.

"Chen Chiu Ta Chheng, Charts Accompanying R. O. Wheeler's Copy of the Chiangsi Edition with Preface of +1680." Needham Research Institute.

Lu, Gwei-Djen. "Pyruvic Acid and Muscle Metabolism in Normal and Vitamin B1-Deficient States." Newnham College, Cambridge, May 1939. Needham Research Institute.

———. "The Inner Elixir (Nei Tan); Chinese Physiological Alchemy." 1971. Needham Research Institute.

———. "Notes on Acupuncture." 1975. Needham Research Institute.

Lu, Gwei-Djen and Joseph Needham. "A Contribution to the History of Chinese Dietetics." *Isis* 42, no. 1 (1951): 13–20.

———. "Medieval Preparations of Urinary Steroid Hormones." *Nature* 200, no. 4911 (December 1963): 1047–1048.

———. "Index Termini Technologici Medic i Sinensio-Anglici Lu-Needhameinsii" [Dictionary of Medical Terms from Chinese to English, Lu and Needham]. 1964. Needham Research Institute.

———. "Notes for Revue Bibliographique de Sinologie." November 1971. Needham Research Institute.

———. *Celestial Lancets: A History and Rationale of Acupuncture and Moxa*. Cambridge: Cambridge University Press, 1980.

———. "The First Half-Life of Joseph Needham." In *Explorations in the History of Science and Technology in China: Compiled in Honour of the Eightieth Birthday of Joseph Needham, FRS, FBA*, 1–38. Shanghai: Shanghai Guji Chubanshe, 1982.

Lu, Gwei-Djen, Joseph Needham, and Phan Chi-Hsing. "The Oldest Representation of a Bombard." *Technology and Culture* 29, no. 3 (1988): 594–605. https://doi.org/10.2307/3105275.

Lu, Gwei-Djen, Joseph Needham, and Nathan Sivin. "Correspondence with Nathan Sivin." November 29, 1972. Needham Research Institute.

Mann, Felix. "Correspondence between Felix Mann and McDonell." October 13, 1975. Needham Research Institute.

Needham, Joseph. Correspondence with the National City Bank in New York City, January 13, 1947. Joseph Needham Papers 1947–48, A.792–A.808, Cambridge University Library Archives.

———. 1958a. "Wellcome Trust Reports," January 10, 1958. Uncatalogued, Needham Research Institute.

———. 1958b. "Wellcome Trust Reports," October 1, 1958. Uncatalogued, Needham Research Institute.

———. Poem by Lu and Needham, Needham Files, A.798, 1969. Cambridge University Library Archives.

———. "Mao and the Dark Aspects." Lecture, Cambridge University Faculty of Divinity, Cambridge, Michaelmas Term, 1970.

———. "Joseph Needham 1972 Notes on China." Acupuncture 4, Needham Research Institute.

———. "Joseph Needham Notes on Meeting with Felix Mann," January 3, 1973. Needham Research Institute.

Wellcome Collection

Head, Henry. 1903a. "Correspondence, Henry Head to Ruth Mayhew." May 15, 1903. PP/HEA D4/16. Wellcome Library Collection.

———. 1903b. "Correspondence, Henry Head to Ruth Mayhew." November 6, 1903. PP/HEA D4/16. Wellcome Library Collection.

———. 1903c. "Correspondence, Henry Head to Ruth Mayhew." November 27, 1903. PP/HEA D4/16. Wellcome Library Collection.

Nichols, Robert. "Correspondence, Robert Nichols to Henry Head." June 26, 1925. PP/HEA/D4/4. Wellcome Library Collection.

Interviews

Beaton Starr, Meredith. Interview by author. August 6, 2012, Harvard Medical School.

Brown, Rolly. Interview by author. July 30, 2021. Zoom.

Carney, Brendan. Interview by author. July 28, 2012, Harvard Medical School.

Cheng, Baixian. Interview by author. August 15, 2009, Tong Ji University.

Domingues, Nilza. Interview by author. March 27, 2013, ATTA, São Paulo, Brazil.

Eduardo, Carlos. Interview by author. March 27, 2013, São Paulo, Brazil.

Fabiana, Strambio. Interview by author. March 27, 2013, São Paulo, Brazil.
Fan-Roche, Jie. Interview by author. July 28, 2012, Harvard Medical School.
Gogate, R. B. Interview by author. March 27, 2012, IASTAM India, Mumbai, India.
Hosmer Kirby, Caitlin. Interview by author. July 30, 2012, Harvard Medical School.
Ji, Hong. Interview by author. July 30, 2014, Chengdu University of Traditional Chinese Medicine.
Kaptchuk, Ted. Interview by author. July 26, 2015.
Kowalski, Matthew H. Interview by author. November 8, 2012, Harvard Medical School.
Langevin, Helene. Interview by author. March 13, 2013, Harvard Medical School.
Levy, Donald B. Interview by author. July 4, 2012, Harvard Medical School.
Li, Cheng. Interview by author. August 18, 2009, Shanghai, China.
Li, Li. Interview by author. August 18, 2009, Shanghai, China.
Li, Liang. Interview by author. April 24, 2015. Institute of Acupuncture and Moxibustion.
Li, Xiuzhi. Interview by author. August 20, 2009, Chengdu Second People's Hospital.
Liang, John Paul. Interview by author. August 4, 2021, Houston, Texas.
Liang, Li. Interview by author. December 28, 2013, Monterey Park, CA.
Liang, Shen Ping. Interview by author. July 30, 2021, Houston, Texas.
Lynn, Mitchell. Interview by author. August 4, 2021. Zoom.
Ma, Bo-Ying. Interview by author. October 25, 2014; March 1, 2015, Cambridge, UK.
Mohite, Swati. Interview by author. March 27, 2012, IASTAM India, Mumbai, India.
Paulsen, Randall. Interview by author. July 12, 2012, Harvard Medical School.
Sivin, Nathan. Interview by author. December 9–10, 2011, Email and Phone.
Strambio, Fabiana. Interview by author. March 27, 2013, São Paulo, Brazil.
Wang, Fang. Interview by author. March 28, 2014, China People's Second Army Hospital.
Wang, Zhiqing. Interview by author. December 27, 2013, Monterey Park, CA.
Wayne, Peter. Interview by author. July 30, 2012, Harvard Medical School.
Yang, Jing. Interview by author. April 28, 2015, Hunan University of Traditional Chinese Medicine.
Zhang, Hui. Interview by author. August 20, 2009, Chengdu Second People's Hospital.

Other Primary Sources

Anon. "*Bu naozhi qianjin yi xiao shuo* (補腦汁淺近易曉說)." *Xinwen Bao* (新聞報), December 6, 1904, first edition, section 8.

Anon. "London Hospital Pathological Institute." *The Times*, July 11, 1901.

Anon. "National Council for Mental Hygiene." *British Medical Journal* 1, no. 3202 (May 13, 1922): 766–767. https://doi.org/10.1136/bmj.1.3202.766.

Anon. "Reston Has Appendectomy in Peking." *Washington Post, Times Herald (1959–1973)*, July 19, 1971, sec. General.

Bell, Charles. *Engravings of the Arteries: Illustrating the Second Volume of the Anatomy of the Human Body, and Serving as an Introduction to the Surgery of the Arteries*. London, 1801.

Boring, Edwin G. "Cutaneous Sensation after Nerve-Division." *Quarterly Journal of Experimental Physiology* 10, no. 1 (1916): 1–95.

———. "Oral History Interviews of Edwin G. Boring." July 5, 1961. rmc K-111-B-3-A. Cornell University Archives.

Bourdon, Amé. *Nouvelles Tables Anatomiques Ou sont représentées au naturel toutes les parties du Corps humain, toutes les nouvelles découvertes, le cours de tolites les humeurs, etc.* Paris: Laurens d'Houry, 1678.

Boym, Michel, and Andreas Cleyer. *Specimen medicinae sinicae sive, Opuscula medica ad mentem sinensium*. Francfort sur le Main: Zubrodt, 1682. Wellcome Collection.

Browe, Peter. *Die Eucharistischen Wunder des Mittelalters* [The Veneration of the Eucharist in the Middle Ages]. Breslauer Studien zur historischen Theologie, NF 4. Breslau: Müller und Seiffert, 1938.

Carpenter, William Benjamin. *Animal Physiology*. London: Wm. S. Orr and Co., 1843.

Carrubba, Robert W., John Z. Bowers, and Willem ten Rhijne. "The Western World's First Detailed Treatise on Acupuncture: Willem ten Rhijne's 'De Acupunctura.'" *Journal of the History of Medicine and Allied Sciences* 29, no. 4 (1974): 371–398.

Chang, Hsiang-tung. "Integrative Action of Thalamus in the Process of Acupuncture for Analgesia." *Scientia Sinica* 16, no. 1 (January 20, 1973): 25–60.

Cheng, Dan'an (承淡安). *Zengding zhongguo zhenjiu zhiliao xue* (增訂中國鍼灸治療學) [Revised Approaches to studying Chinese Acupuncture Moxibustion Therapy]. Shanghai: China Society of Acupuncture Moxibustion Research (中国针灸治疗学), 1933.

———. *Acupuncture and Moxibustion Formulas & Treatments*. Boulder: Blue Poppy Press, 1996.

———. *Cheng Dan'an jianyi jiu zhi—dan fang zhiliao ji* (承淡安簡易灸治 - 丹方

治療集) [Cheng Dan'an's Simple Moxibustion Method—Medicine and Prescription Collection]. Shanghai: Shanghai Science and Technology Press, 2016.

Chiang, Chen-yu, et al. "Studies on Spinal Ascending Pathway for Effect of Acupuncture Analgesia in Rabbits." *Scientia Sinica*, 1975, 651–658.

Dimond, E. Grey. "Acupuncture Anesthesia: When Western Medicine and Chinese Traditional Medicine Meet." *Journal of the American Medical Association* 218, no. 10 (December 6, 1971): 1563–1564. https://doi.org/10.1001/jama.1971.03190230054011.

Dürer, Albrecht. *Hierinn sind begriffen vier Bucher von menschlicher Proportion*. Nürenberg, 1528.

Dürer, Albrecht, David Price, and Warnock Library. *De symmetria partium in rectis formis humanorum corporum; Underweysung der messung*. Palo Alto: Octavo, 2003.

Elliott, T. R. "Wilfred Batten Lewis Trotter, 1872–1939." *Obituary Notices of Fellows of the Royal Society*, 1941, 323–343.

Engell, Shawn, et al. "Differential Displacement of Soft Tissue Layers from Manual Therapy Loading." *Clinical Biomechanics (Bristol, Avon)* 33 (February 23, 2016): 66–72. https://doi.org/10.1016/j.clinbiomech.2016.02.011.

Eriksson, M., and B. Sjölund. "Acupuncturelike Electroanalgesia in TNS-Resistant Chronic Pain." In *Sensory Functions of the Skin in Primates*, 575–581. Oxford and New York: Pergamon Press, 1976.

Galen. *Selected Works*. Edited and translated by P. N. Singer. Oxford: Oxford University Press, 1997.

Garrett, J. J., and F. D. Garrett. "The Segmental Distribution of the Cutaneous Nerves in the Limbs of Man." *The Anatomical Record* 102, no. 4 (1948): 409–437.

Gaskell, W. H. "On the Relation between the Structure, Function, Distribution and Origin of the Cranial Nerves; Together with a Theory of the Origin of the Nervous System of Vertebrata." *Journal of Physiology* 10 (1889): 153–211.

Guo, Yichen (郭一岑). "*Jinian Ba fu luo fu* (纪念巴夫洛夫)" [Remembering Pavlov]. *Journal of Education* (教育杂志), 1936 (6): 1–4.

Halata, Zdenek, Milos Grim, and Klaus I. Bauman. "Friedrich Sigmund Merkel and His 'Merkel Cell,' Morphology, Development, and Physiology: Review and New Results." *Anatomical Record. Part A, Discoveries in Molecular, Cellular, and Evolutionary Biology* 271, no. 1 (March 2003): 225–239. https://doi.org/10.1002/ar.a.10029.

Han, Suyin. "Acupuncture: The Scientific Evidence." *Eastern Horizon* 3, no. 4 (April 1964).

Harvey, William. *On the Motion of the Heart and Blood in Animals*. Translated by Alex Bowie. London: George Bell and Sons, 1889.

Head, Henry. "On Disturbances of Sensation with Especial Reference to the Pain of Visceral Disease." *Brain* 16, no. 1–2 (1893): 1–133.

———. "On Disturbances of Sensation with Especial Reference to the Pain of Visceral Disease. Part II Head and Neck." *Brain* 17, no. 3 (1894): 339–480.

Head, Henry, and William Halse Rivers Rivers. "A Human Experiment in Nerve Division." *Brain* 31, no. 3 (1908): 323–450.

Head, Henry, W. H. R. Rivers, and J. Sherren. "The Afferent Nervous System from a New Aspect." *Brain* 28, no. 2 (1905): 99–115.

Head, Henry, and James Sherren. "The Consequences of Injury to the Peripheral Nerves in Man." *Brain* 28, no. 2 (November 1, 1905): 116–338.

Henson, R. A. "Henry Head's Work on Sensation." *Brain* 84, no. 4 (1961): 529–541.

Hobson, Benjamin (合信), and Xiutang Chen. *Zentai shinron* (全体新论). 3rd ed. Vol. 2. Tokyo: Ansei, 1857.

Hua, Boren (滑伯仁). *Guben jiaozhu, shisi jing fahui* (古本校注 十四經發揮) [Ancient Text Annotation and Elaboration, Treatise on the Fourteen Meridians]. Taipei: Ziyou Publishing, 1969.

Hughes, H. Stuart. *Consciousness and Society; the Reorientation of European Social Thought, 1890–1930*. New York: Knopf, 1958.

Hui, Kathleen K. K. S., et al. "Perception of *Deqi* by Chinese and American Acupuncturists: A Pilot Survey." *Chinese Medicine* 6, no. 1 (January 20, 2011): 2. https://doi.org/10.1186/1749-8546-6-2.

Kim, Bong Han. "On the Kyungrak System." *Journal of the D. P. R. K. Academy of Medical Science*, no. 5 (November 30, 1963): 1–41.

Kulmus, Johann Adam. *Ontleedkundige Tafelen: Benevens De Daar Toe Behoorende Afbeeldingen en Aanmerkingen, Waar in Het Zaamenstel Des Menschelyken Llichaams, en Het Gebruik Van Alle Des Zelfs Deelen Afgebeeld en Geleerd Word*. Amsterdam: de Janssoons van Waesberge, 1734.

Langevin, Helene M., et al. "Connective Tissue Fibroblast Response to Acupuncture: Dose-Dependent Effect of Bidirectional Needle Rotation." *Journal of Alternative and Complementary Medicine* 13, no. 3 (April 2007): 355–360. https://doi.org/10.1089/acm.2007.6351.

———. "Manual and Electrical Needle Stimulation in Acupuncture Research: Pitfalls and Challenges of Heterogeneity." *Journal of Alternative and Complementary Medicine* 21, no. 3 (March 2015): 113–128. https://doi.org/10.1089/acm.2014.0186.

Langevin, Helene M., Nicole A. Bouffard, Gary J. Badger, James C. Iatridis, and Alan K. Howe. "Dynamic Fibroblast Cytoskeletal Response to Subcutaneous Tissue Stretch Ex Vivo and in Vivo." *American Journal of Physiol-*

ogy. Cell Physiology 288, no. 3 (March 2005): C747–756. https://doi.org/10.1152/ajpcell.00420.2004.

Langevin, Helene M., David L. Churchill, Junru Wu, Gary J. Badger, Jason A. Yandow, James R. Fox, and Martin H. Krag. "Evidence of Connective Tissue Involvement in Acupuncture." *FASEB Journal* 16, no. 8 (2002): 872–874. https://doi.org/10.1096/fj.01-0925fje.

Langevin, Helene M., Kirsten N. Storch, Robert R. Snapp, Nicole A. Bouffard, Gary J. Badger, Alan K. Howe, and Douglas J. Taatjes. "Tissue Stretch Induces Nuclear Remodeling in Connective Tissue Fibroblasts." *Histochemistry and Cell Biology* 133, no. 4 (April 2010): 405–415. https://doi.org/10.1007/s00418-010-0680-3.

Lee, Tsung-ying. "Acupuncture Anaesthesia—Its Theory and Practice." *Eastern Horizon* 10, no. 4 (1971): 3–6.

Liu, Shenbin, Zhifu Wang, Yangshuai Su, Lu Qi, Wei Yang, Mingzhou Fu, Xianghong Jing, Yanqing Wang, and Qiufu Ma. "A Neuroanatomical Basis for Electroacupuncture to Drive the Vagal–Adrenal Axis." *Nature* 598, no. 7882 (October 2021): 641–645. https://doi.org/10.1038/s41586-021-04001-4.

Lu, Zhijun (鲁之俊). "Ba (跋)" [Postscript]. In *Xin zhenjiu xue* [New Approaches to Studying Acupuncture-Moxibustion], 277–280. Guangxi: Guangxi People's Publishing House, 1980.

Lewes, George Henry. *The Physiology of Common Life*. Edinburgh: W. Blackwood, 1859. http://archive.org/details/b2146134x_0002.

Lumpkin, Ellen A., and Diana M Bautista. "Feeling the Pressure in Mammalian Somatosensation." *Current Opinion in Neurobiology* 15, no. 4 (August 2005): 382–388. https://doi.org/10.1016/j.conb.2005.06.005.

Lumpkin, Ellen A., Kara L. Marshall, and Aislyn M. Nelson. "The Cell Biology of Touch." *Journal of Cell Biology* 191, no. 2 (October 18, 2010): 237–248. https://doi.org/10.1083/jcb.201006074.

Ma, Jixing (马继兴). "*Zhenjiu zhiliao dian jiepou weizhi cankao tu* (针灸治疗刺激点解剖位里参考图)" [Anatomical Cìjī Chart for Stimulating Acupuncture-Moxibustion Points]. Beijing: Beijing College of Chinese Medicine and Technology at the People's Central Government Ministry of Health, 1952.

Man, S. C., and F. D. Baragar. "Local Skin Sensory Changes after Acupuncture." *CMA Journal* 109 (October 6, 1973): 609–610.

Mann, Felix. *Acupuncture: The Ancient Chinese Art of Healing*. London: W. Heinemann Medical Books, 1962.

———. *The Meridians of Acupuncture*. London: Heinemann, 1964.

———. *Acupuncture: The Ancient Chinese Art of Healing*. London: Heinemann Medical, 1971.
———. "Treatment of Intractable Pain by Acupuncture." July 14, 1973. Acupuncture 6. Needham Research Institute.
———. "Acupuncture Analgesia. Report of 100 Experiments." *British Journal of Anaesthesia* 46, no. 5 (May 1974): 361–364. https://doi.org/10.1093/bja/46.5.361.
———. *Scientific Aspects of Acupuncture*. London: Heinemann Medical Books, 1977.
———. *Scientific Aspects of Acupuncture*. London: Heinemann Medical Books, 1983.
Melzack, Ronald. *The Puzzle of Pain*. New York: Basic Books, 1973.
———. "Shutting the Gate on Pain." In *Science Year: The World Book Science Annual*, 57–67. Chicago: Field Enterprises Educational Corp., 1975.
Melzack, Ronald, and Mary-Ellen Jeans. "Acupuncture Analgesia: A Psychophysiological Explanation." *Minnesota Medicine* 57, no. 3 (March 1974): 161–166.
Melzack, Ronald, and Patrick D. Wall. "Pain Mechanisms: A New Theory." *Science* 150, no. 3699 (November 19, 1965): 971–979. https://doi.org/10.1126/science.150.3699.971.
Morris, E. W. *A History of the London Hospital*. London: Arnold, 1910.
Murray, Elsie. "A Qualitative Analysis of Tickling: Its Relation to Cutaneous and Organic Sensation." *American Journal of Psychology* 19, no. 3 (1908): 299–329.
Nagahama, Yoshio (長濱善夫). *Jingluo zhi yanjiu* (经络之研究) [A Study of Meridians]. Translated by Dan'an Cheng. Shanghai: Qian Qing Tang Shuju, 1955.
Pavlov, Ivan Petrovich. *Lectures on Conditioned Reflexes: Twenty-Five Years of Objective Study of the Higher Nervous Activity (Behaviour) of Animals*. International Publishers, 1941.
Peng, C. K., et al. "Mosaic Organization of DNA Nucleotides." *Physical Review. E, Statistical Physics, Plasmas, Fluids, and Related Interdisciplinary Topics* 49, no. 2 (February 1, 1994): 1685–1689. https://doi.org/10.1103/physreve.49.1685.
Pilla, A. A. "Mechanisms of Electrochemical Phenomena in Tissue Repair and Growth." *Bioelectrochemistry and Bioenergetics* (Uncorrected Proofs), 1973. Acupuncture 2. Needham Research Institute.
Purdue News Service. "Purdue DARPA (ElectRx) Project Focuses on Developing Implantable, Nerve-Stimulating 'Bionode' to Treat Inflammation - Purdue University," n.d. Accessed May 11, 2016. http://www.purdue.edu

/newsroom/releases/2015/Q4/purdue-darpa-project-focuses-on-developing-implantable,-nerve--stimulating-bionode-to-treat-inflammation.html.

Qian, Xinzhong. *Kitayskaya Narodnaya Meditsina* (Китайская Народная Медицина) [Chinese Folk Medicine]. Petrograd: Знание, 1959.

Reston, James. "Now, about My Operation." *New York Times*, July 26, 1971.

———. "Reston Tells of Surgery in Red Chinese Hospital." *Chicago Tribune*, July 26, 1971.

Sarlandière, Jean Baptiste. *Mémoires sur l'électro-puncture: considérée comme moyen nouveau de traiter efficacement la goutte, les rhumatismes et les affections nerveuses, et sur l'emploi du moxa japonaia en France, suivis d'un traité de l'acupuncture et du moxa, principaux moyens curatifs chez les peuples de la Chine, de la Corée et du Japon, ornés de figurés japonaises*. Paris: Chez l'auteur et Chez Mlle Delaunay, Libraire, 1825.

Sechenov, Ivan Mikhaĭlovich. *Reflexes of the Brain*. Cambridge: MIT Press, 1965.

Sima, Qian (司馬遷). *Shijing* (史记) [Records of the Historian]. ca. 91 BCE. In *Dianjiaoben ershisi shi xiudingben: Shiji* (点校本二十四史修订本：史记) [Punctuated and Collated Twenty-Four Histories, Revised Edition: Records of the Historian]. 10 vols. Beijing: Zhonghua shuju, 2014.

Sugita, Genpaku (杉田玄白). *Rangaku Kotohajime* (蘭学事始) [Dawn of Western Science in Japan]. Tokyo: Hokuseido Press, 1969.

Sugita, Genpaku (杉田 玄白), et al. *Kaitai shinsho* (解體新書). Vol. 3. Tōbu [Edo]: Suharaya Ichibē shi, An'ei 3, 1774.

T., C. P. "In Memoriam: H. Morriston Davies." *Annals of the Royal College of Surgeons of England* 36, no. 4 (April 1965): 246–249.

Ten Rhijne, Willem. *Dissertatio de Arthritide*. London: R. Chiswell, Societatis Regalis typographi, 1683.

Trotter, Wilfred, and H. Morriston Davies. "Experimental Studies in the Innervation of the Skin." *Journal of Physiology* 38, no. 2–3 (February 9, 1909): 134–246.

Tsien, Tsuen-hsuin. "Western Impact on China through Translation." *Far Eastern Quarterly* 13, no. 3 (1954): 305–327.

Valverde de Amusco, Juan, et al. *Anatomia del corpo humano*. In Roma: Per Ant. Salamanca, et Antonio Lafrerj, 1560.

Vesalius, Andreas, et al. *De humani corporis fabrica libri septem*. Basileae: Ex officina Joannis Oporini, 1543.

Walshe, F. M. R. "The Anatomy and Physiology of Cutaneous Sensibility: A Critical Review." *Brain* 65, no. 1 (1942): 48–96.

Wang, Haogu (王好古). *Yi yin tang ye zhong jing guang wei dafa* (伊尹汤液仲景广为大法) [Yi Yin's Grand Method of Decoctions Expanded by Zhong-

jing]. 1234. No original publication location. Accessed through the National Archives of Japan. https://jpsearch.go.jp/item/najda-ddEoHPYNhZm1flMZc8vXCERNCBXjuk6A.
———. *Ci shi nan zhi* (此事難知) [Topics of Great Difficulty]. 1308. Taipei: Taiwan shang wu yin shu guan, 1983.
Wang, Weiyi (王惟一). *Xin kan tong ren zhen jiu jing. Xin bian xi fang zi ming tang jiu jing* (新刊銅人鍼灸經 七卷) [Newly Published Bronze Acupuncture-Moxibustion Classic]. 7 volumes. 1515. Ming Dynasty edition, published in Shanxi Pingyang Fu. Accessed at Bodleian Libraries, University of Oxford. Sinica 754.
Wang, Xuetai (王雪苔). *Zhenjiu xue shouce* (针灸学手册) [*The Handbook on Acupuncture and Moxibustion*]. Beijing: People's Medical Publishing House, 1956.
———. *Zhenjiu xue shouce* (针灸学手册) [*The Handbook on Acupuncture and Moxibustion*]. Beijing: People's Medical Publishing House, 1966.
———. "*Maodun faze zai zuguo yixue bianzheng shi zhi zhong de tixian* (矛盾法则在祖国医学辨证施治中的体现)" [Embodying the Law of Contradiction in a Dialectical Treatment of Chinese Medicine]. *Ziran bianzhengfa yanjiu tongxun* [Dialectics of Nature Research Newsletter], March 2, 1959.
———. "*Zhenjiu jiangzuo (xu)* [(針灸講座（續）]" [Acupuncture Moxibustion Seminar (continued)]. *Journal of Chinese Medicine* 6 (June 1960): 55–57.
———. "*Dui jinyibu kaizhan zhenjiu yanjiu de ji dian kanfa* (对进一步开展针灸研究的几点看)" [Perspectives on Improving Acupuncture Moxibustion Research]. *Chinese Medicine Yearbook*, January 1984, 28–31.
———, ed. *Zhonghua zhenjiu tijian* (中华针灸图鉴) [An Illustrated Guide to Chinese Acupuncture Moxibustion]. Beijing: People's Military Medical Press, 2003.
———. Interview. "2006 *Ruhe jianbie zhenjiu de liaoxiao—Wang Xuetai* 2006 (如何鉴别针灸的疗效—王雪苔)" [How to identify the efficacy of acupuncture moxibustion with Wang Xuetai]. http://www.worldtcm.org/131227/15E43950.shtml.
———. *Xuetai zhenlun* (雪苔针论) [Xuetai's Theory on Acupuncture-Moxibustion]. Beijing: People's Health Publishing House, 2008.
Wei, Lifu (韦立富), Xiaoxia Pan (潘小霞), and Zhizhen Mo (莫智珍). "*Zhengwen Zhu Lian laoshi yu zhenjiu* (征文 朱琏老师与针灸)" [Our Teacher Zhu Lian and Acupuncture-Moxibustion]. Number 7 People's Hospital of Nanning, October 24, 2014. http://www.nn7yy.com/notice/?type=detail&id=36.
WHO Scientific Group on International Acupuncture Nomenclature and

World Health Organization. *A Proposed Standard International Acupuncture Nomenclature: Report of a WHO Scientific Group*. Geneva: World Health Organization, 1991. https://iris.who.int/handle/10665/40001.

Willis, Robert. *William Harvey: A History of the Discovery of the Circulation of the Blood*. London: C. Kegan Paul, 1878. http://archive.org/details/b21996404.

Willis, Thomas (1621–1675). *Cerebri anatome: cui accessit nervorum descriptio et usus (Anatomy of the Brain: To which is added a description and use of the nerves)*. London: Printed by James Flesher for John Martyn and James Allestry, 1663. http://archive.org/details/b30342144.

Winchester, Simon. *The Man Who Loved China: The Fantastic Story of the Eccentric Scientist Who Unlocked the Mysteries of the Middle Kingdom*. New York: Harper, 2008.

Woodburne, Russell T. *Essentials of Human Anatomy*. New York: Oxford University Press, 1965.

World Health Organization. "WHO Convenes First High-Level Global Summit on Traditional Medicine to Explore Evidence Base, Opportunities to Accelerate Health for All." Accessed October 4, 2023. https://www.who.int/news/item/10-08-2023-who-convenes-first-high-level-global-summit-on-traditional-medicine-to-explore-evidence-base--opportunities-to-accelerate-health-for-all.

Wotton, William. *Reflections upon Ancient and Modern Learning*. Printed by J. Leake for Peter Buck, 1694.

Xiao, Ying, Jonathan S. Williams, and Isaac Brownell. "Merkel Cells and Touch Domes: More than Mechanosensory Functions?" *Experimental Dermatology* 23, no. 10 (October 2014): 692–696.

Yasuyori, Tanba (丹波康頼). *Ishinpō* (醫心方). 984. Available through the National Diet Library. https://jpsearch.go.jp/item/arc_books-NDL_2555583.

Yoke Ho, Peng. *Reminiscence of a Roving Scholar: Science, Humanities, and Joseph Needham*. Singapore: World Scientific, 2005.

Yu, Yan (余巖). "*Du guoyiguan zhengli xueshu caoan zhi wojian* (讀國醫館整理學術草案之我見)" [My Opinion on the Proposal for Sorting out Chinese Medicine as Issued by the Institute of National Medicine]. *Zhongxi Yiyao* (中西醫藥) [Journal of the Medical Research Society of China] 2, no. 2 (1936): 178–192.

Zhang, Panshi (张磐石). "*Wo yu zhenjiu* (我与针灸)" [My Encounter with Acupuncture-Moxibustion]. In *Xin zhenjiu xue* (新针灸学) [New Approaches to Studying Acupuncture-Moxibustion], 1–18. Shanghai: People's Medical Publishing House, 1954.

Zhao, Yibing (赵以炳). "*Shi nianlai ba fu luo fu xueshuo zai woguo de chengjiu*

(十年来巴甫洛夫学说在我国的成就)" [A Ten-Year Survey of Pavlovian Theory Accomplishment in China]. *Biology Bulletin* (生物学通报) 1959 (10): 468–471, 475.

Zheng, Weigang, and Wei Liu, eds. *Zhenjiu dacheng* (针灸大成). Shanghai: Shanghai Publishing House (上海大文书局发行所), 1936.

Zhenjiu jicheng (針灸集成) [Collected Works on Acupuncture and Moxibustion]. Beijing: People's Health Press, 1956. No author attributed.

Zhu, Lian (朱琏). *Xin zhenjiu xue* (新针灸学) [New Approaches to Studying Acupuncture-Moxibustion]. First edition. Beijing: People's Medical Publishing House, 1950.

———. *Xin zhenjiu xue* (新针灸学) [New Approaches to Studying Acupuncture-Moxibustion]. 2nd ed. Beijing: People's Medical Publishing House, 1954.

———."*Zaiban xu* (再版序)" [Second Preface]. In *Xin zhenjiu xue* (新针灸学) [New Approaches to Studying Acupuncture-Moxibustion], 2nd ed., 23–26. Shanghai: People's Medical Publishing House, 1954.

———. *Xin zhenjiu xue* (新针灸学) [New Approaches to Studying Acupuncture-Moxibustion]. Third edition. Guangxi: Guangxi People's Publishing House, 1980.

Zhu, Bing (朱兵). "*Xitong zhenjiu xue-fuxing 'ti biao yixue'* (系统针灸学—复兴'体表医学')" [Systematic Approach to Acupuncture Moxabustion—Reclaiming 'Medicine at the Periphery']. Beijing: People's Health Publishing House (人民卫生出版社), 2015.

———. "*Zhenjiu de zhendi* (针灸的真谛)" [The True Essence of Acupuncture Moxibustion]. Lecture, China Academy of Traditional Chinese Medicine, Beijing, April 22, 2015.

Zylka, Mark J., Frank L. Rice, and David J. Anderson. "Topographically Distinct Epidermal Nociceptive Circuits Revealed by Axonal Tracers Targeted to *Mrgprd*." *Neuron* 45, no. 1 (January 6, 2005): 17–25. https://doi.org/10.1016/j.neuron.2004.12.015.

Secondary Sources

Allais, Lucia. "Rendering: A History of Experience and Experiments." In *Design Technics: Archaeologies of Architectural Practice*, edited by Zeynep Çelik Alexander and John May, 1–44. 1st ed. Minneapolis: University of Minnesota Press, 2020.

Alpers, Svetlana. *The Art of Describing: Dutch Art in the Seventeenth Century*. Reprint. Chicago: University of Chicago Press, 1984.

Alperson, Philip. "A Topography of Improvisation." *Journal of Aesthetics and*

Art Criticism 68, no. 3 (2010): 273–280. https://doi.org/10.1111/j.1540-6245.2010.01418.x.

Amano, Yosuke (天野陽介). "'Ōtei meidōkyūkei fushin shōshō' kō: Adzuchi-momoyamajidai no keiketsu kenkyū no ichirei to shite (『黄帝明堂灸経不審少々』考: 安土桃山時代の経穴研究の一例として)" [On Kotei-meidokyukyo Fushinshosho: A Look at Early Acupuncture Point Study in Azuchi-Momoyama Era]. *Kampo Medicine* 66, no. 1 (2015): 1–7.

Andrews, Bridie. "Tuberculosis and the Assimilation of Germ Theory in China, 1895–1937." *Journal of the History of Medicine and Allied Sciences* 52, no. 1 (1997): 114–138. https://doi.org/10.1093/jhmas/52.1.114.

———. "From Case Records to Case Histories: The Modernization of a Chinese Medical Genre, 1912–49." In *Innovation in Chinese Medicine*, edited by Elisabeth Hsu, 333–361. Cambridge: Cambridge University Press, 2001.

———. *The Making of Modern Chinese Medicine, 1850–1960*. Honolulu: University of Hawai'i Press, 2015.

Andrews, J. H. "Introduction: Meaning, Knowledge, and Power in the Map Philosophy of J. B. Harley." In *The New Nature of Maps: Essays in the History of Cartography*, edited by J. B. Harley and Paul Laxton, 1–32. Baltimore: Johns Hopkins University Press, 2002.

Andrews, Noam. "Albrecht Dürer's Personal Underweysung Der Messung." *Word & Image* 32, no. 4 (October 1, 2016): 409–429. https://doi.org/10.1080/02666286.2016.1216821.

Arikha, Noga. *Passions and Tempers: A History of the Humors*. Reprint. New York: Ecco, 2008.

Arrington, Robert L., and Hans-Johann Glock, eds. *Wittgenstein's Philosophical Investigations: Text and Context*. London: Routledge, 1991.

Bastian, Heather. "The Genre Effect: Exploring the Unfamiliar." *Composition Studies* 38, no. 1 (2010): 29–51. https://eric.ed.gov/?id=EJ944303.

Bawarshi, Anis S. "The Genre Function." In *Genre and the Invention of the Writer, Reconsidering the Place of Invention in Composition*, 16–48. Denver: University Press of Colorado, 2003.

Bazerman, Charles. "The Life of Genre, the Life in the Classroom." In *Genre and Writing: Issues, Arguments, Alternatives*, edited by Wendy Bishop and Hans Ostrom, 19–26. Portsmouth, NH: Heinemann, 1997.

Beaulieu, A. "Voxels in the Brain: Neuroscience, Informatics and Changing Notions of Objectivity." *Social Studies of Science* 31, no. 5 (October 2001): 635–680. https://doi.org/10.1177/030631201031005001.

Behr, Wolfgang. "Discussion 6: G. Sampson, 'a Chinese Phonological Enigma': Four Comments." *Journal of Chinese Linguistics* 43, no. 2 (2015): 719–732. https://doi.org/10.1353/jcl.2015.0013.

Benická, Jana. "*Xin* as a 'Qualitatively Equal' Co-constituent of Phenomena in Chinese Mahayana Buddhism: Some Remarks on Its Interpretations by Using the Terms of Western Philosophical Discourse." *Monumenta Serica* 54 (2006): 185–194. https://doi.org/10.1179/mon.2006.54.1.007.

Biehl, João, and Peter Locke. "Deleuze and the Anthropology of Becoming." *Current Anthropology* 51, no. 3 (2010): 317–351. https://doi.org/10.1086/651466.

Bildhauer, Bettina, ed. "Medieval European Conceptions of Blood: Truth and Human Integrity." In *Blood Will Out: Essays on Liquid Transfers and Flows*, 56–75. Chichester: Wiley-Blackwell, 2013.

Billeter, Jean François. *The Chinese Art of Writing*. New York: Skira/Rizzoli, 1990.

Bishop, Wendy, and David Starkey. "Genre." In *Keywords in Creative Writing*, 95–99. Denver: University Press of Colorado, 2006.

Bivins, Roberta. *Acupuncture, Expertise and Cross-Cultural Medicine*. New York: Palgrave Macmillan, 2000.

———. "Imagining Acupuncture: Images and the Early Westernization of Asian Medical Expertise." *Asian Medicine* 7, no. 2 (July 2012): 298–318. https://doi.org/10.1163/15734218-12341255.

Blussé, Leonard. "No Boats to China. The Dutch East India Company and the Changing Pattern of the China Sea Trade, 1635–1690." *Modern Asian Studies* 30, no. 1 (1996): 51–76. https://doi.org/10.1017/s0026749x00014086.

Blussé, Leonard, Willem Remmelink, and Ivo Smits. *Bridging the Divide: 400 Years, The Netherlands-Japan*. Leiden: KIT Publishers, 2000.

Bouvier, Valentine, et al. "Merkel Cells Sense Cooling with TRPM8 Channels." *Journal of Investigative Dermatology* 138, no. 4 (April 1, 2018): 946–951. https://doi.org/10.1016/j.jid.2017.11.004.

Bowers, John Z., and Robert W. Carrubba. "The Doctoral Thesis of Engelbert Kaempfer on Tropical Diseases, Oriental Medicine, and Exotic Natural Phenomena." *Journal of the History of Medicine and Allied Sciences* 25, no. 3 (July 1, 1970): 270–310. https://doi.org/10.1093/jhmas/XXV.3.270.

Brain, Robert Michael. *The Pulse of Modernism: Physiological Aesthetics in Fin-de-Siècle Europe*. Reprint. Seattle: University of Washington Press, 2016.

Bray, Francesca, Vera Dorofeeva-Lichtmann, and Georges Métailié, eds. *Graphics and Text in the Production of Technical Knowledge in China: The Warp and the Weft*. Leiden: Brill, 2007.

———. "Introduction: The Powers of Tu." In *Graphics and Text in the Production of Technical Knowledge in China: The Warp and the Weft*, edited by Francesca Bray, Vera Dorofeeva-Lichtmann, and Georges Métailié, 1–81. Leiden: Brill, 2007.

Brook, Timothy, *The Troubled Empire: China in the Yuan and Ming Dynasties*. Reprint. Cambridge: Belknap Press, 2013.

Brown, Marjorie L., George A. Ulett, and John A. Stern. "Acupuncture Loci: Techniques for Location." *American Journal of Chinese Medicine* 2, no. 1 (1974): 67–73.

Bu, Liping. "Social Darwinism, Public Health and Modernization in China, 1895–1925." In *Uneasy Encounters: The Politics of Medicine and Health in China, 1900–1937*, edited by Iris Borowy, 93–124. Frankfurt: Peter Lang, 2009.

Bu, Liping, Darwin H. Stapleton, and Ka-Che Yip, eds. *Science, Public Health and the State in Modern Asia*. 1st ed.. London: Routledge, 2012.

Burton, Antoinette. "Amitav Ghosh's World Histories from Below." *History of the Present* 2, no. 1 (2012): 71–77. https://doi.org/10.5406/historypresent.2.1.0071.

Butler, Judith. *Undoing Gender*. London: Routledge, 2004.

Bynum, Caroline Walker. "Why All the Fuss about the Body? A Medievalist's Perspective." *Critical Inquiry* 22, no. 1 (1995): 1–33. https://www.jstor.org/stable/1344005.

———. *Wonderful Blood: Theology and Practice in Late Medieval Northern Germany and Beyond*. Philadelphia: University of Pennsylvania Press, 2007.

Cambrosio, Alberto, Daniel Jacobi, and Peter Keating. "Ehrlich's 'Beautiful Pictures' and the Controversial Beginnings of Immunological Imagery." *Isis* 84, no. 4 (1993): 662–699. https://www.jstor.org/stable/235104.

Candelise, Lucia, and Elizabeth Guill. "Chinese Medicine Outside of China: The Encounter between Chinese Medical Practices and Conventional Medicine in France and Italy." *China Perspectives*, no. 3 (87) (2011): 43–50. https://doi.org/10.4000/chinaperspectives.5638.

Cao, Dongyi (曹东义). "*Huatuo liubu san fa shanghan xueshuo de lishi yi yi* (華佗六部三法傷寒學說的歷史醫意)" [The Historical Significance of Hua Tuo's 'Six Divisions and Three Methods' Doctrine of Cold Damage]. *Chinese Journal of Medical History Zhonghua Yishi Zazhi* (中華醫史雜誌) 2, no. 2 (2002): 58–60.

Chakrabarty, Dipesh. *Provincializing Europe: Postcolonial Thought and Historical Difference*. First edition. Princeton: Princeton University Press, 2007.

Chan, Man sing. "Sinicizing Western Science: The Case of *Quanti xinlun* 全體新論." *T'oung Pao* 98 (January 1, 2012): 533–573.

Chang, Ning Jennifer. "From Heart to Brain: Ailuo Brain Tonic and the New Concept of the Body in Late Qing China (腦為一身之主:從「艾羅補腦

汁」看近代中國身體觀的變化).” *Bulletin of the Institute of Modern History, Academia Sinica* 74 (December 2011): 1–44.
Chappell, Hilary. “The Romanization Debate.” *Australian Journal of Chinese Affairs*, no. 4 (1980): 105–118.
Chen, Nianfu (陈年福). *Jiaguwen dongci cihui yanjiu* (甲骨文动词词汇研究) [Investigation on Oracle Bone Inscriptions]. Sichuan: Bashu Books (巴蜀书社), 2001.
Chen, Ting. “*Hua shou shengping yu zhushu kao lue* (滑寿生平与著述考略)” [A Survey of Hua Shou’s Life and Writings]. *Beijing Journal of TCM* 23, no. 4 (2004): 242–244.
Chiang, Howard. “Christine Goes to China: Xie Jianshun and the Discourse of Sex Change in Cold War Taiwan.” In *Gender, Health, and History in Modern East Asia*, edited by Angela Ki Che Leung and Izumi Nakayama, 216–243. Hong Kong: Hong Kong University Press, 2017.
———. *Transtopia in the Sinophone Pacific*. New York: Columbia University Press, 2021.
Clarke, Keith C. “What Is the World’s Oldest Map?” *Cartographic Journal* 50, no. 2 (May 1, 2013): 136–143. https://doi.org/10.1179/0008704113z.00000000079.
Clary, Rachel C., Rose Z. Hill, Francis McGlone, Lan A. Li, Molly Kulesz-Martin, and Gil Yosipovitch. “Montagna Symposium 2016—The Skin: Our Sensory Organ for Itch, Pain, Touch, and Pleasure.” *Journal of Investigative Dermatology* 137, no. 7 (July 1, 2017): 1401–1404. https://doi.org/10.1016/j.jid.2017.03.015.
Clunas, Craig. *Pictures and Visuality in Early Modern China*. Reissue. London: Reaktion Books, 2012.
Cook, Harold J. “Boerhaave and the Flight from Reason in Medicine.” *Bulletin of the History of Medicine* 74, no. 2 (June 1, 2000): 221–240. https://doi.org/10.1353/bhm.2000.0062.
———. “Medical Communication in the First Global Age: Willem ten Rhijne in Japan, 1674–1676.” *Disquisitions on the Past & Present* 11 (2004): 16–36.
———. “Medicine.” In *The Cambridge History of Science: Volume 3, Early Modern Science*, edited by Katharine Park and Lorraine Daston, 407–432. New York: Cambridge University Press, 2016.
Cook, Harold John, and Sven Dupré, eds. *Translating Knowledge in the Early Modern Low Countries*. Low Countries Studies on the Circulation of Natural Knowledge. Vol. 3. Zürich: Lit, 2012.
Corner, James. “The Agency of Mapping: Speculation, Critique and Invention.” In *Mappings*, edited by Denis Cosgrove, 213–300. London: Reaktion Books, 1999.
Cosgrove, Denis, ed. *Mappings*. London: Reaktion Books, 1999.

Crampton, Jeremy W. "Maps as Social Constructions: Power, Communication and Visualization." *Progress in Human Geography* 25, no. 2 (June 1, 2001): 235–252. https://doi.org/10.1191/030913201678580494.

Crary, Jonathan. *Techniques of the Observer: On Vision and Modernity in the Nineteenth Century*. Cambridge: MIT Press, 1990.

Csordas, Thomas J. *Embodiment and Experience: The Existential Ground of Culture and Self*. Cambridge: Cambridge University Press, 1994.

———. *Body/Meaning/Healing*. New York: Palgrave Macmillan, 2002.

Cuneo, Pia F. "The Artist, His Horse, a Print, and Its Audience: Producing and Viewing the Ideal in Durer's Knight, Death, and the Devil (1513)." In *The Essential Dürer*, edited by Larry Silver and Jeffrey Chipps Smith, 115–129. Philadelphia: University of Pennsylvania Press, 2011.

Dacome, Lucia. *Malleable Anatomies: Models, Makers, and Material Culture in Eighteenth-Century Italy*. Oxford: Oxford University Press, 2017.

Danzker, Jo-Anne Birnie, et al. *Shanghai Modern, 1919–1945*. Ostfildern-Ruit: Hatje Cantz, 2004.

Daston, Lorraine, and Peter Galison. *Objectivity*. New York: Zone Books, 2010.

Daston, Lorraine, and Elizabeth Lunbeck. 2011. *Histories of Scientific Observation*. Chicago: University of Chicago Press.

Despeux, Catherine. "The System of the Five Circulatory Phases and the Six Seasonal Influences (Wuyun Liuqi), a Source of Innovation in Medicine under the Song (960–1279)." In *Innovation in Chinese Medicine*, edited by Elisabeth Hsu, 121–166. Cambridge: Cambridge University Press, 2001.

———. "From Prognosis to Diagnosis of Illness in Tang China." In *Medieval Chinese Medicine: The Dunhuang Medical Manuscripts*, edited by Christopher Cullen and Vivienne Lo, 176–206. London: Routledge, 2005.

———. "The Body Revealed: The Contribution of Forensic Medicine to Knowledge and Representation of the Skeleton in China." In *Graphics and Text in the Production of Technical Knowledge in China: The Warp and the Weft*, edited by Bray, Dorofeeva-Lichtmann, and Métailié, 635–668. Leiden: Brill, 2007.

———. "Picturing the Body in Chinese Medical and Daoist Texts from the Song to the Qing Period (10th to 19th Centuries)." In *Imagining Chinese Medicine*, edited by Vivienne Lo and Penelope Barrett, 51–68. Leiden: Brill, 2018.

Dodge, Martin, Rob Kitchin, and Chris Perkins, eds. *Rethinking Maps: New Frontiers in Cartographic Theory*. London: Routledge, 2009.

Dubois, Jean Claude. "Revisiting the Medical Work of George Soulié De Morant." *Chinese Medicine and Culture* 2, no. 2 (2019): 53–56. https://doi.org/10.4103/CMAC.CMAC_14_19.

Edney, Matthew H. *Cartography: The Ideal and Its History*. Chicago: University of Chicago Press, 2019.

Efal, Adi. "Generic Classification and Habitual Subject Matter." In *The Making of the Humanities*, edited by Rens Bod, Jaap Maat, and Thijs Weststeijn, 345–358. Vol. 3 of *The Modern Humanities*. Amsterdam: Amsterdam University Press, 2014.

Elkins, James. *How to Use Your Eyes*. London: Routledge, 2000.

Elliott, Mark C. *The Manchu Way: The Eight Banners and Ethnic Identity in Late Imperial China*. Stanford: Stanford University Press, 2006.

Elman, Benjamin A. *On Their Own Terms: Science in China, 1550–1900*. Cambridge: Harvard University Press, 2005.

———. "Sinophiles and Sinophobes in Tokugawa Japan: Politics, Classicism, and Medicine during the Eighteenth Century." *East Asian Science, Technology and Society: An International Journal* 2 (March 1, 2008): 93–121. https://doi.org/10.1215/s12280-008-9042-9.

Espesset, Grégoire. "Traditional Chinese Knowledge before the Japanese Discovery of Western Science in Gabor Lukacs' 'Kaitai Shinsho & Geka Sōden.'" *East Asian Science, Technology, and Medicine* 40 (2014): 113–128. https://shs.hal.science/halshs-01385131/document.

Fall, Juliet. *Drawing the Line: Nature, Hybridity, and Politics in Transboundary Spaces*. Aldershot: Ashgate, 2005.

Fan, Fa-ti. "The Global Turn in the History of Science." *East Asian Science, Technology and Society* 6, no. 2 (June 1, 2012): 249–258. https://doi.org/10.1215/18752160-1626191.

Fan, Ka-wai. "Pavlovian Theory and the Scientification of Acupuncture in 1950s China." In *New Perspectives on the Research of Chinese Culture*, edited by Pei-kai Cheng and Ka Wai Fan, 137–45. Singapore: Springer Singapore, 2013.

———. "The Period of Division and the Tang Period." In *Chinese Medicine and Healing: An Illustrated History*, edited by T. J. Hinrichs and Linda L. Barnes, 71–115. Cambridge: Belknap Press, 2013.

Farquhar, Judith. "Chinese Medicine as Popular Knowledge in Urban China." In *Chinese Medicine and Healing: An Illustrated History*, edited by T. J. Hinrichs and Linda L. Barnes, 272–293. Cambridge: Belknap Press, 2013.

Fishelov, David. *Metaphors of Genre: The Role of Analogies in Genre Theory*. University Park: Penn State University Press, 1993.

Fraley, Jill M. "Images of Force: The Power of Maps in Community Development." *Community Development Journal* 46, no. 4 (2011): 421–435.

Freidin, Betina. "Acupuncture in Argentina." In *Chinese Medicine and Healing:*

An Illustrated History, edited by T. J. Hinrichs and Linda L. Barnes, 294–312. Cambridge: Belknap Press, 2013.
Fu, Jia-Chen. "Houses of Experiment: Making Space for Science in Republican China." *East Asian Science, Technology and Society* 10, no. 3 (September 1, 2016): 269–290. https://doi.org/10.1215/18752160-3595072.
Furth, Charlotte, and Shu-yueh Ch'en. "Chinese Medicine and the Anthropology of Menstruation in Contemporary Taiwan." *Medical Anthropology Quarterly* 6, no. 1 (1992): 27–48.
Galison, Peter, and Caroline A. Jones, eds. *Picturing Science, Producing Art*. New York: Routledge, 1998.
Gavrylenko, Valeria. "The 'Body without Skin' in the Homeric Poems." In *Blood, Sweat and Tears: The Changing Concepts of Physiology from Antiquity Into Early Modern Europe*, edited by Manfred Horstmanshoff, Helen King, and Claus Zittel, 481–502. Leiden: Brill, 2012.
Ghitani, Nima, et al. "Specialized Mechanosensory Nociceptors Mediating Rapid Responses to Hair Pull." *Neuron* 95, no. 4 (August 16, 2017): 944–954. https://doi.org/10.1016/j.neuron.2017.07.024.
Ghosh, Amitav. *The Glass Palace*. New York: Random House, 2001.
Gibson, James J. *The Ecological Approach to Visual Perception*. Boston: Houghton Mifflin, 1979.
Goble, Andrew Edmond. *Confluences of Medicine in Medieval Japan: Buddhist Healing, Chinese Knowledge, Islamic Formulas, and Wounds of War*. Honolulu: University of Hawaii Press, 2011.
Goldman, Andrew J. "Improvisation as a Way of Knowing." *Music Theory Online* 22, no. 4 (December 1, 2016). https://mtosmt.org/issues/mto.16.22.4/mto.16.22.4.goldman.html.
Gole, Andrew Edmund. "Song Printed Medical Works and Medieval Japanese Medicine." In *Chinese Medicine and Healing: An Illustrated History*, edited by T. J. Hinrichs and Linda L. Barnes, 123–128. Cambridge: Belknap Press, 2013.
Gould, Carol S., and Kenneth Keaton. "The Essential Role of Improvisation in Musical Performance." *Journal of Aesthetics and Art Criticism* 58, no. 2 (2000): 143–148. https://doi.org/10.2307/432093.
Guenther, Katja. *Localization and Its Discontents: A Genealogy of Psychoanalysis and the Neuro Disciplines*. Chicago: University of Chicago Press, 2015.
Guerrini, Anita. *Courtiers' Anatomists: Animals and Humans in Louis XIV's Paris*. Chicago: University of Chicago Press, 2015.
Gyatso, Janet. *Being Human in a Buddhist World: An Intellectual History of Medicine in Early Modern Tibet*. New York: Columbia University Press, 2015.

Hacking, Ian. *Representing and Intervening: Introductory Topics in the Philosophy of Natural Science*. Cambridge: Cambridge University Press, 1983.

———. *Historical Ontology*. Cambridge: Harvard University Press, 2002.

Hagberg, Garry. "Foreword: Improvisation in the Arts." *Journal of Aesthetics and Art Criticism* 58, no. 2 (2000): 95–97. https://doi.org/10.1111/1540-6245.jaac58.2.0095.

Hall, Whitney J. *Tanuma Okitsugu, 1719–1788: Forerunner of Modern Japan*. Westport, CT: Praeger, 1982.

Hallam, Elizabeth. *Anatomy Museum: Death and the Body Displayed*. London: Reaktion Books, 2016.

Hansen, Valerie. *The Silk Road: A New History*. Oxford: Oxford University Press, 2015.

Hanson, Marta. "The 'Golden Mirror' in the Imperial Court of the Qianlong Emperor, 1739–1742." *Early Science and Medicine* 8, no. 2 (2003): 101–130. https://doi.org/10.1163/157338203x00035.

———. "Hand Mnemonics in Classical Chinese Medicine: Texts, Earliest Images, and Arts of Memory." *Asia Major* 21, no. 1 (2008): 325–347.

———. *Speaking of Epidemics in Chinese Medicine: Disease and the Geographic Imagination in Late Imperial China*. New York: Routledge, 2011.

Hendriksen, Marieke M. A., *Elegant Anatomy: The Eighteenth-Century Leiden Anatomical Collections, Elegant Anatomy*. Leiden: Brill, 2014.

Herzig, Rebecca M. *Suffering for Science: Reason and Sacrifice in Modern America*. New Brunswick: Rutgers University Press, 2005.

Hinrichs, T. J. "The Song and Jin Periods." In *Chinese Medicine and Healing: An Illustrated History*, edited by Hinrichs and Barnes, 97–128. Cambridge: Belknap Press, 2013.

Hinrichs, T. J., and Linda L. Barnes. *Chinese Medicine and Healing: An Illustrated History*. Cambridge: Belknap Press, 2013.

Hinterwaldner, Inge. "Parallel Lines as Tools for Making Turbulence Visible." *Representations* 124, no. 1 (November 1, 2013): 1–42. https://doi.org/10.1525/rep.2013.124.1.1.

Holmes, Brooke. *The Symptom and the Subject: The Emergence of the Physical Body in Ancient Greece*. Princeton: Princeton University Press, 2014.

Horiuchi, Annick. "When Science Develops Outside State Patronage: Dutch Studies in Japan at the Turn of the Nineteenth Century." *Early Science and Medicine* 8, no. 2 (2003): 131–150. https://doi.org/10.1163/157338203x00044.

House, Michael K. "Beyond the Brain: Sceptical and Satirical Responses to Gall's Organology." In *Neurology and Modernity: A Cultural History of Nervous Systems, 1800–1950*, edited by Laura Salisbury and Andrew Shail, 38–60. New York: Palgrave Macmillan, 2010.

Hsieh, Po Huei. "Paratextual Advantages of Versified Pulse Texts: The Theory and Practice of Wang Shuhe's Rhymed Verses of Pulse-Diagnostics in Chinese Medical History, 11–17th Centuries." PhD dissertation, Johns Hopkins University, 2019.

Hsu, Elisabeth. "Tactility and the Body in Early Chinese Medicine." *Science in Context* 18, no. 1 (2005): 7–34. https://doi.org/10.1017/s0269889705000335.

———. *Pulse Diagnosis in Early Chinese Medicine: The Telling Touch*. Cambridge: Cambridge University Press, 2010.

Huang, Ge. "*Huangdi neijing zhong qi de gainian he fenlei yanjiu* (《黄帝内经》中气的概念和分类研究)" [On the Concept and Classification of Qi in the Yellow Emperor's Inner Classic]. Dissertation, Shandong University of Traditional Chinese Medicine, 2008.

Huang, Longxiang (黄龍祥). *Zhongguo zhenjiu shi tujian* (中国针灸史图鉴) [Graphic History of Chinese Acupuncture-Moxibustion]. Vol. 1. Qingdao: Qingdao Press, 2003.

———. *Zhongguo zhenjiu shi tujian* (中国针灸史图鉴) [Graphic History of Chinese Acupuncture-Moxibustion]. Vol. 2. Qingdao: Qingdao Press, 2003.

Huang, Longxiang, and Fang Wang (王芳). "Acupuncture Illustrations." In *Routledge Handbook of Chinese Medicine*, edited by Vivienne Lo, Michael Stanley-Baker, and Dolly Yang, 189–205. London: Routledge, 2022.

Huang, Wendong. *Skilled Experience from Famous Chinese Medical Experts*. Changsha: Hunan kexue jishu chubanshe, 1983.

Huigen, Siegfried, Jan L. De Jong, and Elmer Kolfin. *The Dutch Trading Companies as Knowledge Networks*. Leiden: Brill, 2010.

Huntington, Eric. *Creating the Universe: Depictions of the Cosmos in Himalayan Buddhism*. Seattle: University of Washington Press, 2019.

Ingold, Tim. *Lines: A Brief History*. New York: Routledge, 2007.

Iwao, Seiichi. "A Dutch Doctor in Old Japan." *Japan Quarterly (Tokyo)* 8 (January 1, 1961): 172–175.

———. "Cultural and Commercial Relations between Japan and the Netherlands in the Tokugawa Period." In *Rangaku and Japanese Culture* (蘭学と日本文化), 30–38. Tokyo: Tokai University Press, 1969.

———. "Nagasaki and Western-Style Paintings." In *Rangaku and Japanese Culture* (蘭学と日本文化), 197–198. Tokyo: Tokai University Press, 1969.

Jackson, Terrence. *Network of Knowledge: Western Science and the Tokugawa Information Revolution*. Honolulu: University of Hawaii Press, 2016.

Jacyna, L. S. *Medicine and Modernism: A Biography of Sir Henry Head*. London: Pickering & Chatto, 2008.

Jiang, Shan. "The 'Ukiyo' of Moxibustion Reflected in the Ukiyo-e." *Chinese*

Medicine and Culture, May 08, 2024. https://doi.org/10.1097/MC9.00000 00000000105.
Jiang, Shan (姜姗), and Jingsheng Zhao (赵京生). *Zhen yu qi: Jingdian zhong de zhenjiu qi lun fa wei* (针与气: 经典中的针灸气论发微) [Acupuncture and Qi: A Study of Acupuncture-Moxabustion and Qi in Classical Texts]. Beijing: People's Medical Publishing House, 2018.
Johnson, Linda C. *Women of the Conquest Dynasties: Gender and Identity in Liao and Jin China*. Honolulu: University of Hawaii Press, 2011.
Joshi, Jayant S., and Rajesh Kumar. "The Dutch Physicians at Dejima or Deshima and the Rise of Western Medicine in Japan." *Proceedings of the Indian History Congress* 63 (2002): 1059–1066.
Kang, Kyung A., Kwang-Sup Soh. "50 Years of Bong-Han Theory and 10 Years of Primo Vascular System." *Evidence-Based Complementary and Alternative Medicine: eCAM* 2013 (2013): 587827.
Kazui, Tashiro, and Susan Downing Videen. "Foreign Relations during the Edo Period: Sakoku Reexamined." *Journal of Japanese Studies* 8, no. 2 (1982): 283–306.
Katz, Joel, and Brittany N Rosenbloom. "The Golden Anniversary of Melzack and Wall's Gate Control Theory of Pain: Celebrating 50 Years of Pain Research and Management." *Pain Research & Management: The Journal of the Canadian Pain Society* 20, no. 6 (2015): 285–286. https://doi.org/10.1155/2015/865487.
Keller, Evelyn Fox. *Making Sense of Life: Explaining Biological Development with Models, Metaphors, and Machines*. Cambridge: Harvard University Press, 2002.
Kitchin, Rob, Justin Gleeson, and Martin Dodge. "Unfolding Mapping Practices: A New Epistemology for Cartography." *Transactions of the Institute of British Geographers* 38 (July 1, 2013): 480–496. https://doi.org/10.1111/j.1475-5661.2012.00540.x.
Knoeff, Rina. *Herman Boerhaave (1668–1738): Calvinist Chemist and Physician*. Amsterdam: Edita, the Publishing House of the Royal, 2003.
———. "Herman Boerhaave's Neurology and the Unchanging Nature of Physiology." In *Blood, Sweat and Tears: The Changing Concepts of Physiology from Antiquity Into Early Modern Europe*, 201–242. Leiden: Brill, 2012.
Knorr-Cetina, K. "Culture in Global Knowledge Societies: Knowledge Cultures and Epistemic Cultures." *Interdisciplinary Science Reviews* 32, no. 4 (2007): 361–375. https://doi.org/10.1179/030801807x163571.
Kubny, Manfred. "*Qi—Lebenskraftkonzepte in China: Defnitionen, Theorien Und Grundlagen*" [Qi—Life Force Concepts in China: Definitions, Theories and Basics]. PhD dissertation, Heidelberg University, 2002.

Kuhn, Dieter. *The Age of Confucian Rule: The Song Transformation of China*. Edited by Timothy Brook. Reprint. Cambridge: Belknap Press, 2011.

Kuzuoğlu, Uluğ. "Telegraphy, Typography, and the Alphabet: The Origins of the Alphabet Revolutions in the Russo-Ottoman Space." *International Journal of Middle East Studies* 52, no. 3 (2020): 413–431. https://doi.org/10.1017/s0020743820000264.

———. "Capital, Empire, Letter: Romanization in Late Qing China." *Twentieth-Century China* 46, no. 3 (2021): 223–246. https://doi.org/10.1353/tcc.2021.0022.

———. "The Chinese Latin Alphabet: A Revolutionary Script in the Global Information Age." *Journal of Asian Studies* 81, no. 1 (2022): 23–42. https://doi.org/10.1017/s0021911821001534.

Kuriyama, Shigehisa. "Between Mind and Eye: Japanese Anatomy in the Eighteenth Century." In *Paths to Asian Medical Knowledge*, edited by Charles Leslie and Allan Young, 21–43. Berkeley: University of California Press, 1992.

———. "The Imagination of the Body and the History of Embodied Experience: Chinese Views of the Viscera." In *The Imagination of the Body and the History of Bodily Experience*, 25–72. Kyoto: International Research Center for Japanese Studies, 2001.

———. *The Expressiveness of the Body and the Divergence of Greek and Chinese Medicine*. New York: Zone Books, 2002.

Kuroda, Genji (黒田源次). *Ki No Kenkyū* (気の研究) [Research on Ki]. Tokyo: Tōkyō bijutsu, 1977.

Kusukawa, Sachiko. *Picturing the Book of Nature: Image, Text, and Argument in Sixteenth-Century Human Anatomy and Medical Botany*. Chicago: University of Chicago Press, 2011.

Langevin, Helene M., et al. "Tissue Stretch Induces Nuclear Remodeling in Connective Tissue Fibroblasts." *Histochemistry and Cell Biology* 133, no. 4 (April 2010): 405–415. https://doi.org/10.1007/s00418-010-0680-3.

Latour, Bruno. *We Have Never Been Modern*. Translated by Catherine Porter. Cambridge: Harvard University Press, 1993.

———. "The More Manipulations the Better." In *Representation in Scientific Practice Revisited*, 331–351. Cambridge: MIT Press, 2014.

Law, John, and Lin Wen-yuan. "Provincializing STS: Postcoloniality, Symmetry, and Method." *East Asian Science, Technology and Society: An International Journal* 11, no. 2 (April 22, 2017): 211–227. https://doi.org/10.1215/18752160-3823859.

Lei, Hanqing (雷汉卿). " '*Shuo wen*' `*shi bu'zi yu shenling jisi kao* (《說文》「示部」字與神靈祭祀考)" [Investigation on "Shi Bu" in *Shuowen*]. Vol. 1. Chengdu: Bashu Books (巴蜀书社), 2000.

Lei, Xiang lin Sean. "How Did Chinese Medicine Become Experiential? The Political Epistemology of Jingyan." *Positions: East Asia Cultures Critique* 10, no. 2 (2002): 333–364.

———. *Neither Donkey nor Horse: Medicine in the Struggle over China's Modernity*. Reprint. Chicago: University of Chicago Press, 2014.

———. "Qi-Transformation and the Steam Engine The Incorporation of Western Anatomy and Re-conceptualisation of the Body in Nineteenth-Century Chinese Medicine." *Asian Medicine* 7, no. 2 (2014): 312–337. https://doi.org/10.1163/15734218-12341256.

Leuenberger, Christine. "Map-Making for Palestinian State-Making." *The Arab World Geographer* 16, no. 1 (April 22, 2013): 54–74. https://doi.org/10.5555/arwg.16.1.nm3625344xr48240.

Li, Chengcheng. "*Quan sou xin xiao de zhuan ze dian zaoqi banben* (《全艘新箫》的撰择典早期版本)" [An Early Version of the Composer's Selection]. *Journal of Chinese Classics and Cultural Studies* (中國典籍與文化論叢) 13 (June 15, 2011): 201–216.

Li, Guangming. "Concerning the English Translation of 脉经 (Mài Jīng)." PULSES, Humanities Research Center, Rice University, July 7, 2020.

———. "Chinese Musical Modes." *Meridian Modes* (blog), May 2021. https://pulses.blogs.rice.edu/2021/05/21/modes.

Li, Guohao, Mengwen Zhang, and Tianqin Cao. *Explorations in the History of Science and Technology in China: A Special Number of the "Collections of Essays on Chinese Literature and History"* [*Zhongguo kejishi tansuo*]. Shanghai: Shanghai Guji Chubanshe, 1982.

Li, Lan A. "The Edge of Expertise: Representing Barefoot Doctors in Cultural Revolution China." *Endeavour* 39, nos. 3–4 (September 2015): 160–167. https://doi.org/10.1016/j.endeavour.2015.05.007.

———. "Invisible Bodies: Lu Gwei-Djen and the Specter of Translation." *Asian Medicine* 13, nos. 1–2 (September 10, 2018): 33–68. https://doi.org/10.1163/15734218-12341407.

———. "Pinpricks: Needling, Numbness, and Temporalities of Pain." In *Imagining the Brain: Episodes in the History of Brain Research*, edited by Chiara Ambrosio and William Maclehose, 205–229. Amsterdam: Academic Press, 2018.

———. "Medical Poetics: Global Health Humanities on Film and the Case of 心." In *The Routledge Companion to Health Humanities*, edited by Paul Crawford, Brian Brown, and Andrea Charise, 163–172. London: Routledge, 2020.

———. "Communist Materialism and Illustrating Medical Textbooks, 1950–1966." In *Making Sense of Medicine: Material Culture and the Reproduction of Medical Knowledge*, edited by John Nott and Anna Harris, 194–204. Bristol, UK: Intellect, 2022.

———. "Measure Your Meridians." Distinguished lecture presented at the Forum for the History of Health, Medicine, and the Life Sciences and History of Science Society Annual Meeting, Portland, November 2023. Video. https://www.youtube.com/watch?v=D5g09q4wJgQ.

Li, Qian (李倩). "*Quanti xin lun yixue shuyu yanjiu* (《全体新论》医学术语研究)" [Investigation of Medical Terms in *Quanti Xinlun*]. Master's thesis, Shandong Normal University, June 6, 2018.

Li, Yanli (李艳丽), and Shuchang Yan (阎书昌). "*Zhou Xiangeng yu Ba fu luo fu xueshuo 1950 niandai de yinjie* (周先庚与巴甫洛夫学说 1950 年代的引介)" [Chou Siegen K. and Introduction of Pavlov's Theory and Works During 1950s]. *Chinese Journal for the History of Science and Technology* 35 (September 15, 2014): 333–343.

Li, Yun (李云). *Zhongyi renming cidian* (中医人名琳典) [Biographical Dictionary of Chinese Medicine]. Beijing: Guoji wenhua chubangongsi, 1988.

Lightman, Bernard V. *The Circulation of Knowledge Between Britain, India and China: The Early-Modern World to the Twentieth Century*. Leiden: Brill, 2013.

Liu, Changhua, Man Gu, Yang Liu, Qi Zhou, and Qiong Luo. "Bamboo Slip Medical Manuscripts Excavated from Tianhui Township, Sichuan: Reflections on Chapter Titles and Textual Sources." *Asian Medicine* 18, nos. 1–2 (January 22, 2024): 38–62.

Liu, Lihong, et al. "The Functioning of Yin and Yang." In *Classical Chinese Medicine*, edited by Heiner Fruehauf, 122–144. Hong Kong: Chinese University Press, 2019.

Liu, Lydia, Rebecca Karl, and Dorothy Ko, eds. *The Birth of Chinese Feminism: Essential Texts in Transnational Theory*. New York: Columbia University Press, 2013.

Livingston, Julie. *Improvising Medicine: An African Oncology Ward in an Emerging Cancer Epidemic*. Durham: Duke University Press, 2012.

Lloyd, Geoffrey. "Pneuma between Body and Soul." *Journal of the Royal Anthropological Institute* 13 (April 20, 2007): S135–S146. https://doi.org/10.1111/j.1467-9655.2007.00409.x.

Lloyd, Geoffrey E., and Nathan Sivin. *The Way and the Word: Science and Medicine in Early China and Greece*. New Haven, CT: Yale University Press, 2002.

Lo, Vivienne (羅維前). "The Influence of Nurturing Life Culture on the Development of Western Han Acumoxa Therapy." In *Innovation in Chinese Medicine*, 19–39. Cambridge: Cambridge University Press, 2001.

———. "Quick and Easy Chinese Medicine: The Dunhuang Moxibustion Charts." In *Medieval Chinese Medicine: The Dunhuang Medical Manuscripts*, 227–251. London: Routledge, 2005.

———. "Imagining Practice: Sense and Sensuality in Early Chinese Medical Illustration." In *Imagining Chinese Medicine*, edited by Lo and Barrett, 69–88. Leiden: Brill, 2018.

———. "Classical Medicine: New Insights from Laoguanshan Cemetery (Second Century BCE), Tianhui Township, Southwest China." *Asian Medicine* 18, nos. 1–2 (January 22, 2024): 1–17. https://doi.org/10.1163/15734218-12341524.

———. "Looms of Life: Weaving a New Medical Imaginary." *Asian Medicine* 18, nos. 1–2 (January 22, 2024): 148–166. http://dx.doi.org/10.1163/15734218-12341531.

Lo, Vivienne, and Penelope Barrett, eds. *Imagining Chinese Medicine*. Leiden: Brill, 2018.

Lo, Vivienne, and Michael Stanley-Baker, eds. *Routledge Handbook of Chinese Medicine*. London: Routledge, 2022.

Lu, Mingxin, "Development of Zhu Gong's Thought on Cold Pathogenic Diseases from Shanghan Baiwen to Huorenshu," *Chinese Journal of Medical History* 41, no. 3 (September 13, 2011): 165–169.

Lukacs, Gabor. *Kaitai Shinsho: The Single Most Famous Japanese Book of Medicine; & Geka Sōden: An Early Very Important Manuscript on Surgery*. Utrecht: Hes & De Graaf, 2008.

Lumpkin, Ellen A., and Diana M. Bautista. "Feeling the Pressure in Mammalian Somatosensation." *Current Opinion in Neurobiology* 15, no. 4 (August 2005): 382–388. https://doi.org/10.1016/j.conb.2005.06.005.

Lumpkin, Ellen A., Kara L. Marshall, and Aislyn M. Nelson. "The Cell Biology of Touch." *Journal of Cell Biology* 191, no. 2 (October 18, 2010): 237–248. https://doi.org/10.1083/jcb.201006074.

Lynch, Michael, and Steve Woolgar. *Representation in Scientific Practice*. Cambridge: MIT Press, 1990.

Macé, Mieko. "La Médecine d'Ishizaka Sōtetsu 石坂宗哲 (1770–1841) En Tant Que Modèle Culturel de L'époque d'Edo:—à Partir de L'exemple du Ei E Chūkei Xu 栄衛中経図 (1825)" [The Medicine of Shizaka Sōtetsu 石坂宗哲 (1770–1841) as a Cultural Model of the Edo Period:—from the Example of Ei E Chūkei Xu 栄衛中経図 (1825)]. *Cahiers d'Extrême-Asie* 8 (1995): 413–438.

———. *Medecins et Medecines dans l'Histoire du Japon: Aventures Intellectuelles Entre La Chine et l'Occident* [Physicians and Medicines in the History of

Japan: Intellectual Adventures between China and the West]. Paris: Les Belles Lettres, 2013.

MacLehose, William. "The Pathological and the Normal: Mapping the Brain in Medieval Medicine." In *Imagining the Brain: Episodes in the History of Brain Research*, edited by Chiara Ambrosio and William MacLehose, 23–54. Amsterdam: Academic Press, 2018.

Maerker, Anna Katharina. *Model Experts: Wax Anatomies and Enlightenment in Florence and Vienna, 1775–1815*. Manchester: Manchester University Press, 2011.

Major, John S., et al., eds. *The Huainanzi*. New York: Columbia University Press, 2010.

Marcon, Federico. *The Knowledge of Nature and the Nature of Knowledge in Early Modern Japan*. Illus. ed. Chicago: University of Chicago Press, 2015.

Martin, Dale B. *The Corinthian Body*. New Haven: Yale University Press, 1999.

Martin, Troy W., "Paul's Pneumatological Statements and Ancient Medical Texts." In *The New Testament and Early Christian Literature in Greco-Roman Context: Studies in Honor of David E. Aune*, edited by David Edward Aune and John Fotopoulos, 105–126. Leiden: Brill, 2006.

Matsumura, Noriaki (松村紀明). "*Kaitai shinsho izen no `shinkei' gainen no juyō ni tsuite* (解体新書以前の「神経」概念の受容について)" [The Japanese Conception of "Nerve" before the Publication of the Kaitai-Shinsho]. *Nihon Ishigaku Zasshi* (日本医史学雑誌), 1998, 85–97. Tokyo: Japanese Society for the History of Medicine.

Mayanagi, Makoto. "The Transmission of Medical Texts and Concepts between Japan, Korea and China - since the 13th Century." Talk presented at the 6th International Congress of Oriental Medicine, Tokyo, October 19–21, 1990.

———. "*Jiang hu qi dulai de zhongguo yishu ji qi riben ban* (江户期渡来的中国医书及其日本版)" [The Import and Reprinting of Chinese Medical Books During the Edo Period]. Talk presented at the Institute of International Asian Studies, Leiden University, Leiden, Netherlands, March 26–28, 1998.

———. "*Gendai-chū-i shinkyū-gaku no keisei ni ataeta Nihon no kōken* (現代中医鍼灸学の形成に与えた日本の貢献)" [The contribution of Japan for the formation of Chinese acupuncture moxibustion studies to the present]. *Zen Nihon Shinkyu Gakkai zasshi* [Journal of the Japan Society of Acupuncture and Moxibustion] 56, no. 4 (2006): 605–615.

———. "*Zhongyi dianji de riben hua* (中医典籍的日本化)" [Japaneseization of Chinese Medicine Canon]. Talk presented at *Huanqiu zhongyiyao* (环球中医药) [World Traditional Chinese Medicine], March 6, 2008.

———. “*Ni Kan Etsu no igaku to chūgoku isho* (日韓越の医学と中国医書)” [Japanese, Korean and Vietnamese Medicine and Chinese Medical Books]. *Nihon ishigaku zasshi* 56, no. 2 (2010): 151–159.
McCloud, Scott. *Understanding Comics: The Invisible Art*. Reprint. New York: William Morrow, 1994.
McKemy, David D. “The Molecular and Cellular Basis of Cold Sensation.” *ACS Chemical Neuroscience* 4, no. 2 (February 20, 2013): 238–247. https://doi.org/10.1021/cn300193h.
Merleau-Ponty, Maurice. *The Primacy of Perception: And Other Essays on Phenomenological Psychology, the Philosophy of Art, History and Politics*. Evanston: Northwestern University Press, 1964.
Messer, Ellen. “Hot/Cold Classifications and Balancing Actions in Mesoamerican Diet and Health: Theory and Ethnography of Practice in Twentieth-Century Mexico.” In *The Body in Balance: Humoral Medicines in Practice*, edited by Peregrine Horden and Elisabeth Hsu, 149–168. New York: Berghahn Books, 2013.
Métailié, Georges. “The Representation of Plants: Engravings and Paintings.” In *Graphics and Text in the Production of Technical Knowledge in China: The Warp and the Weft*, edited by Bray, Dorofeeva-Lichtmann, and Métailié, 485–520. Leiden: Brill, 2007.
Michel-Zaitsu, Wolfgang. “Exploring the ‘Inner Landscapes’—The Kaitai Shinsho (1774) and Its Prehistory.” *Medizinhistorisches Journal* 21 (December 1, 2018): 2018. https://doi.org/10.35276/yjmh.2018.21.2.7.
Michel, Wolfgang. “Willem ten Rhijne und die Japanische Medizin (I).” *Studien Zur Deutschen und Französischen Literatur*, no. 39 (1989): 75–125.
———. “Willem ten Rhijne und die Japanische Medizin (II).” *Studien Zur Deutschen und Französischen Literatur*, no. 40 (1990): 57–103.
Miller, Carolyn R. “Genre as Social Action.” *Quarterly Journal of Speech* 70, no. 2 (May 1, 1984): 151–167.
Ming, Dong Gu. “Sinologism in Language Philosophy: A Critique of the Controversy Over Chinese Language.” *Philosophy East and West* 64, no. 3 (2014): 692–717. https://doi.org/10.1353/pew.2014.0041.
Mishra, Swasti Vardhan. “Critical Cartography and India’s Map Policy.” *Economic and Political Weekly* 50, no. 31 (2015): 22–24. https://www.jstor.org/stable/24482158.
Moehring, Francie, et al. “Uncovering the Cells and Circuits of Touch in Normal and Pathological Settings.” *Neuron* 100, no. 2 (October 24, 2018): 349–360. https://doi.org/10.1016/j.neuron.2018.10.019.
Monmonier, Mark S. *How to Lie with Maps*. Chicago: University of Chicago Press, 1991.

———. "History, Jargon, Privacy and Multiple Vulnerabilities." *Cartographic Journal* 50, no. 2 (May 1, 2013): 171–174. https://doi.org/10.1179/0008704113z.00000000084.

Montello, Daniel R. "Scale and Multiple Psychologies of Space." In *Spatial Information Theory: A Theoretical Basis for GIS*, edited by Andrew U. Frank and Irene Campari, 312–321. Berlin: Springer Science & Business Media, 1993.

Morant, George Soulie De. *Acupuncture Chinoise Atlas*. Illus. ed. Noisy-sur-École: de l'Eveil, 2018.

Mukharji, Projit Bihari. *Doctoring Traditions: Ayurveda, Small Technologies, and Braided Sciences*. Chicago: University of Chicago Press, 2016.

Nagano, Hitoshi (長野仁). *Hari kyū myūjiamu: Dō ningyō meidōzu-hen* (はりきゅう ミュージアム: 銅人形・明堂図篇) [The Museum of Acupuncture and Moxibustion: Volume on Bronze Figures and Numinous Hall Maps]. Osaka: Morinomiya University of Medical Sciences, 2001.

Nakajima, Yoshiaki (中島楽章). "*Seiki no higashiajia kaiiki to kajin chishiki-sō no idō—minamikyūshū no Akito ishi o megutte* (世紀の東アジア海域と華人知識層の移動—南九州の明人医師をめぐって)" [Sixteenth-Century and Seventeenth-Century East Asian Seas and Migration of Chinese Intellectuals: A Study of Dr. Akihito in Southern Kyushu]. *Shigaku Zasshi* (史学雑誌) 113, no. 12 (2004): 1–37.

Nappi, Carla. "Disengaging from 'Asia.'" *East Asian Science, Technology and Society* 6, no. 2 (June 1, 2012): 229–232.

Newman, William R. *Atoms and Alchemy: Chymistry and the Experimental Origins of the Scientific Revolution*. Chicago: University of Chicago Press, 2006.

Ng, On-cho. "An Early Qing Critique of the Philosophy of Mind-Heart (Xin): The Confucian Quest for Doctrinal Purity and the Doxic Role of Chan Buddhism." *Journal of Chinese Philosophy* 26, no. 1 (December 1, 1999): 75–92.

Niu, Yahua. "*Hua shou yixue zhuzuo zai riben de liubo* (滑寿医学著作在日本的流播)" [Spreading of Hua Shou's Medical Books in Japan]. *China Journal of Medical History* 28, no. 3 (1988): 184–189.

Nutton, Vivian. "Physiologia from Galen to Jacob Bording," in *Blood, Sweat and Tears: The Changing Concepts of Physiology from Antiquity Into Early Modern Europe*, 27–40. Leiden: Brill, 2012.

Oak, Sung-Deuk. "Competing Chinese Names for God: The Chinese Term Question and Its Influence upon Korea." *Journal of Korean Religions* 3, no. 2 (2012): 91–122. https://doi.org/10.1353/jkr.2012.0017.

Oliver, Jeff. "On Mapping and Its Afterlife: Unfolding Landscapes in Northwestern North America." *World Archaeology* 43, no. 1 (2011): 66–85. https://doi.org/10.1080/00438243.2011.544899.

Onozawa, Seiichi (小野沢精一), Mitsuji Fukunaga (福永光司), and Yū Yamanoi (山井湧). *Ki No Shisō: Chūgoku Ni Okeru Jinenkan to Nigen-Kan No Tenkai* (気の思想:_中国における自然観と人間観の展開) [Thoughts on Qi: The Emergent Distinction of the Human-Centric Perspective and the Nature-Centric Perspective in China]. Tokyo: Tōkyō Daigaku Shuppanka, 1978.

Orland, Barbara. "White Blood and Red Milk. Analogical Reasoning in Medical Practice and Experimental Physiology (1560–1730)." In *Blood, Sweat and Tears: The Changing Concepts of Physiology from Antiquity Into Early Modern Europe*, 443–480. Leiden: Brill, 2012.

Otori, Ranzaburo. "The Acceptance of Western Medicine in Japan." *Monumenta Nipponica* 19, no. 3/4 (1964): 254–274.

Otsuka, Yasuo (大塚康夫). "Willem Ten Rhyne in Japan." In *Rangaku and Japanese Culture* (蘭学と日本文化), 251–60. Tokyo: Tokai University Press, 1969.

Pacht, Otto. *Van Eyck and the Founders of Early Netherlandish Painting*. Edited by Maria Schmidt Dengler. Translated by David Britt. London: Harvey Miller, 1999.

Pang, Laikwan. "Dialectical Materialism." In *Afterlives of Chinese Communism (LBE): Political Concepts from Mao to Xi*, edited by Christian Sorace, Ivan Franceschini, and Nicholas Loubere, 67–72. Illus. ed. New York: Verso, 2019.

Panzanelli, Roberta, and Julius Schlosser, eds. *Ephemeral Bodies: Wax Sculpture and the Human Figure*. Los Angeles: Getty Research Institute, 2008.

Paranavitana, K. D. "Medical Establishment in Sri Lanka During the Dutch Period (1640–1796)." *Journal of the Royal Asiatic Society of Sri Lanka* 33 (1988): 103–110.

Park, Hijoon, et al. "Does Deqi (Needle Sensation) Exist?" *American Journal of Chinese Medicine* 30, no. 01 (January 2002): 45–50. https://doi.org/10.1142/s0192415x02000053.

Park, Katharine. *Secrets of Women: Gender, Generation, and the Origins of Human Dissection*. New York: Zone Books, 2010.

Pauly, Philip J. "Modernist Practice in American Biology." In *Modernist Impulses in the Human Sciences, 1870–1930*, edited by Dorothy Ross, 272–289. Baltimore: Johns Hopkins University Press, 1994.

Peattie, Mark, Edward Drea, and Hans van de Ven, eds. *The Battle for China: Essays on the Military History of the Sino-Japanese War of 1937–1945*. Stanford: Stanford University Press, 2013.

Prakash, Om. *The Dutch East India Company and the Economy of Bengal, 1630–1720*. Princeton: Princeton University Press, 1985.
Pritchard, James. *In Search of Empire: The French in the Americas, 1670–1730*. Cambridge: Cambridge University Press, 2004.
Qiao, Wenbiao (乔文彪), and Su Ting (苏婷). *Wang Haogu* (王好古). Edited by Guijuan Pan. Academic Research Series of Famous Doctors of Traditional Chinese Medicine through the Ages. Beijing: China Chinese Medicine Publishing House, 2017.
Quarshie, Nana Osei. *An African Pharmakon: Psychiatry and the Mind Politic of Modern Ghana*. Chicago: University of Chicago Press, forthcoming.
———. "Spiritual Pawning: 'Mad Slaves' and Mental Healing in Atlantic-Era West Africa." *Comparative Studies in Society and History* 65, no. 3 (July 2023): 475–499. https://doi.org/10.1017/S0010417523000051.
Rankin, Bill. "Cartography and the Reality of Boundaries." *Perspecta* 42 (2010): 42–45.
———. *After the Map: Cartography, Navigation, and the Transformation of Territory in the Twentieth Century*. London: University of Chicago Press, 2016.
Raphals, Lisa. "The Treatment of Women." In *Chinese Medicine and Healing: An Illustrated History*, 27–48. Cambridge: Belknap Press, 2013.
Ray, Bronson S., Joseph C. Hinsey, and William A. Geohegan. "Observations on the Distribution of the Sympathetic Nerves to the Pupil and Upper Extremity as Determined by Stimulation of the Anterior Roots in Man." *Annals of Surgery* 118, no. 4 (1943): 647–655.
Reill, Peter H. *Vitalizing Nature in the Enlightenment*. Berkeley: University of California Press, 2005.
Rheinberger, Hans-Jörg. *An Epistemology of the Concrete: Twentieth-Century Histories of Life*. Durham, NC: Duke University Press, 2010.
Rivington, Charles A. "Early Printers to the Royal Society 1663–1708." *Notes and Records of the Royal Society of London* 39, no. 1 (1984): 1–27.
Roberts, K. B., and J. D. W. Tomlinson. *Fabric of the Body: European Traditions of Anatomical Illustration*. Oxford: Clarendon Press, 1992.
Rogaski, Ruth. *Hygienic Modernity: Meanings of Health and Disease in Treaty-Port China*. Berkeley: University of California Press, 2004.
Rojas, Carlos. "Contradiction." In *Afterlives of Chinese Communism (LBE): Political Concepts from Mao to Xi*, edited by Christian Sorace, Ivan Franceschini, and Nicholas Loubere, 43–48. Illus. ed. New York: Verso, 2019.
Rosner, Erhard. *Medizingeschichte Japans*. Leiden: Brill, 1988.
Rotman, Brian. *Becoming Beside Ourselves: The Alphabet, Ghosts, and Distributed Human Being*. Durham, NC: Duke University Press, 2008.

Ryan, Judith. *The Vanishing Subject: Early Psychology and Literary Modernism*. Chicago: University of Chicago Press, 1991.

Sadler, Katelyn E., Francie Moehring, and Cheryl L. Stucky. "Keratinocytes Contribute to Normal Cold and Heat Sensation." *eLife* 9 (July 30, 2020): e58625. https://doi.org/10.7554/elife.58625.

Sakade, Yoshinobu (坂出 祥伸), *Taoism, Medicine and Qi in China and Japan*. Osaka: Kansai University Press, 2007.

Salguero, C. Pierce. *A Global History of Buddhism and Medicine*. New York: Columbia University Press, 2022.

Salisbury, Laura, and Andrew Shail. *Neurology and Modernity: A Cultural History of Nervous Systems, 1800–1950*. New York: Palgrave Macmillan, 2010.

Schäfer, Dagmar. *The Crafting of the 10,000 Things: Knowledge and Technology in Seventeenth-Century China*. Chicago: University of Chicago Press, 2015.

Scheid, Volker. "The Globalisation of Chinese Medicine." *The Lancet* 354 (December 1, 1999): SIV10.

———. *Chinese Medicine in Contemporary China: Plurality and Synthesis*. Durham, NC: Duke University Press, 2002.

———. "Remodeling the Arsenal of Chinese Medicine: Shared Pasts, Alternative Futures." *Annals of the American Academy of Political and Social Science* 583, no. 1 (2002): 136–159. https://doi.org/10.1177/000271620258300109.

———. "Traditional Chinese Medicine—What Are We Investigating?: The Case of Menopause " *Complementary Therapies in Medicine* 15, nos. 1–3 (March 2007): 54–68. https://doi.org/10.1016/j.ctim.2005.12.002.

———. "Transmitting Chinese Medicine." *Asian Medicine* (Leiden, Netherlands) 8, no. 2 (2013): 323–349.

Schilling, Tom. "British Columbia Mapped: Geology, Indigeneity, and Land in the Age of Digital Cartography." In *Visualization in the Age of Computerization*, edited by Annemarie Carusi, Aud Sissel Hoel, Timothy Webmoor, and Steven Woolgar, 59–76. London: Routledge, 2014.

Schmalzer, Sigrid. *The People's Peking Man: Popular Science and Human Identity in Twentieth-Century China*. Chicago: University of Chicago Press, 2008.

———. *Red Revolution, Green Revolution: Scientific Farming in Socialist China*. Chicago: University of Chicago Press, 2016.

Schneider, Laurence. *Biology and Revolution in Twentieth-Century China*. Lanham, MD: Rowman & Littlefield. 2003.

Screech, Timon. *The Lens within the Heart: The Western Scientific Gaze and Popular Imagery in Later Edo Japan*. 2nd ed. Honolulu: University of Hawaii Press, 2002.

Seth, Suman. "Putting Knowledge in Its Place: Science, Colonialism, and the Postcolonial." *Postcolonial Studies* 12, no. 4 (December 2009): 373–388. https://doi.org/10.1080/13688790903350633.

———. "Colonial History and Postcolonial Science Studies." *Radical History Review* 2017, no. 127 (January 1, 2017): 63–85. https://doi.org/10.1215/01636545-3690882.

Shackelford, Jole. *William Harvey and the Mechanics of the Heart*. New York: Oxford University Press, 2003.

Shapin, Steven, and Simon Schaffer. *Leviathan and the Air-Pump: Hobbes, Boyle, and the Experimental Life*. Princeton, NJ: Princeton University Press, 2011.

Shapiro, Hugh. "The Puzzle of Spermatorrhea in Republican China." *Positions* 6, no. 3 (1998): 551–596.

Shelton, Tamara Venit. *Herbs and Roots: A History of Chinese Doctors in the American Medical Marketplace*. New Haven, CT: Yale University Press, 2019.

Sherrington, Charles Scott. *The Integrative Action of the Nervous System*. New Haven, CT: Yale University Press, 1961.

Simmons, Richard Vanness. "Whence Came Mandarin? Qīng Guānhuà, the Běijīng Dialect, and the National Language Standard in Early Republican China." *Journal of the American Oriental Society* 137, no. 1 (2017): 63–88. http://dx.doi.org/10.7817/jameroriesoci.137.1.0063.

Siu, Wang-Ngai, and Peter Lovrick. "Using Costumes." In *Chinese Opera: The Actor's Craft*, 131–169. Hong Kong: Hong Kong University Press, 2014.

Smith, Hilary A. *Forgotten Disease: Illnesses Transformed in Chinese Medicine*. Stanford, CA: Stanford University Press, 2017.

Smith, Jeffrey Chipps. *The Northern Renaissance (Art & Ideas)*. London: Phaidon, 2004.

Smith, Pamela H. *The Body of the Artisan: Art and Experience in the Scientific Revolution*. Chicago: University of Chicago Press, 2006.

———. *The Business of Alchemy: Science and Culture in the Holy Roman Empire*. Reprint. Princeton, NJ: Princeton University Press, 2016.

Soh, Kwang-Sup. "Bonghan Duct and Acupuncture Meridian as Optical Channel of Biophoton." *Journal of the Korean Physical Society* 45, no. 5 (November 2004): 1196–1198. http://www.biontology.com/wp-content/uploads/2012/10/bonghan-ducts.pdf.

———. "Bonghan Circulatory System as an Extension of Acupuncture Meridians." *Journal of Acupuncture and Meridian Studies* 2, no. 2 (June 2009): 93–106. https://doi.org/10.1016/S2005-2901(09)60041-8.

Stanley-Baker, Michael. "Qi 氣: A Means for Cohering Natural Knowledge." In *Routledge Handbook of Chinese Medicine*, edited by Vivienne Lo, Michael Stanley-Baker, and Dolly Yang, 23–38. London: Routledge, 2022.

Sterckx, Roel. "Searching for Spirit: Shen and Sacrifice in Warring States and Han Philosophy and Ritual." *Extrême-Orient Extrême-Occident*, no. 29 (2007): 13–47. https://doi.org/10.3406/oroc.2007.1083.

Su, John J. "Amitav Ghosh and the Aesthetic Turn in Postcolonial Studies." *Journal of Modern Literature* 34, no. 3 (2011): 65–86. https://doi.org/10.2979/jmodelite.34.3.65.

Sugimoto, Masayoshi, and David L. Swain. *Science and Culture in Traditional Japan, A.D. 600–1854*. Cambridge: MIT Press, 1978.

Sun, Hai-shu, et al. "Analysis on Cheng Dan-An's Educational Thought in His Book Chinese Acupuncturology." *Journal of Clinical Acupuncture and Moxibustion* 39, no. 5 (October 2014): 410–412.

Suzuki, Tatsuhiko (鈴木 達彦), and Jiro Endo (遠藤 次郎). "*Yakuyō-ryōoyobi fukuyaku-hō kara mita Nihon kanpō no ryūha: Yakuyō-ryōoyobi bun fuku no igi* (薬用量および服薬法から見た日本漢方の流派:-薬用量および分服の意義)" [Schools of Japanese Kampo Medicine from the Perspective of Dosage and Dosage Method: Significance of Dosage and Divided Dosage]. *Journal of Oriental Medicine Japan* 62, no. 3 (2011): 382–391.

Tan, Perciliz L., and Nicholas Katsanis. "Thermosensory and Mechanosensory Perception in Human Genetic Disease." *Human Molecular Genetics* 18, no. R2 (October 15, 2009): 146–153. https://doi.org/10.1093/hmg/ddp412.

Tan, Wei Yu Wayne. "Rediscovering Willem ten Rhijne's *De Acupunctura*: The Transformation of Chinese Acupuncture in Japan." In *Translation at Work: Chinese Medicine in the First Global Age*, 108–133. Leiden: Brill, 2020.

Taylor, Kim. *Chinese Medicine in Early Communist China, 1945–63: A Medicine of Revolution*. London: Routledge Curzon, 2005.

Teng, Emma. "China Comes to Tech: 1877–1931." Exhibit, Massachusetts Institute of Technology, February–November 2017. http://chinacomestomit.org.

Tiquia, Rey. "The Qi That Got Lost in Translation: Traditional Chinese Medicine, Humor and Healing." In *Humor in Chinese Life and Letters*, edited by Jocelyn Chey and Jessica Milner Davis, 39–58. Hong Kong: Hong Kong University Press, 2011.

Toby, Ronald P. *State and Diplomacy in Early Modern Japan: Asia in the Development of the Tokugawa Bakufu*. Stanford: Stanford University Press, 1991.

Todes, Daniel. *Ivan Pavlov: A Russian Life in Science*. Oxford: Oxford University Press, 2015.

Trambaiolo, Daniel. "Diplomatic Journeys and Medical Brush Talks: Eighteenth-Century Dialogues between Korean and Japanese Medicine." In *Motion*

and Knowledge in the Changing Early Modern World: Orbits, Routes and Vessels, edited by Ofer Gal and Yi Zheng, 93–113. Dordrecht: Springer Netherlands, 2014. https://doi.org/10.1007/978-94-007-7383-7_6.
———. "Ancient Texts and New Medical Ideas in Eighteenth-Century Japan." In *Antiquarianism, Language, and Medical Philology: From Early Modern to Modern Sino-Japanese Medical Discourses*, edited by Benjamin A. Elman, 81–104. Sir Henry Wellcome Asian Series 12. Leiden: Brill, 2015.
Turner, Bryan S. *The Body and Society: Explorations in Social Theory*. Oxford: B. Blackwell, 1984.
Unschuld, Paul U., and Hermann Tessenow. *Huang Di Nei Jing Ling Shu: The Ancient Classic on Needle Therapy*. Oakland: University of California Press, 2016.
van de Ven, Hans J. "Public Finance and the Rise of Warlordism." *Modern Asian Studies* 30, no. 4 (1996): 829–868. https://doi.org/10.1017/s0026749x00016814.
van der Eijk, Philip. "Nemesius of Emesa and Early Brain Mapping." *The Lancet* 372, no. 9637 (August 9, 2008): 440–441. https://doi.org/10.1016/s0140-6736(08)61183-6.
Vidal, Fernando. "Brainhood, Anthropological Figure of Modernity." *History of the Human Sciences* 22, no. 1 (February 2009): 5–36. https://doi.org/10.1177/0952695108099133.
Vigouroux, Mathias. "*Commerce Des Livres et Diplomatie: La Transmission de Chine et de Corée Vers Le Japon Des Savoirs Médicaux Liés à La Pratique de l'acuponcture et de La Moxibustion (1603–1868)*" [Book Trade and Diplomacy: The Transmission of Medical Knowledge from China and Korea to Japan Linked to Acupuncture and Moxibustion Practice (1603–1868)]. *Extrême-Orient Extrême-Occident*, no. 36 (2013): 109–154.
———. "The Reception of the Circulation Channels Theory in Japan (1500–1800)." In *Antiquarianism, Language, and Medical Philology: From Early Modern to Modern Sino-Japanese Medical Discourses*, edited by Benjamin A. Elman, 105–132. Leiden: Brill, 2015.
———. "The Surgeon's Acupuncturist: Philipp Franz von Siebold's Encounter with Ishizaka Sôtetsu and Nineteenth Century Japanese Acupuncture." *Revue d'histoire des sciences* 70, no. 1 (2017): 79–108.
———. "*Transmettre Quels Savoirs? Le Rôle des Illustrations des Vaisseaux d'acuponcture dans la Circulation des Savoirs Médicaux entre l'Asie Orientale et l'Europe au Dix-Septième Siècle*" [What Knowledge to Transmit? The Role of Acupuncture Vessels Images in the Circulation of Medical Knowledge between Japan and Europe]. *Eurasie* 28 (2019): 181–185.
Vilaça, Aparecida. "Chronically Unstable Bodies: Reflections on Amazonian

Corporalities." *Journal of the Royal Anthropological Institute* 11, no. 3 (2005): 445–464. https://doi.org/10.1111/j.1467-9655.2005.00245.x.
Wang, Furong (王芙蓉), and Fandong Kong (孔凡栋). "An Aesthetic Study of Song Dynasty Scholarly Dress (宋代士人幅巾的审美研究)." *Fashion Guide* (服饰导刊) 6, no. 4 (August 2017): 10–15.
Wang, Guozhong (王钱国忠著). *Lu Guizhen yu Li Yuese* (鲁桂珍与李约瑟) [Lu Gwei-djen and Joseph Needham]. Guiyang: Guizhou Renmin Chubanshe (贵州人民出版社), 1999.
Wang, Shumin (王淑民), and Gabriel Fuentes. "Chinese Medical Illustration: Chronologies and Categories." In *Imagining Chinese Medicine*, edited by Lo and Barrett, 29–50. Leiden: Brill, 2018.
Wang, Tianping. *Shanghai Sheying Shi* (上海摄影史) [The History of Photography in Shanghai]. Shanghai: Shanghai People's Fine Arts Publishing House (上海人民美术出版社), 2012.
Warner, John Harley, and James M. Edmonson. *Dissection: Photographs of a Rite of Passage in American Medicine: 1880–1930*. New York: Blast Books, 2009.
Waterson, James. *Defending Heaven: China's Mongol Wars, 1209–1370*. London: Frontline Books, 2013.
Wilcox, Lorraine. *Moxibustion: The Power of Mugwort Fire*. Boulder: Blue Poppy Press, 2008.
———. *Moxibustion: A Modern Clinical Handbook*. Boulder: Blue Poppy Press, 2009.
Winslett, Justin T. "Deities and the Extrahuman in Pre-Qin China: Lesser Deities in the 'Zuozhuan' and the 'Guoyu.'" *Journal of the American Academy of Religion* 82, no. 4 (2014): 939–967. https://doi.org/10.1093/jaarel/lfu035.
Winter, Alison. *Mesmerized: Powers of Mind in Victorian Britain*. Chicago: University of Chicago Press, 1998.
Wittgenstein, Ludwig. *Philosophical Investigations*, edited by P. M. S. Hacker and Joachim Schulte. 4th ed. Chichester: Wiley-Blackwell, 2009.
Wool, Zoë. "Homunculus Revolts: Re-Figuring the Neurological Subject." *Somatosphere* (blog), August 21, 2019. http://somatosphere.net/2019/homunculus-revolts-re-figuring-the-neurological-subject.html.
Wragge-Morley, Alexander. "Imagining the Soul: Thomas Willis (1621–1675) on the Anatomy of the Brain and Nerves." In *Imagining the Brain: Episodes in the History of Brain Research*, edited by Chiara Ambrosio and William Maclehose, 55–74. Amsterdam: Academic Press, 2018.
———. *Aesthetic Science: Representing Nature in the Royal Society of London, 1650–1720*. Chicago: University of Chicago Press, 2020.

Wright, David. "The Translation of Modern Western Science in Nineteenth-Century China, 1840–1895." *Isis* 89, no. 4 (December 1, 1998): 653–673. https://doi.org/10.1086/384159.
Wright, Jessica. "Ventricular Localization in Late Antiquity: The Philosophical and Theological Roots of an Enduring Model of Brain Function." *Progress in Brain Research* 243 (2018): 3–22. https://doi.org/10.1016/bs.pbr.2018.10.007.
Wright, Thomas Edward. *William Harvey: A Life in Circulation*. Oxford: Oxford University Press, 2013.
Wu, Yi-Li. "The Gendered Medical Iconography of the Golden Mirror (Yuzuan Yizong Jinjian, 1742)." *Asian Medicine* 4, no. 2 (July 2008): 452–491.
———. *Reproducing Women: Medicine, Metaphor, and Childbirth in Late Imperial China*. Berkeley: University of California Press, 2010.
Xiao, Ying, Jonathan S. Williams, and Isaac Brownell. "Merkel Cells and Touch Domes: More than Mechanosensory Functions?" *Experimental Dermatology* 23, no. 10 (October 2014): 692–695. https://doi.org/10.1111/exd.12456.
Xu, Xiaoqun. *Chinese Professionals and the Republican State: The Rise of Professional Associations in Shanghai, 1912–1937*. Cambridge: Cambridge University Press, 2001.
Yakazu, Domei. "On the Medicine Established by Dōsan Manase." *Kampo Medicine* 42, no. 2 (1991): 189–203.
Yee, Cordell D. K. "Reinterpreting Traditional Chinese Geographical Maps." In *Cartography in the Traditional East and Southeast Asian Societies, The History of Cartography*. Vol. 2, Book 2, edited by J. B. Harley and David Woodward, 35–70. Chicago: University of Chicago Press, 1994.
———. "Taking the World's Measure: Chinese Maps between Observation and Text." In *Cartography in the Traditional East and Southeast Asian Societies, The History of Cartography*, Vol. 2, Book 2, edited by Harley and Woodward, 96–127. Chicago: University of Chicago Press, 1994.
Yeh, Wen-hsin. *Shanghai Splendor: A Cultural History, 1843–1949*. Berkeley: University of California Press, 2007.
Yoeli-Tlalim, Ronit. *ReOrienting Histories of Medicine: Encounters along the Silk Roads*. New York: Bloomsbury Academic, 2021.
Yood, Jessica. "Writing the Discipline: A Generic History of English Studies." *College English* 65, no. 5 (2003): 526–540. https://doi.org/10.2307/3594251.
Yu, Ning. *The Chinese HEART in a Cognitive Perspective*. Berlin: De Gruyter Mouton, 2009.
Zhan, Mei. *Other-Worldly: Making Chinese Medicine through Transnational Frames*. Durham: Duke University Press Books, 2009.
Zhang, Cheng (张程). "*20 Shiji 50 niandai xuexi sulian xianjin yixue yanjiu*

(20 世纪 50 年代学习苏联先进医学研究)" [Review of Soviet Union Research in Advancing Medicine in the 1950s]. *Journal of Nanjing Medical University (Social Sciences)* 4, no. 93 (August 2019): 299–302.

Zhang, Lijian (张立剑). *Zhu lian yu zhenjiu* (朱琏与针灸) [Zhu Lian and Acupuncture Moxibustion]. Beijing: People's Health Publishing House, 2015.

Zhang, Qicheng (張其成), and David Dear. "Embodying Animal Spirits in the Vital Organs: Daoist Alchemy in Chinese Medicine." In *Imagining Chinese Medicine*, edited by Lo and Barrett, 389–396. Leiden: Brill, 2018.

Zhu, Pingyi (祝平一). "*Shenti, Linghun Yu Tianzhu: Ming Mo Qing Chu Xixue Zhong de Renti Shengli Zhishi* (身體、靈魂與天主: 明末清初西學中的人體生理知識)" [Body, Soul, and God: Knowledge of Human Physiology in Western Studies from the End of Ming Dynasty to the Early Qing Dynasty]. *Xin Lishi* (新史學) 7, no. 2 (June 1996): 47–98.

Zürn, Tobias Benedikt. "The Han *Imaginaire* of Writing as Weaving: Intertextuality and the *Huainanzi*'s Self-Fashioning as an Embodiment of the Way." *Journal of Asian Studies* 79, no. 2 (May 2020): 367–402. https://doi.org/10.1017/s0021911819001906.

INDEX

Page numbers in *italics* indicate illustrations.